Material Literacy in
Eighteenth-Century Britain

Material Culture of Art and Design

Material Culture of Art and Design is devoted to scholarship that brings art history into dialogue with interdisciplinary material culture studies. The material components of an object—its medium and physicality—are key to understanding its cultural significance. Material culture has stretched the boundaries of art history and emphasized new points of contact with other disciplines, including anthropology, archaeology, consumer and mass culture studies, the literary movement called "Thing Theory," and materialist philosophy. **Material Culture of Art and Design** seeks to publish studies that explore the relationship between art and material culture in all of its complexity. The series is a venue for scholars to explore specific object histories (or object biographies, as the term has developed), studies of medium and the procedures for making works of art, and investigations of art's relationship to the broader material world that comprises society. It seeks to be the premiere venue for publishing scholarship about works of art as exemplifications of material culture.

The series encompasses material culture in its broadest dimensions, including the decorative arts (furniture, ceramics, metalwork, textiles), everyday objects of all kinds (toys, machines, musical instruments), and studies of the familiar high arts of painting and sculpture. The series welcomes proposals for monographs, thematic studies, and edited collections.

Series Editor:
Michael Yonan, University of Missouri, USA

Advisory Board:
Wendy Bellion, University of Delaware, USA
Claire Jones, University of Birmingham, UK
Stephen McDowall, University of Edinburgh, UK
Amanda Phillips, University of Virginia, USA
John Potvin, Concordia University, Canada
Olaya Sanfuentes, Pontificia Universidad Católica de Chile, Chile
Stacey Sloboda, University of Massachusetts Boston, USA
Kristel Smentek, Massachusetts Institute of Technology, USA
Robert Wellington, Australian National University, Australia

Volumes in the Series

Forthcoming Books

Material Literacy in Eighteenth-Century Britain

A Nation of Makers

Edited by
Serena Dyer and Chloe Wigston Smith

BLOOMSBURY VISUAL ARTS
LONDON • NEW YORK • OXFORD • NEW DELHI • SYDNEY

BLOOMSBURY VISUAL ARTS
Bloomsbury Publishing Plc
50 Bedford Square, London, WC1B 3DP, UK
1385 Broadway, New York, NY 10018, USA

BLOOMSBURY, BLOOMSBURY VISUAL ARTS and the Diana logo are trademarks of
Bloomsbury Publishing Plc

First published in Great Britain 2020
Reprinted 2020
This paperback edition first published 2022

Series design: Irene Martinez Costa
Cover image: *The Ladies Waldegrave*, 1780 (oil on canvas), Joshua Reynolds (1723–92) /
National Galleries of Scotland, Edinburgh / Bridgeman Images

A catalogue record for this book is available from the British Library.

Library of Congress Cataloguing-in-Publication Data
Names: Dyer, Serena, editor. | Smith, Chloe Wigston, editor.
Title: Material literacy in eighteenth-century Britain: a nation
of makers / edited by Serena Dyer and Chloe Wigston Smith.
Description: New York : Bloomsbury Visual Arts, [2020] | Series: Material
culture of art and design
LC record available at https://lccn.loc.gov/2020032492
LC ebook record available at https://lccn.loc.gov/2020032493

ISBN: HB: 978-1-5013-4961-4
          PB: 978-1-3502-8241-4
          ePDF: 978-1-5013-4963-8
          eBook: 978-1-5013-4962-1

Series: Material Culture of Art and Design

Typeset by Integra Software Services Pvt. Ltd.,
Printed and bound in Great Britain

To find out more about our authors and books visit www.bloomsbury.com
and sign up for our newsletters.

Contents

List of Figures

List of Plates

List of Tables

Notes on Contributors

Hilary Davidson is a dress historian and curator. She is Honorary Associate at the University of Sydney, and is completing her PhD at La Trobe, Melbourne. She was previously curator of fashion and decorative art at the Museum of London, and has published, taught, lectured and broadcast extensively on fashion history, archaeological textiles, material culture and haptic knowledge. Her book *Dress in the Age of Jane Austen* (Yale University Press, 2019) examines clothing and fashion across the British Regency world.

Serena Dyer is Lecturer in History of Design and Material Culture at De Montfort University. She has taught at the University of Warwick and the University of Hertfordshire, and was Postdoctoral Fellow at the Paul Mellon Centre for Studies in British Art. She was previously Curator of the Museum of Domestic Design and Architecture. She has published on albums, wallpaper, consumer culture and childhood in the eighteenth century. She is author of *Material Lives: Women Makers and Consumer Culture in the 18th Century* (Bloomsbury, 2021) and editor of *Shopping and the Senses, 1800–1970: A Sensory History of Retail and Consumption* (Palgrave, 2022).

Laura Engel is Professor of English at Duquesne University, where she specializes in eighteenth-century British literature and theatre. She is the author of *Women, Performance, and the Material of Memory: The Archival Tourist, 1780–1915* (Palgrave, 2019), *Actresses, Accessories, and Austen: Much Ado About Muffs* (Palgrave Pivot, 2014) and *Fashioning Celebrity: Eighteenth-Century British Actresses and Strategies for Image Making* (Ohio State University Press, 2011), and co-editor of *Stage Mothers: Women, Work, and the Theater 1660–1830* (2015). She recently co-curated an exhibition entitled 'Artful Nature: Fashion and Theatricality' at the Lewis Walpole Library and is working on a new digital book project, *The Art of the Actress*.

Ariane Fennetaux is Associate Professor of Eighteenth-Century History at the University of Paris. In addition to essays and articles on women, material culture, textiles and dress, she co-edited *The Afterlife of Used Things: Recycling in the Long Eighteenth Century* (Routledge, 2015). She is the author, with Barbara Burman, of *The Pocket: A Hidden History of Women's Lives 1660–1900* (Yale University Press, 2019).

Elisabeth Gernerd is the Deidre Murphy Postdoctoral Fellow at Historic Royal Palaces and a former fellow at the Paul Mellon Centre for Studies in British Art and the UCLA Centre for 17th- and 18th-Century Studies. Her current book project, based on her PhD research, focuses on eighteenth-century dress, underwear and accessories, as well as the relationship between dress and visual representation. She has published on men's stockings and masculinity in *Textile History* (2015).

Sarah Howard is a freelance conservation consultant and textile conservator who has worked with a variety of dress collections in the UK, including those in independent and national museums and the National Trust. Her conservation projects on items of eighteenth-century men's fashion in the collections of Hampshire Culture Trust have advanced her research into the accounts and archives of the era's tailoring trades.

Alicia Kerfoot is Associate Professor of English at SUNY Brockport. Her published research includes articles on Jane Austen and Frances Burney, and on fashion and dress in eighteenth-century literature. Her current book project considers the role of footwear in literature of the long eighteenth century.

Crystal B. Lake is Professor of English at Wright State University and a co-editor of the-rambling.com. Her research has appeared in journals such as *ELH, Modern Philology, The Review of English Studies* and *Word & Image*. Lake's first book, *Artifacts: How We Think and Write about Found Objects* (Johns Hopkins University Press, 2020), recovers the broken, dirty, rusty, mouldy objects that were dug up in England during the long eighteenth century. Her current book studies the relationships between popular reading practices and maker cultures.

Nicole Pohl is Professor of English at Oxford Brookes University, Oxford. She has published and edited books on women's utopian writing in the seventeenth and eighteenth centuries, eighteenth-century European salons, epistolarity and the Bluestockings. She edited *The Letters of Sarah Scott* (2013) and is the Editor-in-Chief of *Elizabeth Montagu Correspondence Online* (*EMCO*).

Robbie Richardson is Assistant Professor of English at Princeton University. Previously he was Senior Lecturer in Eighteenth-Century Literature at the University of Kent. He is the author of *The Savage and Modern Self: North American Indians in Eighteenth-Century Literature and Culture* (University of Toronto, 2018). He is a member of Pabineau First Nation in New Brunswick, Canada.

Chloe Wigston Smith is Senior Lecturer in the Department of English and Related Literature and the Centre for Eighteenth Century Studies at the University of York. She is the author of *Women, Work, and Clothes in the Eighteenth-Century Novel* (Cambridge University Press, 2013), as well as co-editor of the collection, *Small Things in the Eighteenth Century: The Personal and Political Value of the Miniature* (Cambridge University Press). Her current British Academy–funded project looks at domestic crafts in the Atlantic world.

Jon Stobart is Professor of History at Manchester Metropolitan University. His research explores various aspects of retailing and consumption in the long eighteenth century, with a particular focus on shops and shopping, and on the supply of material goods to the country house. He is currently working on a project which examines comfort in the European country house. Recent publications include an edited collection, *The Comforts of Home in Western Europe, 1700–1900* (Bloomsbury, 2020).

Emily Taylor is Assistant Curator, European Decorative Arts at National Museums Scotland in Edinburgh. She specializes in fashion history and her current research centres on British garment construction, material culture and identity, *c.* 1700–1850. She completed her PhD on Scottish women's dress in the eighteenth century at the University of Glasgow in 2013 and has published research from this project in *Fashion Practice*.

Beth Fowkes Tobin, Professor Emerita, Department of English, Arizona State University, is author of four monographs, most recently *The Duchess's Shells: Natural History Collecting in the Age of Cook's Voyages* (Yale University Press, 2014), a project for which she received a Scholars Award from the National Science Foundation. She is the co-editor of a series of books on women and material culture as well as *The Materiality of Color: The Production, Circulation, and Application of Dyes and Pigments, 1400–1800* (Ashgate, 2012). Her current research concerns the visual and material culture of Enlightenment entomology.

Acknowledgements

The idea for this book originated from a conference at the University of Warwick in May 2017. Organized by Serena, and with Chloe as the keynote, 'Fashioning Dress: Sewing and Skill, 1500–1800' brought together colleagues working on making and embodied knowledge throughout the early modern period. First and foremost, our thanks go to those speakers, delegates and makers. Their enthusiasm, knowledge and innovative thinking have shaped our approach here to the study of material culture. We thank the University of Warwick's Institute of Advanced Studies for supporting the conference financially, and for their support of illustrations, the Department of English and Related Literature at the University of York's F. R. Leavis Fund, the Institute of History at De Monfort University and the York Georgian Society. We thank our anonymous readers for their thoughtful and thorough advice. At Bloomsbury, we benefitted from the support and assistance of Frances Arnold, Barbara Cohen Bastos, Erin Duffy, April Peake, Shanmathi Priya Sampath, Margaret Michniewicz and James Thompson, and we remain grateful for the encouragement of series editor, Michael Yonan. Our contributors pushed forward our thinking on material literacy in new and exciting ways, and we feel fortunate to have learned from their research, writing and maker's knowledge. Finally, we thank our families and each other for listening to and supporting our successes and our blunders – pin pricks and all.

1

Introduction

Serena Dyer and Chloe Wigston Smith

All goods are made or touched by someone. Processes of production were vital to the flourishing eighteenth-century 'world of goods', its textiles, ceramics, wood and metal objects.[1] The skills and knowledge of the makers who transformed lengths of wool into fashionable coats, or intricately hand-carved wooden bedframes from pieces of mahogany, have often been minimized in narratives of the Industrial Revolution and its technological innovations. The looming figure of the factory and the roar of its machines, has diminished the manual work of human hands and positioned production as a masculine, professional and mechanized process.[2] Yet, as this volume argues, making practice, skill and knowledge were not only widespread, they were fundamental to men, women and children alike. They mattered to individuals and also to communities, provincial and urban. Manufacturing and making took place not only in workshops and factories, but in the drawing rooms, shops, parlours and backrooms of Britain. This collection upsets the familiar delineation of women's craft and men's production and makes the case for a flexible and inclusive approach to material practices. Crucially, the chapters in this volume contend that maker's knowledge was integral to the ways in which eighteenth-century people navigated the material world.

This volume sees making knowledge and skill as evidence of 'material literacy' and calls attention to the competence, knowledge and understanding of the material world which coursed across eighteenth-century society. Material literacy, we suggest, incorporates three categories of maker and making activity. Firstly, those who held the needle or chisel, and acted as active producers, physically moulding and shaping the material properties of the goods they produced. Secondly, those who guided and advised, acting in partnership with other professional or amateur makers. Finally, those who did not directly contribute to production, but who mobilized their knowledge of making to comment upon, judge and inform their own activities as consumers and owners of material objects. In recognizing these pathways by which individuals encountered and engaged with making, the term 'material literacy' encompasses the diversity of eighteenth-century material knowledge. Similarly, it celebrates the fluidity with which individuals moved between distinct types of material activity, highlighting the ways in which different making practices coexisted. Our attention to the range of eighteenth-century making is reflected in the interdisciplinary chapters gathered here. The array of scholars represented, from the disciplines of history, art history and literature, as well as museum professionals, demonstrates the multiple ways in which making knowledge could be accessed through image, practice, object, sociability, text and instruction. Material literacy cuts across conventional disciplinary boundaries, bringing a range of perspectives, sources and practices into view.

Making occupied not only professional workshops and factories, but also those spaces traditionally associated with commerce and sociability, such as shops and homes. The painting *The Ladies Waldegrave* (1780) by Joshua Reynolds, the cover image to our collection, shows a group of elite sisters within the domestic interior. Occupied by a variety of material practices, the women are intently focused on their manual work. Lady Charlotte, on the left, holds a skein of silk taut between her hands, offering it up to her sister, Lady Elizabeth, who fixes her steady look on the silk threads that she is winding onto the small card in her left hand. Lady Anna, on the right, turns her head away from us and towards the tambour frame nestled in her left arm, intently focused on the silver hook in her right hand. Further evidence of the sisters' material and textual literacy rests on the table, where a book balances on a pair of scissors and is partially covered by an embroidered silk workbag, possibly also the handiwork of the sisters. Diverging from the usual tea-tables and outdoor scenes of many conversation pieces, these women perform their femininity not through conventions of sociability, but through their manual material labour.[3] Their making practices, framed as female accomplishments, position material literacy as a domestic and artistic pursuit.[4] Reynolds's portrait invites us inside the domestic interior of making, as the women's gazes look either inwards or downwards to their work, and away from the glimpse of greenery and blue sky on the right.

Women's decorative efforts and amateur handicraft – often considered as the domain of the leisured and genteel – have frequently ringfenced making as the preserve of elite, leisured and distinctly domestic feminine hands. Scenes of feminine accomplishment, captured by *The Ladies Waldegrave*, have long stood as a synecdoche for the handiwork of eighteenth-century Britons. This collection, however, broadens the horizons of not only the age, gender and class identities of makers, but also the settings in which they started, pursued, completed or set aside their projects and products. It shows how making was experienced by eighteenth-century Britons, elite, middling and labouring. Many of our makers did indeed complete their work in refined domestic settings, by the comfort of a fire or in lushly appointed drawing rooms, as the group portrait above illustrates. Other professional makers, however, relied on workrooms and built environments for their respective trades. But, crucially, divisions between the home and the workspace were never steady nor consistent for makers, as our contributors show. Crafts, and the toxic chemicals required for them, could be fashioned at home in the drawing room or kitchen, as well as in purpose-made laboratories, ateliers and work rooms. Dressmakers stitched garments and accessories in the windows of their shops, turning their making to the full view of customers and passers-by, rather than secreting the creative process to backrooms. Makers from across the social spectrum took their work outdoors, from chairmakers and wheelwrights to genteel ladies. Lady Salisbury's watercolour of the Misses Van, for instance, depicts one sister intently focused on her needlework, her gaze cast downwards to the cloth in her hands, her bright red housewife of tools and materials resting in her lap. The sisters sit in a garden: industrious feminine activity is framed by the tree and foliage that surround them, fastening manual skill to the pleasures of nature (Figure 1.1). Examples from literature reaffirm these practices. In Charlotte Lennox's novel, *Euphemia* (1790), the eponymous heroine recounts an excursion into the woods of colonial New York, in which she and her friends take tea in 'a little valley, surrounded with lofty trees'; there 'Miss Bellenden produced her netting, Louisa her flower-piece, Mrs. Benson and I our plain-work, and Clara her book.'[5] Material literacy is distributed across age and social status: the younger, unmarried women pursue decorative work, while the married women execute the practical plain-work of household management. The bookish Clara has brought Frances Burney's *Cecilia* (1782), which she reads out loud to her industrious friends.[6]

Just as our collection unfolds around makers working in a variety of spaces, it considers also the decorative handicrafts – so synonymous with feminine making – as part of a broader rubric of making knowledge and skill. This comprehensive approach encompasses skills which enabled women and men not only to make, but to alter and mend existing material possessions. Maintaining objects – in terms of both individual things and bodies of objects – required a knowledge of making beyond the initial process of production. While women maintained the bulk of household linens, men also altered and adjusted garments.[7] Material literacy ensured the maintenance and preservation of objects throughout the lifecycle, and not only at the key moments of production and consumption, as Ariane Fennetaux deftly demonstrates in her work on pockets in this volume. Reworking and recycling added layers of meaning to objects which demanded ongoing and active material literacy.[8]

Figure 1.1 Mary Emilia Cecil, Marchioness of Salisbury, *The Misses Van*, 1791, Yale Center for British Art, Paul Mellon Collection (see Colour Plate 1).

The diverse varieties of material literacy can also be observed in the experience and skill which makers and consumers used to apprehend and understand both domestic and imported goods. Material literacy was essential to eyeing the bright, rich dyes that saturated Indian textiles, especially calicoes, in the early eighteenth century and to feeling the paper-thin muslins of the later eighteenth century. Did a piece of porcelain hail from China or was it manufactured in Chelsea? What makers' marks or patterns would help to identify the real from the fake? Which techniques could amateur makers acquire to imitate imported goods, as Hannah Robertson advised in her craft manual (discussed in Chapter 4)? Consumers had to exercise careful discernment and close scrutiny to navigate the world of eighteenth-century goods, especially when shopping by correspondence. As Robbie Richardson examines in his chapter on steel and perceptions of savagery, gaps and weaknesses in material literacy could lead consumers and collectors to misidentify objects under inspection.

We know that material literacy, like textual literacy, was never a zero-sum game, but instead ranged up, down and across a person's age, class, experience and dexterity. A person's material literacy could change over the course of a lifetime and it could surpass, sit comfortably alongside or trail behind their textual literacy. Far more women could make their marks in thread than in ink.[9] Material literacy involved a combination of manual know-how that depended on several senses; hands, eyes, ears all contributed to the various stages of transforming raw materials into finished objects.[10] Such skills often benefitted from literacy and numeracy yet reading and writing were not prerequisites for manual praxis. Histories of the reading public have documented the gender gaps for reading and writing, where rates, in England, for men (40 per cent in 1700) far exceeded those for women (25 per cent in 1700).[11] Over the eighteenth century, literacy rates for English men rose steadily (to 60 per cent in 1795), but expanded rapidly for women, almost doubling by mid-century to around 40 per cent and rising to 50 per cent by 1830.[12] As J. Paul Hunter notes, some women could read but neither write nor sign their names, and 'men were more likely to be engaged in occupations and tasks that put a premium on the ability to sign'.[13] Material literacy, however, was equally available to women and men and for some types of material literacy, women's knowledge and skills exceeded those of men. At the same time, maker's knowledge was shaped by amateur and professional practices: a male tailor would have possessed techniques and training in drafting collars and pad-stitching lapels that would not have fallen within the remit of even the most accomplished domestic needlewoman. Still she would know things he did not. The rolled hems and whipped gathers which went into the construction of caps and handkerchiefs were undoubtedly feminine practice.

It would be all too easy in stressing the literacy and close reading skills of eighteenth-century Britons to overplay analogies to reading that turn on the legibility and illegibility of objects and maker's knowledge. In addressing this range of manual practices, we are keen to resist the temptation to set material literacy against reading practices. Skill in one area did not overwrite expertise in the other. Busy hands did not lead to idle minds. Material endeavours offered opportunities for 'covert power and … radical expression' in ways not dissimilar to reading and writing.[14] In fact reading was often a crucial part of material literacy, as the chapters here show, communicating the advantages of knowledge to novices and experts alike, and providing sources for craft inspiration. Women, in particular, often engaged in literary pursuits during making: reading aloud and literary conversations were frequent partners to making; this is the case especially for those elite enough to possess the leisure time to engage in non-commercial making. In Lady Salisbury's watercolour (Figure 1.1), one sister stitches while tilting her head towards the other sister, who holds an open book in her hands. Making could be combined with other leisure pursuits, such as reading, visiting and tea-taking, that maximized and mixed the activities of women's time. It was also subject to the stresses of the time-poor: making contributed to the numerous accomplishments that women, in particular, were expected to excel in, but that competed with other domestic and sociable duties. We see this in the unfinished and abandoned projects that many makers left behind, but that were nonetheless preserved in their partially completed conditions. These projects bear the traces of their finishing points, with beginning and middles that indicate what was to come, but that were permanently interrupted by other duties and distractions. In this sense, making, particularly its temporal conditions and constraints, evokes Christina Lupton's insights into the divided time of eighteenth-century readers, where women 'in particular complained of the buzz of social obligation and communication from which they find it impossible to protect themselves'.[15]

At the same time professional makers found moments to combine making with reading, and several of our contributors address the increasing availability of print sources that supported and expanded access to maker's knowledge. From recipe books for the kitchen to voluminous, illustrated encyclopaedias, eighteenth-century Britons could improve their material literacy not only through verbal, practical training but also through text and image. Reading held out the potential to distract the attention of makers; too much print, especially the kind of stories that diverted makers from the task at hand, held the potential to damage profits and productivity.[16] In William Hogarth's 1747 series, *Industry and Idleness*, plate 1 shows 'The fellow 'prentices at their looms', with ballads stuck on the walls of the weaver's workshop (Figure 1.2). Tom Idle's shiftless and dismal future finds expression in a ballad of Daniel Defoe's 1722 novel *Moll Flanders*, which has been pinned to the loom's back beam, against which Tom Idle dozes, mouth agape. By contrast, the diligent Francis Goodchild centres his gaze downwards to his loom, ignoring the ballads, 'The London Prentice' and 'Whitington, Ld Mayor' stuck to the wall behind him. In such scenes, texts appear as part of the worker's day, a feature of their workspace. The printed pages' content and placement serve as further moralizing emblems of how the apprentices will approach their acquisition of material literacy and the degree to which they will, or won't, in Tom Idle's case, absorb knowledge and manual expertise from the master weaver.

Many of our historical makers were well-versed in multiple approaches to handiwork, even if they didn't follow the bourgeoning print market for aesthetic theory in the period. If treatises by the third Earl of Shaftesbury, Joseph Addison and Edmund Burke sought variously to explain how we see and why we respond to visual phenomena, then our makers expressed their commitment to the visual and material world through manual practice. They thus align more closely with Ruth Mack's analysis of William Hogarth's 'practical aesthetics' in *The Analysis of Beauty* (1753), where the artisan-artist articulates 'his embrace of an unusually bodily empiricism'.[17] In their acquisition of

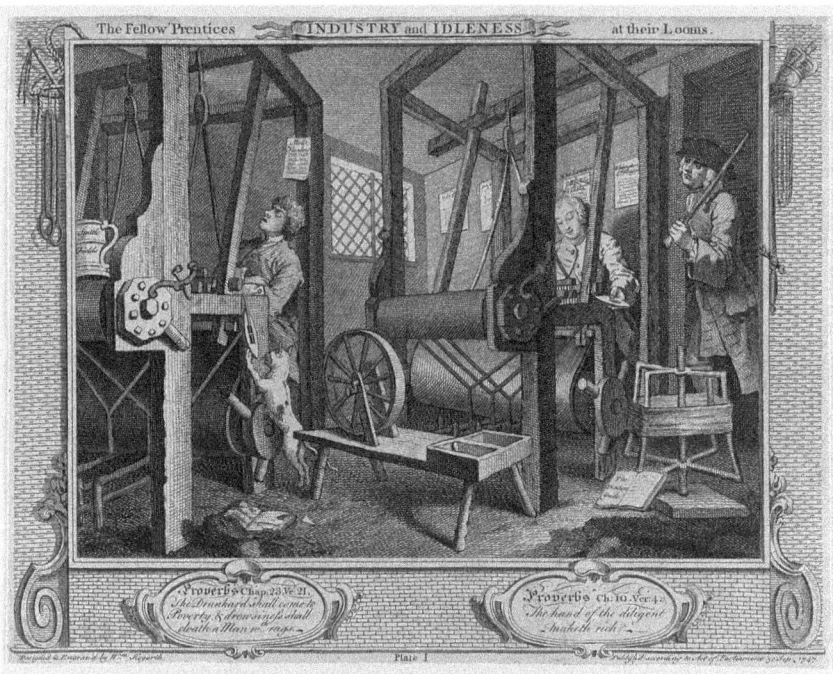

Figure 1.2 William Hogarth, 'The fellow 'prentices at their looms', Plate 1, *Industry and Idleness*, 1747. Courtesy of The Lewis Walpole Library, Yale University.

material literacy, professional and amateur makers would all have encountered the challenges of translating bodily empiricism into finished objects; bodily empiricism demanded experience, training and repetition. Many of our contributors profitably turn to Richard Sennett's work on the gaps between the language of explication and the manual demonstration of technique.[18] Our collection thus attends to the moments where ambition and skill manage not to meet, where inexperience shows or where text fails to express what the hand manages to elucidate. Material literacy was not always a satisfying experience or even always welcome to the individuals who exercised it.

In addition to book learning, eighteenth-century Britons could acquire material literacy through formal apprenticeships, whose agreements linked the language of contracts and obligation to the acquisition of manual techniques. Apprenticeship contracts for weavers, mantua-makers, furniture-makers, potters and metal-workers codified the transfer of material literacy from experienced authorities to novices.[19] Teaching and learning within the apprenticeship model relied upon the transferability of material literacy, showing, rather than telling, the apprentice how to acquire experience with the loom, potter's wheel, forge and patterned paper. Such formal agreements ran in parallel to numerous verbal and textual outlets for acquiring making skills, which superseded the often-gendered classifications of professional and amateur making. Eighteenth-century Britons found they could increasingly turn to print culture to improve their material literacy, by studying or thumbing through collections of recipes and instructions that unveiled the secrets of making. From Chippendale's *The Gentleman and Cabinet-maker's Director* (1754) to *The Female Instructor; or, Young Woman's Companion* (1811), print culture helped to bridge amateur and professional access to material literacies.[20] Such publications, far from delineating consumers from producers, facilitated the exchange of information and technique between professional and non-professional makers.[21] Aspiring makers could access such instructions in books, periodicals and individual sheets, in addition to seeking tailored tutorials in specific shops, as Beth Fowkes Tobin details in her chapter. These instructional models constitute the more formal ways that Britons could improve their material literacy, yet informal instruction, especially the transmission of making practices within households and between generations, often dominated the teaching of children, especially girls. In their chapters, Crystal B. Lake and Chloe Wigston Smith examine how the material arts found a firm place in domestic instruction, showing also that material literacy didn't always square with the obedience demanded by pious needlework.

For Serena Dyer, too, both the home and school functioned as key sites of material instruction in childhood. However, as Dyer notes, such instruction concerned not only the attainment of specific material techniques but also the development of a judicious economic eye that could assess, in adulthood, the execution of professional makers and the pricing of their wares. Several chapters in this volume, including those by Jon Stobart, Richardson, Elisabeth Gernerd and Laura Engel, also draw attention to the material literacy of the consumer and collector finding moments of mastery, but also missed opportunities for connection and comprehension. Our efforts to understand material literacy as linked to both making and consumption echo Tobin's key emphasis on the 'meaning-making potential of consuming practices' that spilled over into the making practices which buttressed them.[22] The study of material literacy offers a rich opportunity to dismantle the binary distinctions so often made between those who made things and those who bought things. This producer and consumer paradigm, frequently gendered to set masculine industrious production against feminine frivolous consumption, has dominated cultural depictions of making both in the eighteenth century and in scholarly work. At the same time, focus on the moment of purchase has emphasized Britain's characterization as a 'nation of shopkeepers'.[23] Placed on either side of the shop counter, neatly ordered into roles seen to contribute to a healthy national economy, participation in Britain's consumer culture has been built firmly around commercial transactions. Work such as Helen Berry's influential browse-bargain model has reinstated the materiality of consumer activity within the shopping process, but this model continues to position the material knowledge of browsing and consumption as disconnected from the knowledge of making.[24] Yet making and buying were always inherently interconnected, however small the project. In order to make a small work bag, silk fabric, sewing threads, embroidery silks, scissors, needles and ribbon would all need to be purchased. Most makers needed to be consumers in order to produce. The rise of sensory history and the acknowledgement of the haptic skills required to shop have demonstrated that shopping skills extended beyond economic bartering and objective assessments, but relied upon complex schemas of haptic and sensory understanding.[25] Our collection engages the full material experience of both producer and consumer practices, and recasts Britain as 'a nation of makers'.

The interconnected nature of making and buying is epitomized in a fashion plate from the 1812 edition of *The Lady's Magazine; or, Entertaining Companion for the Fair Sex*. The subject of the image, dressed in 'London fashionable full Dress', handles a decorated straw bonnet (Figure 1.3). Her white, un-gloved hand grasps the bonnet, which she intently inspects. Her haptically informed scrutiny of the bonnet conveys not a culture of casual browsing, but a process of material scrutiny so embedded in the consumption of dress that it was depicted in a vehicle of fashion dissemination. The woman's fingers, in feeling the ridges of the straw, might detect the type of straw, or the method of plaiting. She might consider the stitching, and the neatness of the application of the pleated blue ribbon which adorns the bonnet, or whether the sprig of foliage is stitched on sufficiently to prevent it detaching during wear. This consumer is materially engaged, and materially literate. Understanding the intricacies of the fibre content of a textile, the reliability of a stitching method or the cultural implications of an object all constituted material literacy.

In addressing the making practices of eighteenth-century Britons, we are also forced to contend with our own material literacy as scholars. As experienced makers, curators or material culture scholars, our contributors have approached historical making from viewpoints which are, undeniably, influenced by their engagement with material practices. A key challenge in compiling this volume has been negotiating the points at which our own embodied knowledge might aid, or even hinder, an exploration of eighteenth-century making. This volume thus sits within a movement in material culture studies which turns towards experiential research and research-as-practice.[26] Pamela H. Smith's *Making and Knowing* project at Columbia University explores the ways in which artistic making and scientific knowledge intersected in the early modern period, through continuous and methodical experimentation in the process of making, replicating recipes and processes found in contemporary texts.[27] In a similar vein, Aalto University's *Refashioning the Renaissance*, led by Paula Hohti, includes scientific experiments and reconstructions as part of the project's core activities. Indeed, *Fashioning Fashion*, the conference held at the University of Warwick in 2017 and at which this collection was conceived, included experiential making as a key facet of the programme. Within this collection, Emily Taylor's and Hilary Davidson's chapters consider the ways in which practical knowledge of construction and making methods might shape how we study the social and cultural implications of manual labour. Other contributors, including Dyer and Nicole Pohl, are experienced in the period hand-stitching that their subjects exercised. Together our contributors take seriously Leonie Hannan and Kate Smith's model of 'return and repetition', in which repeated, detailed and careful examination of the very material objects created by eighteenth-century makers will lead us towards an informed and nuanced knowledge of material labour.[28] As Fennetaux states in her opening chapter, 'If materiality carries the past, it is by becoming ourselves literate in objects … that we may hope to unpack some of the meanings they carried for the people who made and used them.'

Placing necessary editorial constraints upon these expansive notions of making and material literacy has proven no easy task. Overall our collection addresses material literacy in the long eighteenth century and several chapters cluster around the late eighteenth and early nineteenth centuries – a moment of key change for fashions in

N.º 9.   *Lady's Magazine.— September, 1812.*

London fashionable full Dress.

Figure 1.3 London fashionable full Dress, *The Lady's Magazine; or, Entertaining Companion for the Fair Sex*, September 1812, Private Collection.

clothes and the domestic interior, but also of transition in the very methods of making. Sidestepping the 'common misstep' of 'the instinctive convergence on print media to the exclusion of other sources', our contributors source their evidence from visual culture, archival records and, crucially, the material objects themselves in assembling a diverse and complex history of making.[29] Our collection is structured to move broadly from the domestic interior to public places. This flow from the private to the public enables us to trace how makers and consumers acquired, improved and exercised various forms of material literacy in different settings, and how material literacy both intersected with and diverged from print literacy. Throughout, our contributors show that material literacy and kinetic learning united, rather than separated, private and public spaces, as well as practices of production and consumption. Material literacy in the home could involve a range of haptic practices and consumer habits, such as shopping by correspondence and domestic manufacture. In the marketplace, whether at the cobbler's, tailor's, dressmaker's or printer's shop, consumers trained their skilled eyes and hands on goods and services for sale, implementing maker's knowledge acquired at home or in workshops. But these commercial spaces were also sites of learning, and material instruction occurred both in the domestic interior and on the shop floor. While this book begins in perhaps the most interior of spaces – women's tie-on pockets – and then moves to public spaces such as shops and museums, its chapters illuminate the shared material practices that cut across varied settings. These overlapping material literacies and strategies inform also our book's movement from British makers to the global commodities addressed in later chapters. Our book shows how material literacy was central to the way that producers shaped imported natural materials from around the globe and to how Britons interpreted objects that were marketed as exotic, foreign and novel. The maker who held her own hand-made pocket or the consumer who anticipated the feel and touch of a sofa might also find themselves purchasing feathers from the African continent or touching Indigenous objects from the American continent. In the role of maker, consumer or museum-goer, eighteenth-century Britons applied their material literacy to assess and interpret objects before them, whether or not they always did so accurately. Throughout, our chapters engage with the haptics skills and sensory competencies required of makers and consumers to navigate the eighteenth-century world of goods.

We begin with women's tie-on pockets in '"Work'd pocketts to my intire sattisfaction": Women and the multiple literacies of making' by Ariane Fennetaux. Tie-on pockets were discreet accessories of women's dress that were frequently hand-made. Whereas the pen and the needle are sometimes seen in opposition to each other, Fennetaux shows how eighteenth-century needlework required its own material literacy that often complemented rather than opposed text-based literacy. Little print and manuscript discussion of pockets survives, yet the making, care and use of pockets were routine features in women's lives, and sat alongside female education and print literacy, as women adapted their manual techniques to the pocket's affordances. 'Needlework verse' by Crystal B. Lake also centres on the convergences between material and textual literacies by exploring both how eighteenth-century embroidered works engaged with printed texts and how printed texts engaged with metaphors and practices of embroidery. Rather than read women's needlework as symptomatic of patriarchal oppression, Lake considers how embroidery became a means of engaging with a poetics of absorption and free association. In samplers and embroidered pictures made by known and unknown women, including Mary Linwood's uncanny museum of embroidered paintings, Lake finds evidence of a material literacy that is both aesthetically productive and politically provocative.

The following two chapters move from the pockets and needlework of individual women to consider communities of craftswomen and the economic contours of material literacy. 'Domestic crafts at the *School of Arts*' by Chloe Wigston Smith tracks the growing place of craft instruction in the numerous editions of Hannah Robertson's manual, *The Young Ladies School of Arts*. Robertson steadily prioritized the place of handicrafts over cookery recipes. For Robertson, material literacy does more than signal the virtues of female accomplishments, but rather functions as an essential economic tool, allowing women to establish themselves as professional artisans. As Smith shows, Robertson returns to material literacy as a boundless source for feminine artistic expression, moral redemption and economic survival. Nicole Pohl's '"To Embroider What Is Wanting": Making, consuming and mending textiles in the lives of the Bluestockings' addresses the different skill sets and attitudes towards the making, consumption and recycling of textiles that were shaped by costs and circumstances, drawing on the unpublished correspondence of the sisters Elizabeth Montagu and Sarah Scott, both leading members of the Bluestocking circle. Montagu employed teams of needlewomen to execute her ambitious and decorative fibre arts, where Scott maintained, mended and updated her clothes and accessories herself, relying on her sister and family for fashion advice and materials.

Next 'Material literacies of home comfort in Georgian England' by Jon Stobart brings consumer practices into the home to examine how furniture was promoted and purchased. Stobart investigates the ways in which producers and retailers sold the idea and ideal of comfort through trade cards and puff pieces in periodicals and pattern books. Their rhetoric of comfort was tied to bodily experience: how it might feel to sit in the sofa, lie on the bed or write at the desk. It depended upon a shared material literacy between producer and consumer that drew on a common interest in understanding and imagining how objects might feel to their users. Serena Dyer extends these ties between home-making and consumerism in 'Stitching and shopping: The material literacy of the consumer'. From childhood, the material literacy of girls was cultivated through didactic texts and the sewing of miniature garments for their dolls. This formative training equipped adult women consumers with the knowledge and skills required to navigate the increasingly sophisticated world of goods. Such consumer knowledge contributed to a shared material literacy between consumers and professional makers of fashionable dress. The links between production and consumption continue to be central to Alicia Kerfoot's 'Stitching the it-narrative in *The History and Adventures of a Lady's Slippers and Shoes*'. As Kerfoot demonstrates, the authors of it-narratives explicitly display their material literacy through intricate descriptions of production and consumption. In particular, *The History and Adventures of a Lady's Slippers and Shoes* attends to the workshop hierarchies in footwear production and the use of women's needlework as part of that production. In coupling the history of the production process with the literary representation of shoe-making,

Kerfoot conjures up the rich narratives around the stitched lives of shoes, re-centring the place of artisanal production and material intertexts within narratives about object-protagonists.

The following three chapters study makers of fashion, from tailors to mantua-makers and dressmakers. In 'Making, measuring and selling in Hampshire: The provincial tailor's accounts of George and Benjamin Ferrey', Sarah Howard focusses on the tailor's craft, underlining the degree of specialism and skill retained by professional makers. Provincial tailors adapted metropolitan fashions to suit the needs and lifestyles of their clients, contributing to provincial fashion, and also provided a surprising range of ready-made clothes. Tailor account books, as Howard shows, unlock tailors' comprehensive understanding of how cloth, seams and stitches would be worn, stretched and pressed by their customers' bodies and limbs. Emily Taylor, in 'Gendered making and material knowledge: Tailors and mantua-makers, *c.* 1760–1820' uncovers the material traces of now largely anonymous garment makers, examining how clothes memorialize the knowledge and material literacy of tailors and mantua-makers. Taylor studies the materials, alterations to and construction of garments to reveal the trials, errors and expertise of eighteenth-century makers. Late eighteenth-century garments illuminate the material literacy of mantua-makers and tailors, showing how their hands left evidence of experimentation and experience. Also attuned to the theme of adaptation in women's dress, Hilary Davidson's 'Dress and dressmaking: Material evolution in Regency dress construction' demonstrates how evolving women's dress styles during the late eighteenth and early nineteenth centuries were inherently connected to the material skills and material literacy required to produce them. The technical demands of the era's bust, in particular, created challenges for professional and amateur makers, as did the introduction of new textiles and trimmings. The Regency period's fashions and silhouettes emerged from makers adapting and establishing innovative techniques that transformed long-established concepts about gown construction.

The following two chapters move outward geographically to consider imported natural materials and transculturated objects. First Elisabeth Gernerd's 'Fancy feathers: The feather trade in Britain and the Atlantic world' looks at the complex circulation, trade and manufacture of feather accessories. The ostrich feather has long been seen as a symbol of late-century high fashion, prominently portrayed in print culture and painted portraiture. Yet fashionable plumes were the product of the eighteenth-century global commodity trade, skilled makers and the specialized sartorial business that developed to deal exclusively with feathers. Gernerd follows the journey of the feather, tracing it from raw material to finished fashionable accessory. In 'Tomahawks and scalping knives: Manufacturing savagery in Britain', Robbie Richardson turns to the global and cross-cultural dimensions of material literacy, drawing attention to how tomahawks and scalping knives were mass-manufactured by Europeans yet subsequently understood and represented as uniquely 'savage' objects. Richardson underlines how collectors, consumers and museum visitors understood, and misunderstood, these transculturated objects. His entangled objects illuminate the complex links between material literacy and colonialism.

The following chapter, 'The lady vanishes: Madame Tussaud's self-portrait and material legacies', by Laura Engel, turns to a different type of museum setting, the wax museum of celebrity culture, but one that was also a foreign import. Marie Tussaud is most prominently associated with her work in wax, yet Engel illustrates the range of material skills required to create her models, including her maker's knowledge of textiles, wigs, ribbons, necklaces, hair and paint. Engel locates the visual connections between Madame Tussaud's early self-portrait in wax and the work of Elisabeth Vigée Le Brun. For Engel, Tussaud's waxworks, and later photographs of her creations, produce both material and maternal legacies that give rise to a material literacy that looks across time and artistic form. Our final chapter, 'Learning to craft' by Beth Fowkes Tobin, revisits multiple themes and topics which appear across the entire volume, thus serving as a capstone to the varied forms of making that occurred at home, in the commercial setting and contributed to fashionable practices that joined manual skill to sociability. Tobin considers the popular production of shellwork, taxidermy and the watercolour paint for flower drawings to interrogate the sites of material learning where skills were obtained and refined. Weaving together the amateur activities of the maker at home, the material literacy articulated in shops and within the marketplace, and fashionable decorative arts manuals, Tobin's chapter revitalizes the overlooked communities of making practice in eighteenth-century Britain. Her emphasis on the transmission of craft knowledge centres attention on the key place of embodied learning and sensory experience to material literacy.

Similar to the other chapters in this volume, Tobin draws on diverse source material to display the pleasures of making, for both practitioners and instructors. Whether training their eyes on miniscule beads or wielding large pieces of mahogany and cumbersome lengths of fabric, eighteenth-century makers brought their bodies, their eyes, ears and hands to the manual manufacture of objects. Making could be taxing, laborious and frustrating, but it also involved sociability and learning. Eighteenth-century makers took pleasure and felt a sense of accomplishment in how their hands might turn their taste and imagination into physical objects. Their enjoyment of making limns this volume and our efforts to recover, identify and grapple with the objects and traces of eighteenth-century material literacy, in all its contours, forms and mysteries. Our collection exhibits making as a comprehensive practice – at turns complex, codified and creative – in which eighteenth-century Britons forged collective vocabularies of manual labour. Material literacy joined individuals to the broader community of creativity and production that underscored how this nation of shopkeepers was as much a nation of makers.

Notes

1   Neil McKendrick, John Brewer and John Harold Plumb, *The Birth of a Consumer Society: The Commercialization of Eighteenth-Century England* (Bloomington: Indiana University Press, 1982); John Brewer and Roy Porter, ed., *Consumption and the World of Goods* (London: Routledge, 1993).

2   On industrial manufacture in Britain, see Maxine Berg, *The Age of Manufactures, 1700–1870* (London: Routledge, 1985).

3 On activities depicted in conversation pieces, see Kate Retford, *The Conversation Piece: Making Modern Art in Eighteenth-Century Britain* (London: Yale University Press, 2017), 67.

4 Women's making and female accomplishment are extensively dealt with in Maureen Daly Goggin and Beth Fowkes Tobin, ed., *Women and the Material Culture of Needlework and Textiles, 1750-1950* (Farnham: Ashgate, 2009); Goggin and Tobin, ed., *Material Women, 1750-1950: Consuming Desires and Collecting Practices* (Farnham: Ashgate, 2009); Goggin and Tobin, ed., *Women and Things, 1750-1950* (Farnham: Ashgate, 2009).

5 Charlotte Lennox, *Euphemia*, ed. Susan Kubica Howard (Toronto: Broadview, 2008), 255-6.

6 *Euphemia* was published after *Cecilia* but set several decades prior to the American Revolution.

7 For a discussion of women's contributions to the maintenance of household linens, see Amanda Vickery, *The Gentleman's Daughter* (London: Yale University Press, 1998), 150.

8 Ariane Fennetaux, Amélie Junqua and Sophie Vasset, ed., *The Afterlife of Used Things: Recycling in the Long Eighteenth Century* (London: Routledge, 2015).

9 On the topic of non-elite women making their marks in thread, see John Styles, *Threads of Feeling: The London Foundling Hospital's Textile Tokens, 1740-1770* (London: Foundling Museum, 2010), 56-62.

10 In this sense our understanding of making squares with other interdisciplinary treatments that prioritize 'the shared, collective nature of knowledge making: the communications between different modes of cognition and between different strata of society' (Pamela H. Smith, Amy R. W. Meyers and Harold J. Cook, *Ways of Making and Knowing: The Material Culture of Empirical Knowledge* (New York: Bard Graduate Center, 2017), 7).

11 For a breakdown of literacy rates, see David Mitch, 'Education and Skill of the British Labour Force', in *The Cambridge Economic History of Modern Britain, Vol. I: Industrialisation, 1700-1860*, ed. Roderick Floud and Paul Johnson (Cambridge: Cambridge University Press, 2004), 332-56, 344. David Cressy addresses the challenges of measuring literacy rates in *Literacy & the Social Order* (Cambridge: Cambridge University Press, 1980), 42-61. For an overview of responses to Cressy's sources and methodology, see Heidi Brayman Hackel, *Reading Material in Early Modern England: Print, Gender, and Literacy* (Cambridge: Cambridge University Press, 2007), 56-62.

12 These data are drawn from Michael Suarez, 'Introduction', in *The Cambridge History of the Book in Britain*, Vol. V, 1695-1830, ed. Michael Suarez, S. J. and Michael L. Turner (Cambridge: Cambridge University Press, 2009), 1-36, 11. Suarez here relies on research into English marriage registers by R. S. Schofield, 'Dimensions of Illiteracy, 1750-1850', *Explorations in Economic History* 10, no. 4 (1973): 437-54, 445, 446. Suarez includes a detailed summary of research into literacy in England, Scotland and Wales.

13 J. Paul Hunter, *Before Novels: The Cultural Contexts of Eighteenth Century English Fiction* (New York: Norton, 1990), 74, 73. See also his extended discussion of literacy and reading practices, 62-85.

14 Amanda Vickery, 'The Theory and Practice of Female Accomplishment', in *Mrs Delany and Her Circle*, ed. Mark Laird and Alicia Weisberg-Roberts (London: Yale University Press, 2009), 99. See also discussions of surface reading, close reading

and the descriptive turn in material culture studies in Eugenia Zuroski and Michael Yonan, 'Material Fictions: A Dialogue as Introduction', *Eighteenth-Century Fiction* 31, no. 1 (2018): 1–18.

15 Christina Lupton, *Reading and the Making of Time in the Eighteenth Century* (Baltimore: Johns Hopkins University Press, 2018), 34.

16 David H. Solkin, *Painting Out of the Ordinary: Modernity and the Art of Everyday Life in Early Nineteenth-Century Britain* (London: Yale University Press, 2008), 90–1.

17 Ruth Mack, 'Hogarth's Practical Aesthetics', in *Mind, Body, Motion, Matter: Eighteenth-Century British and French Literary Perspectives*, ed. Mary Helen McMurran and Alison Conway (Toronto: University of Toronto Press, 2016), 21–46, 27. See also the comprehensive treatment of Hogarth in Ronald Paulson, *Hogarth*, 3 vols. (New Brunswick: Rutgers University Press, 1991–3).

18 Richard Sennett, *The Craftsman* (Harmondsworth: Penguin, 2009).

19 See Joan Lane, *Apprenticeship In England, 1600–1914* (London: UCL Press, 2009), especially chapter 3.

20 For discussion of Chippendale's *Director*, see Akiko Shimbo, *Furniture-Makers and Consumers in England, 1754–1851: Design as Interaction* (London: Routledge, 2016), 34.

21 Kate Smith, *Material Goods, Moving Hands: Perceiving Production in England, 1700–1830* (Manchester: Manchester University Press, 2014).

22 Beth Fowkes Tobin, 'Introduction: Consumption as a Gendered Social Practice', in *Material Women, 1750–1950: Consuming Desires and Collecting Practices*, ed. Maureen Daly Goggin and Beth Fowkes Tobin (London: Routledge, 2009), 1.

23 Adam Smith, *An Inquiry into the Nature and Causes of the Wealth of Nations*, ed. Kathryn Sutherland (Oxford: Oxford University Press, 1998), 358. See also John Benson and Laura Ugolini, ed., *A Nation of Shopkeepers: Five Centuries of British Retailing* (London: I.B. Tauris, 2003); Maxine Berg and Elizabeth Eger, ed., *Luxury in the Eighteenth Century: Debates, Desires and Delectable Goods* (Basingstoke: Palgrave, 2003); Maxine Berg, *Luxury and Pleasure in Eighteenth-Century Britain* (Oxford: Oxford University Press, 2005).

24 Helen Berry, 'Polite Consumption: Shopping in Eighteenth-Century England', *Transactions of the Royal Historical Society* 6, no. 12 (2002): 375–94.

25 Kate Smith, 'Sensing Design and Workmanship: The Haptic Skills of Shoppers in Eighteenth-Century London', *Journal of Design History* 25, no. 1 (2012): 1–10; Serena Dyer, 'Shopping and the Senses: Retail, Browsing and Consumption in Eighteenth-Century England', *History Compass* 12, no. 9 (2014): 694–703; Smith, *Material Goods, Moving Hands*.

26 For a collection of contemporary perspectives on hand-work and embodied research, see Alice Kettle and Jane McKeating, ed., *Hand Stitch: Perspectives* (London: Bloomsbury, 2011). See also Tim Ingold's emphasis on 'knowing from the inside', in *Making: Anthropology, Archaeology, Art and Architecture* (New York: Routledge, 2013), 6–11.

27 Smith, Meyers and Cook, *Ways of Making and Knowing*.

28 Kate Smith and Leonie Hannan stress the crucial place of understanding materiality to the historical examinations of objects. See Smith and Hannan, 'Return and Repetition: Methods for Material Culture Studies', *The Journal of Interdisciplinary History* 48, no. 1 (2017): 1–17.

29 Smith, Meyers and Cook, *Ways of Making and Knowing*, 4.

2

'Work'd pocketts to my intire sattisfaction': Women and the multiple literacies of making

Ariane Fennetaux

'Drank coffee and work'd pocketts to my intire sattisfaction', Gertrude Savile wrote in her diary in August 1727[1], a few days before she had cut the pockets out before making and embroidering them in the summer house of her London residence. This is a very rare reference, in writing, to a woman making pockets. Yet, for over 200 years, from at least the end of the seventeenth century to the very end of the nineteenth century, women routinely made pockets, either domestically or professionally, at a time when pockets were discreet items of clothing worn tied around the waist independently of the rest of clothing.[2] Although all women wore them, and many made them, very little written evidence has survived that testifies to the making of this object, or even its existence. Pockets were so much part of the everyday that the minority of women who had the ability and time to write rarely bothered to mention them, whilst the largely illiterate majority left no written record at all. If we were to rely on written sources alone, we would risk overlooking entirely the existence of this mundane but ubiquitous item of clothing. Whether or not women could read and write, all knew how to make a pocket, from the humblest market stall holder to the educated gentry, the landed aristocrat or the published author. These women might not have all been literate in the traditional sense of the word, but all understood the advantages and purposes of a pocket and how it should best be made to serve them. Not possessing textual literacy didn't deprive women of understanding the subtleties of material literacy when it came to pockets – or to any other home-made textile item for that matter.

Building on extensive research in UK textile collections, the chapter is underpinned by a vast body of surviving pockets that forms a critical mass of unparalleled scope and diversity ranging from the elite to the plebeian and registers evidence that cannot be easily garnered from other sources.[3] Through the particular example of tie-on pockets and the striking discrepancy between the scarcity of written sources about them and the rich material testimony embedded in the surviving artefacts themselves, the chapter underlines eighteenth- and nineteenth-century women's multiple literacies. In the process it seeks to establish the crucial importance of materiality as a source for history. It argues that scholars cannot rely on written documents alone if they are to study not only this particular object and how it was made, but more generally speaking, if they are to understand a period when material literacy was much more widespread than textual literacy. This seems to be all the more important if one is to grasp something of the life and experience of women of the period who often could not read or write and therefore did not always leave a trail in the traditional written archives. Laurel Thatcher Ulrich has called for scholars interested in women's history to extend the remit of their sources and move beyond merely 'sifting through hundreds of pages of court records, sermons, diaries, family letters, and account books, almost literally looking for needles (that is any evidence of women's lives) in haystacks of male prose'.[4] Embracing the fact that women authored poetry as well as 'stitchery' she urges:

> We might begin with Anne Bradstreet's famous line: 'I am obnoxious to each carping tongue/Who says my hand a needle better fits.' That sentence establishes a creative tension between pens and needles, hands and tongues, written and non-written forms of female expression, inviting us not only to take oral traditions and material sources more seriously … but also to examine the roots of the written documents we take so much for granted.[5]

In her footsteps, this chapter looks at one particular body of textile artefacts, namely pockets, to show how they preserve evidence about women's lives that might have no exact equivalents in traditional written archives.

As Gertrude Savile's example demonstrates, however, women sometimes clearly combined different types of literacy – textual and material. The chapter will start by looking at women as both writers and makers – whether of pockets or other accessories – by examining the many instances where women described their needlework endeavours in writing in their diaries and letters. These literate engagements with the practices of making on the part of women usefully run against the grain of too quickly opposing the pen and the needle and help to contextualize women's multiple literacies. Honing in on pockets, the chapter moves on to explore how different types of literacies sometimes cohered in the making of pockets such as when a pocket bore an inscription or was made after a pattern obtained from a magazine. But even when women could not read or did not mark their pockets, when they made pockets they still left their trace and made a mark manifesting a different type of literacy, one upon which textual sources are mostly silent. Some of that silent history can be retrieved by looking at surviving artefacts in detail. But in order to do so we need, ourselves, to become literate and learn to read material sources, to combine academia's traditional and necessary reliance on textual literacy with a level of material literacy.

Women, the pen *and* the needle

In the 1790s Anna Margareta Larpent (1758–1832), an educated ambassador's daughter married to a senior civil servant, recorded her extensive reading and theatre going in her diary. Literate enough to keep a diary, she read widely, using the pages of her journal to reflect on the novels and essays she was reading. In May 1796, as she was half way through Edward Gibbon's recently published *Autobiography* (1796), she recorded:

> I know most of the books he mentions at least generally – it brings back a train of ideas pleasant to me – the library of old books my father sat in, the volumes I used to climb up & ask him the contents of – many that at 15 or 16 I took down, tried to understand and understood enough to ask questions which used to make my father wish me a boy that I might with propriety pursue the studies they pointed out. Then I journalize & feel a sort of interest in journal writing. After tea I took my X stitch.[6]

Larpent's diaries are filled with such juxtapositions of notations on reading books of literature, philosophy or history and making petticoats, neckcloths, waistcoats or mending stockings and handkerchiefs or working aprons, chairs and carpets in embroidery, cross or tent stitch.[7] She constantly – and effortlessly – moved from one to the other, evidently as comfortable with words as she was with stitches. At one point she muses on her own versatility:

> This day and Saturday occupied in mending up shirts for Georges. I often smile at my self one hour studying with the delights such pursuits give me the next no Mrs Notable more eagerly patching up work … &c but uniting the idea of a manly with that of a feminine character has from my earliest age been my ambition. It arose in my mind from the perusal of Mrs Cockburns memoirs, a book I found in my father's library. She was a woman of great learning, a metaphysician &c yet when her domestic duties drew her from these pursuits she was most domestic & useful.[8]

Larpent may have commented on her versatility as a distinctive trait of her personality, yet the same multiple literacies were mastered by many of her contemporaries.

For many elite women, accomplished literacy and needlecraft went hand in hand, as Nicole Pohl addresses in this volume for sisters Elizabeth Montagu and Sarah Scott. Montagu and Scott's friend, Mary Delany (1700–1788), a member of elite circles including the Court and the Bluestockings, worked profusely with her needle, from spectacular embroidery to plain sewing.[9] And her letters to friends and family record her endeavours, discuss materials and patterns in detail, and manifest how the pen and the needle were not incompatible appendages in an eighteenth-century woman's hand.[10] Similarly Catherine Hutton (1756–1846), in her autobiography, aged eighty-nine, took a retrospective glance at her life listing her achievements, starting with: 'I have made shirts for my father, and brother, and all sorts of wearing apparel for myself … I have made furniture for beds, with window curtains, and chair and sofa covers; these included a complete drawing rooms set. I have quilted counterpanes and chest covers in fine white linen, in various patterns of my own inventions.' Following on from her needlework achievements, she then lists her reading and writing track record with equal, rather than superior, pride: 'I have been a reader from three years old to the present day, and I have read innumerable English books and many French.

... I have written nine volumes which have been published by Longman and Co. and three which have been published by Baldwin and Cradock and I have written sixty papers which have been published in different periodicals.'[11] Looking back, Hutton lists alongside each other her quilted counterpanes and publications, both equally shaping her sense of self and identity.

Countless such examples of the harmonious combination of literary and needlework achievements may be found in the letters and diaries of literate women over the course of the long eighteenth century and, to the surprise of some, even keenly beloved literary authors such as Jane Austen or Charlotte Brontë did needlework.[12] 'For the last three or four weeks have had weakness in my eyes; it was well for you it didn't come sooner for I could not now make petticoats, pockets and dressing gowns for any bride expectant', wrote Jane Austen to her niece Anna Austen, in 1814, as she was suffering from poor eyesight.[13] While Austen might have already been a published author of several novels, she still made petticoats, pockets and dressing gowns for her beloved niece's trousseau when the latter was preparing to marry Benjamin Lefroy. Austen was indeed an assiduous needleworker as we see from several notations in her letters – and even seems to have excelled at her needle if we are to believe her words. 'We are very busy making Edward's shirts and I am proud to say I am the neatest worker of the party', she boasted in a letter written to her sister on 1 September 1796.[14] A muslin shawl reputedly worked by her own hand, now kept in Chawton House, the house where she completed several novels, confirms that she was as skilful working embroidered delicate patterns onto the translucent fabric, as she was crafting words on the page. Her nephew, who famously wrote about his aunt's writing practices, also commented on her proficiency with her needle 'her needlework both plain and ornamental was excellent ... She was considered especially great in satin stitch.'[15] That many today marvel at or are astounded by the fact that acclaimed literary authors also relished something so seemingly menial as needlework speaks volumes about the epistemological chasm that now seems to separate the hand and the mind. Yet, as the rest of the chapter will show through the particular example of women as makers of pockets there were many ways in which making and the skills it involved actually resonated with different types of literacies – some text-based, others more material.

The multiple literacies of pocket making

Women learnt and practised needlework from a very young age. In London in the 1820s, Mary Young (c. 1790–1876) drew up plans for the education of her children including her two daughters, Lucy Jane and Emma, who, aged five, were to practise needlework for half an hour every day until they turned seven when the daily regime was increased to a full hour.[16] Tie-on pockets, which integrated most of the basic needlework techniques necessary to complete many clothing essentials such as shirts, caps and under petticoats, constituted a perfect learning tool for girls. In 1786, Charlotte Papendiek (1765–1839), a diarist and assistant keeper to Queen Charlotte's Wardrobe, recalled proudly the development of her four-year old daughter: 'She could stitch a pocket, she read prettily, and now began to write.'[17] On a par with reading and writing, knowing how to stitch a pocket was part of the rudiments of female education expected of an elite girl of four. Smaller and more manageable still was making a pocket for a doll.[18] In a 1780 pedagogical book intended for young girls written by Dorothy Kilner (1755–1836), making a pocket for a doll provides an early exercise of good domestic practice. It anticipates motherhood whilst representing selflessness and care.[19]

The association between making pockets and female education carried on into the nineteenth century. In her pedagogical novel entitled *Grandmamma's Pockets* (1849), Anna Maria Hall depicts little Annie's reformation from slipshod and scatty to industrious and provident, presided over by her grandmother and her pockets. Apart from virtuous lessons into the balance between preservation and benevolence materialized by the use of pockets, their very making was part of Annie's education. At one point she is 'intrusted with the task of stitching a pair of little watch-pockets to go inside' her grandmother's large pair of pockets, whilst she says she 'learnt to backstitch on her [mother's] dimity pockets'.[20] A vehicle for the cross-generation transmission of skills and values, pockets and their making took part in the progression from girlhood into adulthood. Undoubtedly pockets as objects to be made by girls did help the transfer and acquisition of material skills. Their presence in didactic books as tropes for good housewifery and virtuous femininity shows the lessons taught by pockets were also conveyed, at least in part, through words.

Just as pockets had a literary presence in novels for girls, patterns for pockets were available to the readers of the burgeoning female press in the eighteenth century.[21] Amidst the advertisements placed in the *London Evening Post* on 25 July 1772, readers found the promise of an 'exceeding pretty Pattern of a Lady's Pocket' to be included in the next issue of Wheble's *Lady's Magazine*, a competitor to the original *Lady's Magazine*.[22] The printed advertisement manifests how the making of material things and traditional text-based literacy could often combine, with an advertisement in one publication alerting female readers and makers to an upcoming needlework pattern in another. In November 1786, another embroidery pattern for a pocket was published by London bookseller Alexander Hogg in his own rival version of the *Lady's Magazine* entitled *The New Lady's Magazine* (Figure 2.1). Hogg was a well-established bookseller and print and map seller on Paternoster Row, the epicentre of the book trade in the late eighteenth century. Alongside a sound stockpile of classics such as John Foxe's *Book of Martyrs*, family Bibles or John Bunyan's *Pilgrim's Progress*, Hogg also carried books on geography and recent explorations, publishing, for instance, James Cook's voyage around the world that came complete with several copperplate engravings. He also published fashionable novels and magazines including *The New Lady's Magazine*, which he launched in February 1786 'embellished with no less than 3 most elegant copper plates, among which is a fashionable pattern for some useful article of dress'.[23] The 'pattern for lady's pocket' representing Vincent Lunardi's balloon ascent, which is now detached from the magazine that published it, should be recontextualized within the vibrant print culture targeted at women which was awash with novels and essays, reports of current events and exotic explorations. Clearly these patterns for making

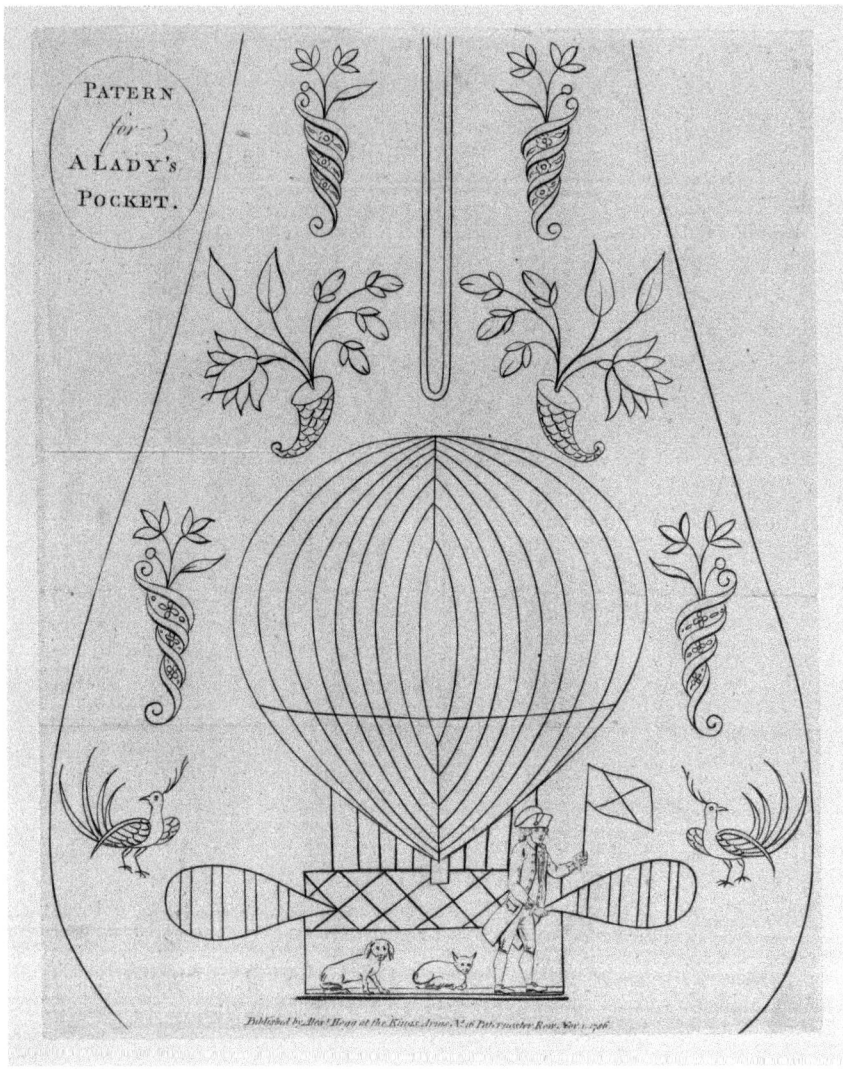

Figure 2.1 Pattern for a pocket front, 1786, Huntington Library, San Marino, California, Box 1, Folder 25b.

pockets circulated along the same pathways as printed texts, illustrating how female material and textual literacies might not be so easily disentangled. Conversely actual pockets sometimes integrated letters or words to be read on them, a reverse instance of how material literacy could dovetail with textual literacy.

If we turn to our group of extant pockets, we see many bearing an intentional mark on their surface.[24] Marks on linen, made in cross-stitch or ink, served to indicate ownership of particular items. Composed sometimes of a full name, initials or, at their most minimal, of a single letter, marks were particularly necessary when linen was put to wash in big households or when linen was entrusted with a washerwoman who took in linen from different families. Such marks also played important roles in identifying and recovering belongings after loss or theft. The practice of marking linen obviously relied on some level at least of literacy on the part of women who applied the marks but also of numeracy because marking systems frequently combined letters and numbers – with sometimes full years inscribed, as well as, in better-off households where linen was kept in multiples, isolated digits meant to identify the item in a series of similar items.

Sometimes whole names were inscribed on the front of pockets, working as joint assertions of individuality and literacy on the part of the maker.[25] Here literacy is compound in nature as it signals both a text-based type of literacy and one that is needle-based: the ability to write one's name and the ability to do so in stitch. Such is the dated and signed pocket (Figure 2.2) made by Mary Hibberd (or Hebbert) after the pattern published in the *New Lady's Magazine* in 1786 discussed above. Some names marked on pockets have clear connections to the sampler tradition and are thus a kind of *me fecit* in thread, identifying the maker and commemorating accomplishment and good work. From an early age, girls learnt to stitch and mark through the making of samplers where they practised their cross-stitch, sewing and sometimes their darning techniques.[26] These samplers, made as school or domestic exercises, were signed and dated and sometimes featured the age of the girl upon completion, a commemoration of effort, progress and accomplishment at a given point in the development of a girl's needle skills. Samplers also often carried religious overtones, with the inclusion of biblical motifs and religious mottos, as Crystal B. Lake also notes in this volume.[27] A pocket (Figure 2.3) decorated in cross-stich in the collections of the Fitzwilliam Museum clearly partakes of this tradition. Dated 1844 and signed Sarah Roberts, it carries floral cross-stitched motifs and a pious motto in a roundel that reads 'prepare to meet thy God'. It bears strong similarities to a little workbag made by the same needlewoman a year later, showing how pockets could fit into the tradition of samplers or sampler-like items recording needlework progress at regular intervals.

Despite the many instances in which textual and textile skills met in pockets, one area where they didn't overlap was that of written instructions. While written instructions on how to make other types of garments or accessories appear over the course of the eighteenth century, pockets remained with no real 'instruction manual' until 1838. Yet, pockets had been made by girls and women for at least 150 years before that.[28] Richard Sennett has highlighted the disjunction between gestures and words, between making practices and written language.[29] Words somehow always fail to accurately describe and account for making practices; they fall short of gestures and are inadequate to convey things that pertain more to a tacit understanding between the hand and the eye. The making of pockets – and the transmission of the skills

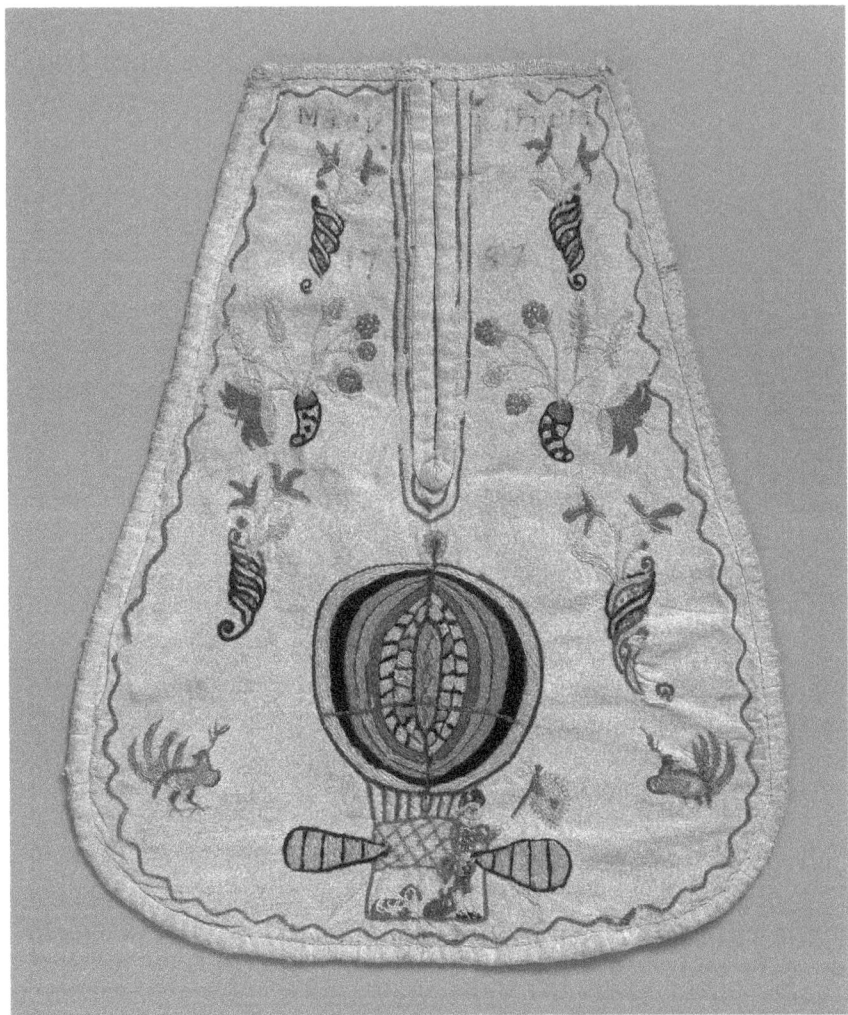

Figure 2.2 Embroidered pocket made by Mary Hibberd (or Hebbert), 1787, Museum of Fine Arts, Boston, 40.80. Gift of Mrs Samuel Cabot (see Colour Plate 2).

required to do so – mostly occurred in familiar settings such as at a mother's knee, or in informal interactions between women. To grasp something of these material practices however we can turn to the pockets themselves. The surviving pockets offer a striking illustration of making practices and their subtleties, the consistency in their shape and form for over two centuries, a clear manifestation that women did not need the written word to teach or learn how to make a pocket.

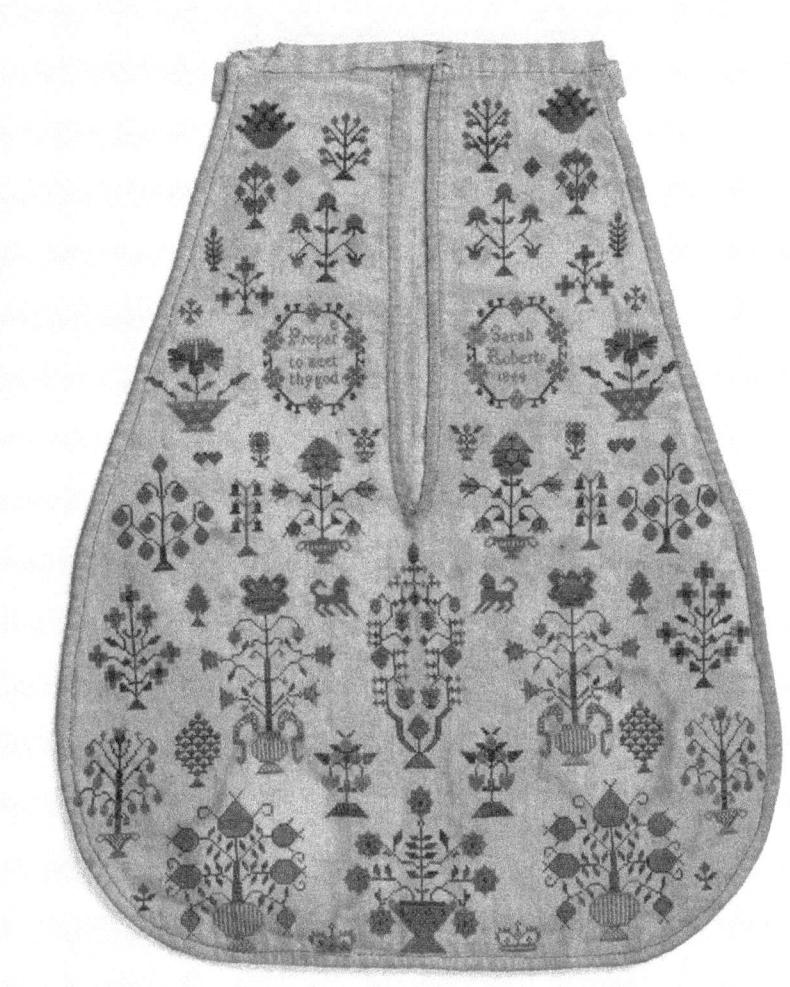

Figure 2.3 Sarah Roberts sampler-style pocket, 1844 Fitzwilliam Museum, Cambridge, T.67–1938.

Making a mark

The range of materials and techniques used on surviving specimens evidences careful, knowing choices on the part of their makers who have excluded some materials from their repertoire whilst designating others clear favourites. In the making of a pocket and their choice of cloth, technique and decoration, women had to balance practicality, aesthetics and economy. A pocket had to accommodate substantial weight and volume, so it made sense to choose a durable cloth to sustain the strain and wear. The overall majority of surviving pockets from the eighteenth and nineteenth centuries is made of woven cloth with a smooth surface with linen and cotton being the fibres of choice.[30] Striped or figured dimity lent additional strength to pockets whilst the back of pockets was sometimes made of twilled fabric instead of plain toile, a sensible choice when considering the mix of objects carried in the pocket.[31] Such sturdy fabric also offered additional protection against the pickpocketing method that consisted of cutting a whole in the bottom of the pocket with a knife or razor. If the knife was sharp and the thief agile, even strong fabric was no guarantee against theft. When Jane Searle's pocket was cut as she was watching the fountains with her children in Sadler's Wells in 1806, she was shocked to find 'the bottom [of her pocket] cut in a slit' and could only protest how provident she had been: 'it was a new pocket made with double cotton' she declared in court, evidence 'double cotton' was thought appropriate if not infallible protection against such crime.[32]

In terms of cut and construction, too, makers made shrewd choices. Weight meant that ties had to be preferably strong and firmly attached to the pocket whilst the opening, an area that sustained particular wear from the hand constantly reaching in and out for objects, was often bound for extra strength. As Linda Baumgarten reminds us, eighteenth-century needleworkers carefully adapted their stitches, materials and tools to the projected use of objects, using long loose stitches for garments expected to be recycled and refashioned, typically gowns made from silk, and 'small, tight stitches and all raw edges of the fabric turned inside' for articles worn close to the body and sustaining the strain of 'boiling, scrubbing and pounding'.[33] Falling into the latter category, pockets were closely stitched, using simple sewing techniques that were well fitted to the work done by them. So-called French, double seams or extra bindings around the side or around the opening are thus common features. Undecorated pockets form the majority of surviving pockets, but those that *are* decorated illustrate how there too women made knowing choices that balanced economy, appearance and practicality. Saving larger pieces of the fashionable printed cottons and linens that were so popular in the eighteenth and nineteenth centuries for other purposes, women seem to have used them for pockets principally as scraps in patchwork for instance.[34] This thrifty and yet decorative use of fabric shows clever deployment of resources on the part of makers, reflecting the fine balance of design and economy discussed by Serena Dyer in this collection. A pocket (Figure 2.4) made with tiny appliqué pieces of colourful cotton is a neat illustration of how even small scraps could be used to maximum effect by a skilled craftswoman. In another pocket, which is constructed from colourful bits of printed cotton tastefully arranged in a traditional pinwheel pattern, the back is made from a stamped selvedge, its maker using bits of normally unusable cloth.[35]

In all these making practices women manifested what design historians and anthropologists call 'affordance' and 'fittingness', a neat adaptation of technique and purpose.[36] Although fittingness for purpose meant there was commonality in the way pockets were made, the needlework used on them is often distinctive. As material objects, pockets embody and preserve not only a wide range of needle crafts including plain sewing, embroidery, quilting, patchwork and leatherwork, knitting and crochet

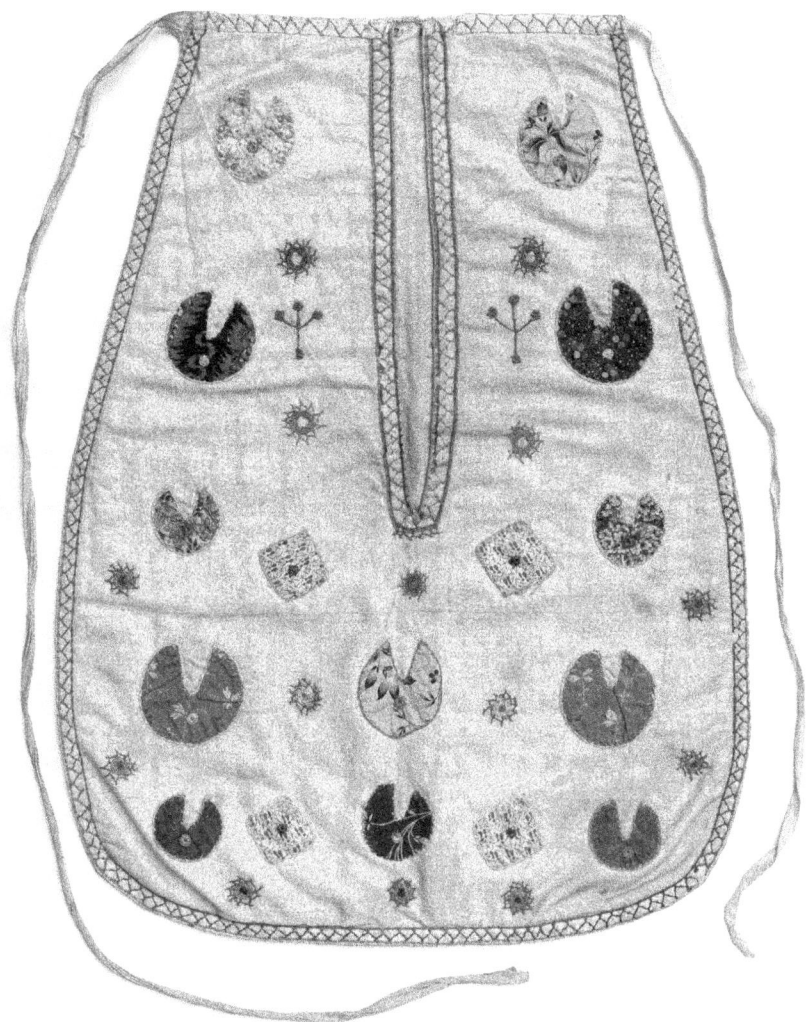

Figure 2.4 Pocket front with appliqué decoration, early to mid-nineteenth century, Worthing Museum and Art Gallery, WMAG.2029. Photo credit: Mike Halliwell.

as well as darning, patching, mending or marking, but also distinctive renditions of these. Following on from Rozsika Parker's *The Subversive Stitch*, scholars interested in women's domestic needlework have tended to see it as an expression of tension between domestic duties and self-expression, sometimes almost viewing it as an activity born of a diabolical pact between patriarchy and zealous disciplinary processes bent on breaking girls into obedient women.[37] Doubtless, domestic needlework was sometimes employed for such implicit or explicit purposes and there is evidence of pockets being made by little girls as part of their education into dutiful housewifery, Sarah Robert's pocket being a case in point. On the other hand, some pockets exemplify women's self-expressive capacity through needlework. Others bear the names of their makers or very distinctive motifs or decorations suggesting references to personal or family memory.[38] They show the needle being used by women to 'write something of their life story' or 'stitch themselves' although the narratives that these encode are sometimes difficult for us to access.[39] Alert to how physical objects, processes, meanings, memory and feelings can be experienced as inseparable, George Eliot in *The Mill on the Floss* (published in 1860, but set in the 1820s) pictures Mrs Tulliver, fearing her stock of linen will be sold at auction and weeping 'over the mark "Elizabeth Dodson" on the corner of some table cloths she held in her lap'. Mrs Tulliver looks back on the day when the weaver delivered the cloth made from yarn she had spun herself: 'And the pattern as I chose myself – and bleached so beautiful – and I marked 'em so as nobody ever saw such marking – they must cut the cloth to get it out, for it's a particular stitch.'[40] Eliot reminds us of the intimate bond between things and their makers and the evocative power that even a small stitched initial or name could hold for its creator. Much more than a mere indication of textual or needle craft literacy, the name 'Elizabeth Dodson' holds deep personal meanings for Mrs Tulliver. Interwoven with her sense of self and memory is her sensory appreciation of the material properties of the linen she has spent hours making and marking. The whiteness of the cloth, the minuteness of the 'particular stitch' she used to mark her linen all reveal that her sense of self is as much embodied in the actual name spelt out on its surface as in the act of making the mark, a specific type of intimacy fostered by the hand and eye working together. As Richard Sennett reminds us, any maker's mark, even small, even possibly invisible to or undecipherable by all but its maker, claims a presence. And even if it didn't take the shape of letters, it says 'I was there', 'I exist'.[41]

To many people today, one stitch or one sewn article might look much like another, but a closer look reveals there is more to a stitch than meets the untrained eye. Stitch by stitch, the surviving pockets reveal many differences in the skill, taste, resources and even the mood or temperament of their makers. Some pockets are assembled with botched and wild stabs of the needle, signs of an impatient, unschooled or unwilling hand. Others appear to be made by an expert but hurried hand that flew along as fast as possible. And others still show the discipline of a steady hand forming stitches of impeccable regularity in length and tension. The differences that appear when one looks closely manifest how needlework far from being uniform is, on the contrary, so deeply idiosyncratic and personal that it could in some cases act as a signature even when no name or initial was used.

Extant pockets bring to life what those women meant when they said, repeatedly in court trials for instance, they could recognize not only their mark but also their own work or indeed their own mending on stolen items. In 1794, Elizabeth Horde whose house had been broken into could positively swear in court to a pair of pockets saying 'they are mine, they are patchwork of my own doing'.[42] Mary Butler also recognized her things, telling the court in 1797: 'These things are all my property … here is an apron and petticoat of my own mending. I have no doubt that they are all mine.'[43] Women's hands-on engagement with the material nurtured a specific relation to objects where

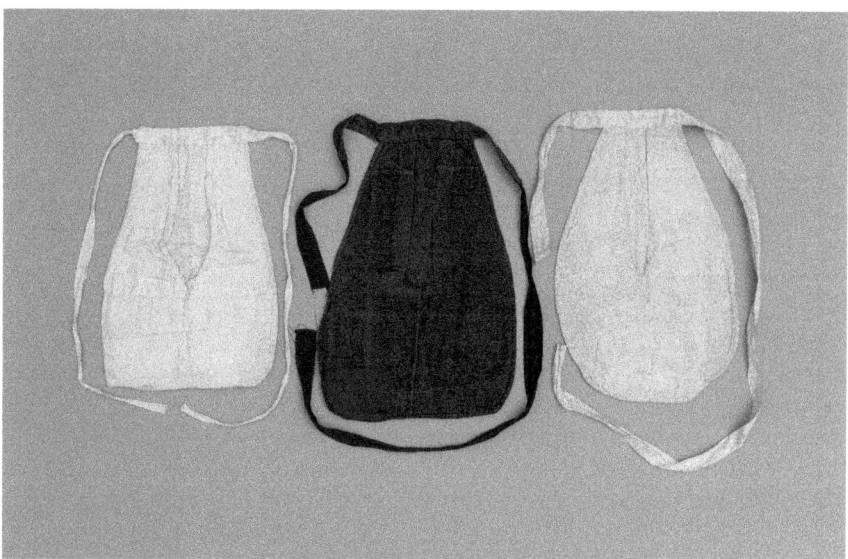

Figure 2.5 Group of late nineteenth-century pockets from the same household showing similar mending habits, National Museum Wales (left to right) 59.245.30, 59.285.9, 59.285.8.

the smallest detail of construction in something they had made, the peculiarity of a stitch, became meaningful carriers not only of identification but also of identity.

That intimate bond fostered by hours of silent labour spent on making, repairing or marking things is difficult to grasp. In the absence of words, surviving objects might be the only source which registers something of this bond and, in the case of less privileged women, something of the existence of those who left no written record behind. Claude Lévi-Strauss's *bricoleur* we are told speaks 'through the medium of things: giving an account of his personality and life by the choices he makes'.[44] Women with no formal education or access to textual self-expression may similarly be said to express something of their lives through the medium of the things they made. In the case of heavily patched up pockets, the material strategies at work to keep pockets in use similarly express individual choices that evince a presence. A group of late nineteenth-century pockets (Figure 2.5) coming from the same household all show signs of heavy repair. Despite their subtle differences, they share a similar 'rhetoric of mending' and in the similarities of their patches what can be read is the singularity of a hand, a unique presence, a signature. Whoever made and repaired these pockets is unknown to us but for her patched up pockets. Yet undoubtedly, in the palimpsest of patches made on the pockets she has left her mark.

If materiality carries the past, it is by becoming ourselves literate in objects, by attempting to 'read artefacts' and understand their specificities, that we may hope to unpack some of the meanings they carried for the people who made and used them. A stain, a darn or a stitch are as expressive and evocative as any kind of textual evidence if one cares to look. They, too, can be read. These small details speak volumes not only about making and technology but also about the sociality of the object, its life as a material artefact which was made, used, worn, damaged, repaired, discarded or preserved, restored, altered. The artefact, rips and all, forms a material text that calls for interpretation. It demands new skills be acquired or developed in order to be able to attempt what French textile conservator Emmanuelle Garcin has called a 'philology of materiality' – the study of material objects as texts to be deciphered and interpreted.[45] We must attempt as scholars to become as literate with objects as we are with texts and images. The task is not easy and demands that we be ready to adopt a 'slow approach to seeing'.[46] And just as with textual sources, reading can differ in style. Literal readings of objects – the merely descriptive rather than interpretative and contextualized – have sometimes dominated interpretations, misleading scholars into thinking that materiality held little epistemological value. It is high time, however, we faced the challenges of material literacy and embraced what it can bring to our understanding of the past. Objects are 'not only alternative, but imperative sources' for the writing of history.[47] If we agree with Sennett that 'anonymous workers can leave traces of themselves in inanimate things', being able to engage with these traces is crucial to excavating the lives of those women usually silenced in the traditional written sources.[48]

Notes

1   Gertrude Savile, *Secret Comment: The Diaries of Gertrude Savile 1721–1757*, ed. Alan Saville (Nottingham: Kingsbridge History Society, 1997).

2   The appearance of tie-on pockets of the kind discussed in this chapter is difficult to date precisely. The study carried out by this author has focused on the 1660–1900 period but there is evidence to suggest that at least in continental Europe, they might have existed before, see Rebecca Unsworth, 'Hands Deep in History: Pockets in Men and Women's Dress in Western Europe', *Costume* 51, no. 2 (2017): 148–70.

3   The research project, led by Barbara Burman and funded by the AHRC, has resulted in an exhibition, catalogue and monograph. Barbara Burman and Ariane Fennetaux, *The Pocket: A Hidden History of Women's Lives 1660–1900* (London: Yale University Press, 2019).

4   Laurel Thatcher Ulrich, 'Of Pens and Needles: Sources in Early American Women's History', *The Journal of American History* 77, no. 1 (1990): 200.

5   Ibid., 202. See also Crystal B. Lake's discussion of verse and embroidery in this collection.

6   Anna Margaretta Larpent, Diaries, Huntington Library, San Marino, MS HM 31201, Vol. 2, 33.

7   Mary Anne Garry, '"After They Went I Worked": Mrs Larpent and Her Needlework, 1790–1800', *Costume* 39, no. 1 (2005): 91–9.

8   Anna Margaretta Larpent, Diaries, Vol. 2, 21.

9   L. Llanover, *Specimens of Rare and Beautiful Needlework Designed and Executed by Mrs Delany and Her Friend, the Hon Mrs Hamilton Photographed from the Originals in the Possession of Lady Hall of Llanover* (London: Dickinson, 1860); Ruth Hayden,

Mrs Delany, Her Life and Her Flowers (London: British Museum Publication, 1992); Mark Laird and Alicia Weisberg-Roberts, ed., *Mrs Delany and Her Circle* (London: Yale University Press, 2009).

10   Mary Delany, *Autobiography and Correspondence of Mary Granville, Mrs Delany, with Interesting Reminiscences of King George III and Queen Charlotte*, ed. Lady Llanover, 3 vols. (London: Richard Bentley, 1861); Mary Delany, *Autobiography and Correspondence of Mary Granville, Mrs Delany*, Second Series, ed. Lady Llanover, 3 vols. (London: Richard Bentley, 1862). For a discussion of the pen and the needle trope, see Ulrich, 'Of Pens and Needles'; Rozsika Parker, *The Subversive Stitch: Embroidery and the Feminine* (New York: Routledge, 1984). See also Maureen Daly Goggin, 'Stitching a Life in "Pen of Steele and Silken Inke": Elizabeth Parker's circa 1830 Sampler', in *Women and the Material Culture of Needlework and Textiles, 1750–1950*, ed. Maureen Daly Goggin and Beth Fowkes Tobin (London: Routledge, 2009), 36–48; Maureen Daly Goggin, 'One English Woman's Story in Silken Ink: Filling in the Missing Strands in Elizabeth Parker's circa 1830 Sampler', *Sampler and Antique Needlework Quarterly* 8, no. 4 (2002): 8–49.

11   Catherine Hutton, *Reminiscences of a Gentlewoman of the Last Century*, ed. C. H. Beale (Birmingham: Cornish Bros., 1891), 213.

12   For references to Jane Austen as a needleworker, see Deirdre Le Faye, *Jane Austen's Letters*, 3rd ed. (Oxford: Oxford University Press, 1995). See also Hilary Davidson, 'Reconstructing Jane Austen's Silk Pelisse', *Costume* 49, no. 2 (2015): 198–223; Serena Dyer, 'Trained to Consume: Dress and the Female Consumer in England 1720–1820' (PhD Thesis, University of Warwick, 2016), 249–50. On the Brontës and needlework, see Sally Hesketh, 'Needlework in the Lives of the Brontë Sisters', *Brontë Society Transactions* 22, no. 1 (1997): 72–85.

13   Letter from Jane Austen to Anna Austen Lefroy, July 1814, in *Jane Austen's Letters*, ed. Deirdre Le Faye (Oxford: Oxford University Press, 1995), 643.

14   Letter to Cassandra, *Jane Austen's Letters*, 7.

15   J. E. Austen-Leigh, *A Memoir of Jane Austen and Other Family Recollections*, ed. Kathryn Sutherland (1871; Oxford: Oxford University Press, 2002), 77. See also Jane Taylor, '"Important Trifles": Jane Austen, The Fashion Magazine, and the Inter-Textual Consumer Experience', *History of Retailing and Consumption* 2, no. 2 (2016): 113–28.

16   Mary Young Papers, 'School Book', Museum of London, 48.85/6.

17   Mrs Papendiek, *Court and Private Life in the Time of Queen Charlotte: Being the Journals of Mrs Papendiek, Assistant Keeper of the Wardrobe and Reader to Her Majesty*, 3 vols., ed. Mrs Vernon Delves Broughton (London: Richard Bentley & Son, 1887), Vol. I: 280.

18   Making garments for dolls had much in common with needlework exercises that took the form of small-scale garments, as Serena Dyer notes in her chapter in this collection. See also Vivienne Richmond, 'Stitching the Self: Eliza Kenniff's Drawers and the Materialization of Identity in Late-Nineteenth-Century London', in *Women and Things, 1750–1950: Gendered Material Strategies*, ed. Maureen Daly Goggin and Beth Fowkes Tobin (Farnham: Ashgate, 2009), 43–54; Vivienne Richmond, 'Stitching Women: Unpicking Histories of Victorian Clothes', in *Gender and Material Culture in Britain since 1600*, ed. Hannah Greig, Jane Hamlett and Leonie Hannan (Basingstoke: Palgrave, 2016), 90–103. See also Ariane Fennetaux, 'Transitional Pandoras: Dolls in the Long Eighteenth Century', in *Childhood by Design: Toys and the Material Culture of Childhood*, ed. Megan Brandow-Faller (London: Bloomsbury, 2018), 47–66.

19   Dorothy Kilner, *Dialogues and Letters on Morality, Oeconomy and Politeness* (London, 1780), 21.

20   Anna Maria Hall, *Grandmamma's Pockets* (Edinburgh: William and Robert Chambers, 1849), 44, 156.

21   On women's magazines, see Jennie Batchelor and Manushag N. Powell, ed., *Women's Periodicals and Print Culture in Britain, 1690–1820s: The Long Eighteenth Century* (Edinburgh: Edinburgh University Press, 2018). See also Alison Adburgham, *Women in Print, Writing Women and Women's Magazines from the Restoration to the Accession of Victoria* (London: Allen and Unwin, 1972). On patterns in the publication, see Jennie Batchelor, 'Patterns and Posterity: Or What Not in the Lady's Magazine', *The Lady's Magazine (1770–1818): Understanding the Emergence of a Genre* (blog), 27 April 2015, https://blogs.kent.ac.uk/lady's-magazine/2015/04/27/patterns-and-posterity-or-whats-not-in-the-ladys-magazine, accessed 22 May 2020; Chloe Wigston Smith, 'Fast Fashion: Style, Text, and Image in Late Eighteenth-Century Women's Periodicals', *in Women's Periodicals and Print Culture in Britain 1690–1820s*, ed. Jennie Batchelor and Manushag N. Powell (Edinburgh: Edinburgh University Press, 2018), 440–57.

22   Advertisement in *General Evening Post* (London), Saturday 25 July 1772. This short-lived monthly magazine was a pirated version of the original *The Lady's Magazine* published by John Wheble, one of the original publishers of the magazine, after John Coote sold his interests to Robinson and Roberts. Despite losing his case in court against Robinson and Roberts in July 1771, John Wheble carried on his own version of the magazine until December 1772. Jennie Batchelor, 'Robinson and Roberts vs Wheble: Periodicals and Piracy', *The Lady's Magazine (1770–1818): Understanding the Emergence of a Genre* (blog), 24 November 2014, https://blogs.kent.ac.uk/ladys-magazine/2014/11/24/robinson-and-roberts-vs-wheble-periodicals-and-piracy/, accessed 22 May 2020. See also Adburgham, *Women in Print*, 142–4; Jan Fergus, *Provincial Readers in Eighteenth-Century England* (Oxford: Oxford University Press, 2006), 200, no. 5, 201. The pattern has just been found by Jennie Batchelor in the Bayerische Staatsbibliothek, inv: Per. 123 m-3.

23   *A New Catalogue of Valuable, Useful, Instructive and Entertaining Books* now publishing by Alex. Hogg, at the King's Arms, n°16 Paternoster Row (London, 1786).

24   For further analysis of these marks, see Burman and Fennetaux, *The Pocket*, chapter 3. See also Barbara Burman, '"A Linnen Pockett, a Prayer Book & Five Keys": Approaches to a History of Women's Tie-on Pockets', in *Textiles and Text: Re-Establishing the Links between Archival and Object-Based Research*, ed. Maria Hayward and Elizabeth Kramer (London: Archetype Publications, 2007), 157–63.

25   See, for instance, Pair of sampler-style pockets, mid-nineteenth century, Carmarthenshire County Museum, CAASG.1976.3668 or Elizabeth Reynolds's knitted and embroidered pocket, mid-nineteenth century, York Castle Museum, YORCM: 1953.284.9.

26   On darning samples see Vivienne Richmond, *A Remedy for Rents: Darning Samplers and Other Needlework from the Whitelands College Collection*, exhibition booklet (London: Constance Howard Gallery, Goldsmiths, 2016).

27   On samplers see Edwina Ehrman, *The Judith Hayle Samplers* (London: Needleprint, 2007); Carol Humphrey, *Sampled Lives: Samplers from the Fitzwilliam Museum* (Cambridge: Fitzwilliam Museum Publications, 2017); Helen Wyld, *Embroidered Stories. Scottish Samplers* (Edinburgh: National Museums Scotland, Edinburgh, 2018); Chloe Wigston Smith, 'Gender and the Material Turn', in *Women's Writing*

1660–1830: Feminisms and Futures, ed. Jennie Batchelor and Gillian Dow (Basingstoke: Palgrave Macmillan, 2016), 159–78.

28 A Lady, *The Workwoman's Guide*, facsimile of the 1840 edition (London: Bloomfield Books, 1975).

29 Sennett, *The Craftsman*, 179–80.

30 For a more extensive analysis of the cloths chosen for pockets, see Burman and Fennetaux, *The Pocket*, chapter 2.

31 Money, boxes or tools are examples of such contents. For a full exploration of contents, see Burman and Fennetaux, *The Pocket*, chapter 4.

32 *The Old Bailey Proceedings Online*, 1674–1913, www.oldbaileyonline.org, version 7.0. 24 March 2012, hereafter OBP, OBP t18060917-16.

33 Linda Baumgarten, *What Clothes Reveal: The Language of Clothing in Colonial and Federal America, the Colonial Williamsburg Collection*, The Colonial Williamsburg Foundation (London: Yale University Press, 2002), 40.

34 Our survey, conducted in British collections, contains only six patchwork pockets in public museums: Bowes Museum Barnard Castle County Durham, inv. CST.2.523.1981.25.5; Charles Paget Wade Collection, Berrington Hall, National Trust, inv. SNO 1454; Manchester Art Gallery, inv. MCAG.1947.1262 and inv. MCAG.1947.1250 and National Museums Scotland, inv. H.UF 91. One is in a private collection.

35 National Museums Scotland, inv. H.UF 91.

36 Ian Hodder, *Entangled: An Archaeology of the Relationships between Humans and Things* (Chichester: Wiley-Blackwell, 2012), 114–15. On affordance see also James Jerome Gibson, *The Ecological Approach to Visual Perception* (Dallas: Houghton Mifflin, 1979); Donald Norman, *The Psychology of Everyday Things* (New York: Basic Books, 1988).

37 Parker, *The Subversive Stitch*. See also Lake on Parker in this volume.

38 See, for instance, Nottingham City Museums and Galleries, NCM 1964–35.

39 For a particularly striking example of a woman 'stitching her life', see Elizabeth Parker's autobiographical sampler in the V&A, inv.T.6–1956. For an analysis of this extraordinary piece, see Goggin, 'One English Woman's Story in Silken Ink: Filling in the Missing Strands in Elizabeth Parker's circa 1830 Sampler'; Goggin, 'Stitching a Life in "Pen of Steele and Silken Inke": Elizabeth Parker's circa 1830 Sampler'; Nigel Llewellyn, 'Elizabeth Parker's Sampler: Memory, Suicide and the Presence of the Artist', in *Material Memories: Design and Evocation*, ed. Marius Kwint, Christopher Breward and Jeremy Aynsley (Oxford: Berg, 1999), 59–71.

40 George Eliot, *The Mill on the Floss* (1860; Harmondsworth: Penguin, 1985), 282.

41 Sennett, *The Craftsman*, 130, 134.

42 OBP t17941208-22.

43 OBP t17971025-5. Another example is Sarah Smith who swears to a shift being her property 'by the mending', see OBP t17900915-51.

44 Claude Lévi-Strauss, *The Savage Mind* (London: Weidenfield and Nicolson, 1966), 21.

45 Emmanuelle Garcin, 'Le Costume comme source historique: Pour une philologie de la matière' in *Fabrique de l'habit: Artisans, techniques et production du vêtement (fin du Moyen Âge-XVIIIe siècle)*, ed. Astrid Castres and Tiphaine Gaumy (Paris: Publications de l'École nationale des chartes, forthcoming). See also Garcin, 'La Restauration des matériaux textiles: la fibre du métier', *Patrimoines, Revue de l'Institut National du Patrimoine* 6 (2010): 78–85, 78.

46   Ingrid Mida and Alexandra Kim, *The Dress Detective: A Practical Guide to Object-Based Research in Fashion* (London: Bloomsbury, 2015), 33.
47   This evocative phrase was used by Josephine Tierney in 'The Design and Trade of British Textiles to West Africa, *c.* 1830–1914', paper given at the Pasold Conference, Museum of London, 16 November 2018.
48   Sennett, *The Craftsman*, 10.

3

Needlework verse

Crystal B. Lake

In 1974, the artists Kate Walker and Sally Gallop were friendly neighbours in London. When Gallop moved to the Isle of Wight, the two women began exchanging small, homemade artworks through the post. In 1975, Walker spoke at the Women's Art History Conference where she invited other women to join her and Gallop's art exchange, and more than twenty women joined their Postal Art Collective.[1] Much of what the women exchanged were needleworks, and the artists' experiments with needlework were the most notable and influential art objects that their collective produced; in particular, the Postal Art Collective's second-wave feminist experiments in embroidery especially shaped contemporary interpretations of seventeenth- and eighteenth-century needlework.[2]

The historical practices of embroidery have since been recovered, but only aslant. Although women artists and art historians began by celebrating the technical proficiency that early works of embroidery entailed and, occasionally, praising the aesthetic judgment they exhibited, second-wave feminists also established pre-twentieth-century needlework as a symptom of women's oppression. Contrasting the needle with the pen has been central to this seemingly inescapable interpretation of historical embroidery's compromised availability for feminist art. Where needlework has been taken to have disciplined women by confining their artistry to the domestic sphere, women's writing has been taken to signify a rejection of such disciplining and to have enabled a more uncompromised aesthetic mode for advancing women's resistance to patriarchal oppression as a result.[3]

Opposing pens to the needles, however, threatens to erase the writing that appeared on women's embroidery as well as the important role that embroidery played in literacy instruction, as Ariane Fennetaux also notes in her chapter for this volume. Embroidered works produced during the long eighteenth century remind us that learning to read and write frequently entailed handling not only pens, ink and paper but also needles, thread and fabric.

As Chloe Wigston Smith and Serena Dyer explain in their introduction, however, 'skills' like embroidery 'often benefitted from literacy and numeracy yet reading and writing were not prerequisites for manual praxis'. Embroiderers could allude to printed texts without necessarily reading them. Letters, after all, could be copied as shapes. Embroiderers could also illustrate texts that had been read to them out loud and craft narratives or poetic expositions – either of their own or of an adaptive, allusive nature – out of images and combinations of patterns.

In this way, the material literacy on display in the embroidered works that were produced during the long eighteenth century leads us to discover what we might describe as a poetics of needlework. More specifically, embroidery from the period can be seen to prefigure aesthetic modes that we tend to characterize today as a twentieth-century phenomenon: variations of surrealist or avant-garde sensibilities that find political as well as artistic purchase in dreamlike states of absorption. The sensibilities that emerge in the entanglements of texts and textiles in the long eighteenth century make it possible to see how embroidery's feminist critiques, in the past and still today, can operate via form as well as content – despite the patriarchal imperatives that have previously been assigned to women's needlework. The material literacy embodied in the embroidered works from the period, in other words, foregrounds a material poetics where it was possible to prioritize the associative relationships between objects, images and ideas over the conscious, reasoned logics we often associate with the period's popular texts.

Pen and needles

As Amy Tobin documents, the works of art that Walker and Gallop exchanged in the 1970s were consciousness-raising meditations on their everyday experiences as women, housewives, mothers and artists.[4] Their art was small and quickly made, usually out of ready-to-hand, recycled materials.[5] Walker's *Sampler* has come to stand as a representative example of what the *Postal Art* Collective exchanged: a small piece of quilted embroidery that features a miniature, erect, disembodied penis sprouting five sunrays or petals in the upper left-hand corner, followed by the phrase 'WIFE IS A FOUR LETTER WORD' stitched in big, bubble, capital letters.[6] Needleworks like *Sampler* were commonly exchanged by Walker and Gallop as well as by Beryl Weaver and Sue Richardson.[7]

Rozsika Parker, an art historian who founded the Women's Art History Collective, personally knew many of the artists who were involved in the *Postal Art Event*, and their work inspired Parker's important 1984 study, *The Subversive Stitch: Embroidering and the Making of the Feminine*. In the *Subversive Stitch*, Parker recovered the long history of needlework's association with femininity in order to assess the feminist turn to embroidery that occurred in the 1970s. Now taken as an exemplary work of feminist art history, *The Subversive Stitch* grew directly out of Parker's association with artists like Walker, whose *Sampler* Parker reproduced and singled out for analysis.

Although Parker admired the skill that embroidery historically entailed, she found little else to celebrate about women's needlework before the twentieth century. Accordingly, Parker locates the origins of second-wave experiments in feminist embroidery in the twentieth century. For Parker, the feminist needlework artists of the 1970s were inspired by the handstitching of Suffragettes and the textile experiments of

Russian Constructivist, Dadaist and Surrealist artists, more so than by the needlework that had come before. As Parker explains, 'embroidery' had become 'entirely fused' by the nineteenth century with an ideal of 'femininity' characterized by 'docility, obedience, love of home, and a life without work' (11). Hence, feminist artists of the 1970s aspired to reposition their needleworks as artworks defined by technical, aesthetic and political acumen and influenced by twentieth-century activists and artists. Their artworks were not the products of leisurely time, nor were they expressions of 'docility', 'obedience' or a 'love of home'. Rather, feminist's needlework art staged a stark contrast between the medium and the message. By featuring an embroidered penis alongside the motto, 'WIFE IS A FOUR LETTER WORD', a work like *Sampler* seized on the medium of needlework – rife with the history of women's oppression – as an uncanny and therefore especially effective vehicle for its message of feminist resistance.

In *The Subversive Stitch*, however, Parker suggests that conceiving of needlework as resistance before the twentieth century largely meant resisting doing needlework in the first place. Parker notes, for example, that Lucy Hutchinson 'hated' her needle (84), that Ann Fanshaw reported she would have much rather played outside (83), and that Mary Wollstonecraft declared that needlework 'contract[ed] [women's] faculties' (139). 'Embroiderers', she writes, 'employed the needle, not the pen,' and so 'they left no record of their attitudes toward their subject matter' (102). Parker maintains that because we have no written record that early embroiderers worked 'in *conscious* opposition to their ordained role, in rebellion against the inactivity, immobility and obedience enforced by embroidery itself', that we cannot say that they were 'proto-feminists' (102). Feminist histories of needlework that appeared in the wake of *The Subversive Stitch* likewise 'link[ed]' the pen to the needle 'oppositionally'. As Kathryn King notes: '[T]he needle conveys metonymically the constraints, frustrations, and sheer tedium of the domestic role. And when set against the activities of the pen, needlework ... always assumes negative values.'[8] In 2000, Ann Rosalind Jones and Peter Stallybrass read the relationship between needles and pens in the early modern period a little more optimistically than King had in 1995, but only a little. For some women, the needle '*could* be a pen'.[9]

More recent histories of needlework have reassessed the relationships between early women's texts and textiles. Susan Frye observes in *Pens and Needles* that 'the insistence' second-wave feminists placed 'on an absolute split between the pen and the needle' doesn't quite hold.[10] Frye reconsiders early modern women who conflated their needles with their pens, turning to Margaret Cavendish as one representative example. In the first dedication to her *Poems and Fancies* (1653), Cavendish declared that the 'studying or writing Poetry' is 'Spinning with the Braine'.[11] In another dedication, one she wrote for her female readers, Cavendish argued something similar: that women are especially adept poets because poetry is 'built upon the fancy', and women's 'brains work usually in a fantastical motion' as evidenced by the many '*Curious*' things they make out of 'divers sorts of *Stitches*'.[12] If Cavendish here appears to confirm that needlework could be used to discipline women's minds as well as their bodies, her rendering of the poetic imagination as so much thread to be spun out was also part of a larger conversation about the minds of poets and the nature of poetry.[13] In other words, Cavendish represents poets as needleworkers who spin the threads of fancy into poems not only because she was a woman but also because the materials commonly used in needlework were also commonly used as metaphors for the mind and the process of poetic composition and comprehension.

Embroidered minds and embroidered verse

John Locke celebrated the disciplining power needlework could have on women's minds. In his *Thoughts Concerning Education* (1693), he describes a scene where women chose to 'employe themselves in Needle-work' – rather than spend their time gambling. As the women stitched they also 'read, in their turns, such an entertaining and instructive piece, as may give rise to Reflections, on which everyone passes her Judgment'.[14] Locke goes on to devote an entire, albeit brief, section of the *Thoughts* to needlework, explaining that he does not 'expatiate' on needlework's virtues because it is taken for granted as a necessary component of women's moral as well as practical education.[15] Readers of Locke's philosophy would have readily recognized that Locke also understood cognition as metaphorically akin to needlework insofar as the mind combined discrete sensory information like so many stitches into a tapestry of ideas and a network of associations. In the section of his *Essay Concerning Human Understanding* (1690) devoted to parsing 'simple modes', Locke considers 'how from simple *Ideas* taken in by Sensation, the Mind comes to extend itself even to Infinity'.[16] Famously, for Locke, our senses apprehend basic ideas from external objects; these are the 'Materials of all our Knowledge'.[17] We build more complex ideas by reflecting on and combining simple ideas.[18] This process of idea-making was difficult, however, to delineate because, as Locke himself realized, our senses work in tandem, and we often perceive multiple phenomena at once. Locke explains, then, how 'single modes' of sense perception can also function as 'mixed modes'. Locke takes one colour composed of many shades as an analogy to illustrate what he means. We comprehend multiple shades of a colour simultaneously as types of a single-colour kind, and we rarely encounter a single colour itself as an isolated phenomenon.[19] Moreover, when our senses take in colour, 'Figure is taken in also' and contributes to our understanding of the colour we think we see – as happens, Locke explains, when we encounter examples of 'Painting, Weaving, Needle-works, etc.'[20] For Locke, this proves 'how remote soever [an idea] may seem to be from any Object of sense or Operation of our mind', the idea itself 'nevertheless' originates in an external object; we perceive many things at once, record them as distinct phenomena and then combine them into increasingly complex ideas.[21]

Locke famously preferred to use a blank sheet of paper and the malleable substance of metal for his metaphors of mind, but his brief allusion to needlework as a way of explaining simple modes, mixed modes and perception reflects the conventional uses of needlework and its materials as metaphors in the period's philosophy. In the *Principles of Philosophy* (1644), for example, René Descartes imagined the brain as a 'twisted' skein and the nerves as 'threads' that extended to the body's limbs.[22] In the *Cyclopaedia* (1728), Ephraim Chambers declared that 'all of the Organs' of the body 'consist of little Threads' which originate 'in the Middle of the Brain' and 'terminate' in the body's 'exterior' and were the conduits for the transmission of images to the imagination.[23] Benjamin Martin's *Philosophical Grammar* (1735) likewise attested that 'the innumerable Divisions, Windings, serpentine Progressions, and frequent Inoculations of Veins with Veins, and Arteries with Arteries' in the body 'make a most agreeable Embroidery'.[24] In the second volume of *The Fable of the Bees* (1729), Bernard Mandeville wrapped together the needlework metaphors used to characterize anatomy, 'the close Embroidery of Veins and Arteries that environ the Brain', with Locke's representation of the mind as a 'slate to cypher' – 'or', as Mandeville puts it: 'a Sampler to work upon'.[25]

Two episodes in *The Spectator* – one depicting the dissection of a woman and the other the dissection of a beau – suggest that such metaphors were so common as to be readily accessible in allegorical satires. Addison makes quick work of using needlework in both of these episodes to characterize minds and bodies. In the first episode, Mr Spectator reports that he and a friend are eager to 'dissect a Woman's Tongue' in order to determine 'whether the Fibres of it' are made of 'Thread' or not.[26] A year later, Addison offers a second, similar episode. He narrates a dream in which he attends the dissection of a 'Beau's Head'. Inside, Addison finds 'a Heap of strange Materials', including a 'Cavity … filled with Ribbons, Lace, and Embroidery, wrought together in a most curious piece of Network' (2:570–1).

Addison here shows that Cavendish likely seized on the image of poetry as a spinning of the brain in order to rebrand stereotypical femininity as philosophy and women's writing as smart artistry. In this regard, Addison's allegories and Cavendish's turn of phrasing both illustrate the claim that Parker and other second-wave feminist scholars have made: that by the eighteenth century, needlework had become thoroughly associated with femininity and concordantly denigrated as a mere craft, well-suited for the lesser sex's lesser mind. Another essay from *The Spectator* is frequently cited to illustrate this point. A beleaguered aunt writes to Mr Spectator, begging him to extol the virtues of needlework. The aunt herself has 'plied [her] needle these fifty years, and by [her] good will would never have it out of her [hand]'; she would like Mr Spectator to convince her 'flirt[y]' nieces to do the same (5:71). Mr Spectator obliges, and the essay (which is attributed to Thomas Tickell and not Addison) notes that embroidery is good for women because it keeps them busy. Women who do needlework are not bored and so they do not gossip or meddle in politics. Above all Tickell praises the moral education – if not the education in reading and writing praised by Locke – that needlework offered to girls by suggesting that a woman might do worse than spend her whole life embroidering scenes from the Bible.[27]

Such associations between femininity and needlework were rendered, however, with more complexity than they might at first seem. The aunt asks Mr Spectator to meditate on the 'laudable *mystery* of embroidery', and Tickell concordantly invokes the close relationships between needlework and the 'fancy' that led Cavendish to invoke philosophies of mind in order to claim that poetry and needlework entailed similar types of cognitive work. Tickell, for example, describes embroideries as 'works of fancy' and the hours passed in embroidery as time spent not only 'imitating Fruits and Flowers' (5:72) but also creating new images. Some embroiderers 'transplant all the Beauties of nature into her own Dress' while others '[raise] a new Creation in [their] Closets and Apartments' (5:72). Tickell imagines that embroiderers experience pleasure in such clothes and in such rooms, 'walking among the Shades of groves planted by themselves' and 'surveying' the 'heroes slain by their needle, or little Cupids which they have brought into the world without pain!' (5:72).

Despite relegating women to their needles – even wishing that women writers 'had chosen to apply themselves rather to tapestry than rhyme' because the former is a better medium in which a 'lady can shew a fine genius' (5:72) – Tickell conflates embroideries with the genres of romance, the pastoral and the epic. He characterizes women who embroider as 'poetess[es]' who 'vent their fancy in rural landscapes, and place despairing shepherds under silken willows, or drown them in a stream of mohair' (5:72). Tickell grants that women who embroider 'work up battles as successfully [as pastoral poets]', using metal thread to 'enflame them with gold' or red thread to 'stain them' with blood (5:72). More importantly, Tickell imagines embroidered works in ways that reveal a depth to Cavendish's invocation of the mind that spins its fancy into poetry; the embroidery celebrated by Tickell transforms into three-dimensional poems, and the aesthetic experience of 'reading' such 'texts' is revealed to be both embodied and immersive. This was a spinning of the brain, too, whereby embroidery produced waking poetic dreamscapes.

By associating embroidery with poetry and mental states of dreamy absorption and association, Tickell invoked a long-running debate about poetry itself. John Dryden summed up the debate this way in 1664: critics dismissed poetry as the 'embroidery of sense'.[28] They maintained that poetry did little more than transform what 'is ordinary' into something that 'pass[es] for excellent with less examination' (1:179). Dryden neither critiques nor refutes the denigration of needlework that inheres in the critics' complaint that poetry is merely the 'embroidery of sense'. His purpose, after all, is to establish that poetry is not like embroidery. Nevertheless, Dryden elevates embroidery by emphasizing the concentration and skill that rhyming requires. Rhyming forces poets to discipline their minds as well as their hand in acts of composition as they simultaneously attempt to copy what they have observed in nature and what they have imagined. The poet's imagination is particularly liable to be 'wild and lawless', according to Dryden, and so rhyming usefully 'bounds and circumscribes' a poet's fancy (1:179). Constructing rhymes takes effort, which takes time, which likewise introduces opportunities for the poet to exercise their 'judgment' (1:179). This process of copying, imagining and rhyming ensures that poets make 'the richest and clearest thoughts' available to their readers, ever mindful of the 'help' that poetry 'brings to memory, which rhyme so knits up' (1:178).

Defenders of poetry not only continued to counter those critics who denigrated verse as mere embroidery, they also turned needlework metaphors of mind to their advantage. In 1709, Richard Steele recommended the study of poetry to the aspirational gentleman: 'The graceful Sentences, and the manly Sentiments so frequently to be met with in every great and sublime Writer are ... the most ornamental and valuable Furniture that can be a for a young Gentleman's head'.[29] 'Methinks', he continued, that such poetry is 'like so much rich Embroidery upon the Brain'.[30] In his *Discourse on the Education of Children and Youth* (1753), Isaac Watts insisted that poetry was not 'a mere Amusement, or useless Embroidery of the mind'; Watts admires poetry, in fact, for its needlework-like qualities.[31] '[O]ur senses and our souls', Watts writes, are enraptured by the sounds or language of poetic verse, its 'harmony' as well as by its 'colours'.[32] For his part, Blaise Pascal laid out his case bluntly: '[P]ersons of true Sense and Judgment' agree that there is no 'Difference between the trade of the Poet, and that of an Embroiderer'.[33] Dismissals of poetry as embroidery were met with rebuttals that emphasized the poet as a skilful artist who copied from nature, indulged their imagination and then painstakingly stitched the two worlds together so that the reader could encounter a poem as not only text but also texture.

Elsewhere, Addison himself used textile metaphors to explain what made for good poetry. In his 'Essay on Virgil's Georgics' (1697), for example, Addison admired how Virgil offered his readers 'precepts' based on the unlikely content proffered by 'the science of Husbandry'.[34] Addison pronounces Virgil to be the master of cloaking the 'natural simplicity and nakedness of [the poem's] subject' in 'the pleasantest *dress* that poetry can bestow' (1: 250). The precepts of Virgil's *Georgics* affect readers more powerfully than those offered by other writers because Virgil 'loves' to use descriptions, discreet actions and poetic 'pomp' 'to suggest a truth indirectly' (1:251). Virgil never gives his reader 'a full and open view' of the moral (1: 252). Instead, he allows the reader to 'see just so much as will naturally lead the imagination into all the parts that lie concealed' (1:252). Addison appreciates both the affect and the reasoned insight that result. It 'is wonderfully diverting to the understanding', Addison writes, 'to receive a precept that enters as it were through a by-way, and to apprehend an idea that draws a whole train after it. For here the mind, which is always delighted with its own discoveries, only takes the hint from the poet, and seems to work out the rest by the strength of her own faculties' (1:252).

All of this makes reading Virgil an engrossing experience for Addison, and the pleasures of discovery and understanding that Virgil provides are presaged by an experience of reading that is like entering into another world or a dream state. Virgil's precepts naturally 'fall in after each other by a natural, unforced method' (1:250), as if by a thread of association. There are no 'digression[s]' in Virgil's Georgics, even when the poet appears to digress. Every part of the poem is 'something of a piece' and can be shown to 'have a remote alliance, at the least, to the subject' (1:252). Everything, in short, that Virgil describes 'immediately present[s] itself, and rise[s] up to the reader's view' (1:253). Readers see the farmer toiling naked in the summer's heat, shovelling 'dung about with an air of gracefulness' (1:256). The reader sweats with Virgil as he sweats recounting the farmer's labour, and then 'shiver[s]' (1:255) reading Virgil's depiction of winter. The meter and the rhyme of the poet's lines about beekeeping produce a veritable auditory hallucination; readers hear the 'noise and hurry' of the bees. When Virgil describes the process of grafting trees, he considers the perspective of the trees themselves, which leads us to share with the trees an uncanny experience of 'oblivion, ignorance, wonder, [and] desire' (1:256). Virgil's *Georgics*, Addison concludes, are 'perfection' (1:258).

The encomiums that Addison heaps on Virgil translate the *Georgics* from texts into texture and then textile. Virgil gives a 'pleasant dress' to the coarse experience of the everyday. So thoroughly does Virgil wrap the reader in the poem's world, the reader who sweats with the farmer and feels cold when the winter pinches is moved to take off an article of clothing or reach for warmer fabric. Like Ariadne, Virgil unravels threads for readers to follow; they 'apprehend an idea that draws a whole train after it' and discover that every episode of the poem is 'finely wrought together in the same piece' (1:250). As Addison phrased it: Virgil's poem is like a 'curious brede [*sic*] of needle-work', in which 'one colour falls away by such just degrees, and another rises so insensibly, that we see the variety, without being able to distinguish the total vanishing of the one from the first appearance of the other' (1:250).

Throughout the previous century, the metaphors of mind that invoked needlework had similarly turned to thread to construe not only the relationships between the body and the brain but also the relationships between different ideas, as Locke's parsing of 'mixed modes' implied. Hume, for example, used thread to name the 'principle of connexion between the different thoughts or ideas of the mind'.[35] Hume argued that 'even in our wildest and most wandering reveries, nay in our very dreams, we shall find, if we reflect ... that there was still a connection upheld among the different ideas, which succeed each other'.[36] He continues: '[N]otwithstanding the empire of the imagination, there is a secret tie or union among particular ideas, which causes the mind to conjoin them more frequently together, and makes the one, upon its appearance, introduce the other. Hence arises what we call the *apropos* of discourse: hence the connection of writing: and hence that thread, or chain of thought, which a man naturally supports even in the loosest *reverie*.'[37] Hume uses a thread metaphor to suggest that the links between thoughts in 'reveries' or 'dreams' are similar to the chain-like connections that characterize reason, but they are not quite the same. Threads connect associations that can be rationalized after the fact; chains connect conscious thoughts. The quick slip that Hume makes from thread to chain risks being straightened into lines that are familiar as those that characterize the shape of narrative – a plot composed of causes and effects, a story organized chronologically – rather than the shape of poetry.[38]

However, Hume's suggestion that threads are like 'secret tie' between disparate ideas and Addison's observation that Virgil delivers his precepts through a 'by-way' both preserve a pre-novelistic sense of needlework's less linear qualities. *The Needle's Excellency* (1631) by the water poet John Taylor gestures to the variety of shapes as well as literary forms that needlework engaged. Needlework was a combination of 'true history', 'poesie rare' and 'various pleasant fiction', and it gave 'dimension' to that combination by making use of 'Ovals, Squares, and Rounds'.[39] In 'forming shapes so Geometricall', needlework makes 'Art seemeth merely natural'.[40] Taylor's use of 'seemeth' here is key. Although thread might be unstitched and straightened to impose narrative order, needleworks still retained their affiliation with a poetics that disavowed the strictly representational and the consciously reasonable. Ben Jonson reminded his readers that the 'Soule of any Poeticall worke, or *Poeme*' is 'Fable and Fiction'.[41] Edmund Burke similarly insisted that poetry did not 'present a clear idea of things themselves'; rather, it 'display[ed] the effect[s]' that things themselves had already had on 'the mind'.[42] 'The essence of poetry', as Samuel Johnson similarly reminded his readers, was 'invention'; poetry 'produc[es] something unexpected' and expresses 'an idea' that 'things themselves' cannot 'afford'.[43]

Needlework metaphors could be used, as Dryden did, to defend poetry as the conscious control the poet executed over her imagination as she shaped ideas into rhymed arrangements, thereby enhancing the ability of the poem's reader to memorize and rationalize what she read. Yet poetry as embroidery and embroidery as poetry both remained not only entangled but also capable of entangling in ways that productively confused what was real and what was imagined. Addison, for example, dreamed in embroidery. The dissection scenes I discussed above were dream visions. Addison also dreamed about the 'Regions of False Wit' and the figure of 'Fancy'. In the first of these dreams, the 'Trees' in the Region of False Wit sprouted 'Bone-Lace' and the Flowers 'grew up in Pieces of Embroidery' around 'Fountains that bubbled in an Opera tune, and were filled with Stags, Wild-Boars, and Mermaids' while 'Birds' with 'golden Beaks' spoke in 'human Voices' (1:271). In the next dream, 'Fancy' wears a 'flowing Robe, embroidered with several Figures of Fiends and Spectres, that discovered themselves in a thousand chimerical shapes' (4:507). Addison was both intrigued and disturbed.

Entangled poetics

As entangled, entangling poetics, the period's needlework samplers – which prominently feature stitched verses – appear less easily consigned to a history in which women laboured under the strictures of domestic ideology by laying down their pens to take up their needles. At first glance, however, the poetry featured on samplers does not bode well for a history either of poetry or of women. On the whole, the verses are aphoristic; their metres are qualitative, their lines are short-footed and strict couplets or alternate rhymes proliferate. Thematically, the verses on needlework samplers are Protestant in their tenor, offering pat bon mots about faith and labour.[44] Thus, Bible verses commonly feature on samplers. So, too, do poems by the period's best-selling authors. In the more than three hundred samplers that were made between 1650 and 1850 and which are currently available in the Victoria and Albert Museum's digital collection, Thomas Dyche's *A Guide to the English Tongue* appears, however, to have been embroiderers' preferred source for poetry to copy. Dyche's *Guide* is a literacy primer.[45] First published in 1707, more than a hundred editions of Dyche's text – some in print runs of 250,000 – were in circulation by 1800.[46] After providing his young readers with instructions in handwriting, Dyche offers three series of alphabet acrostics for them to practise copying so that they learn by physical action and repetition how to align the shape of letters with their respective sounds. Series one includes one-line verses; series two includes two-line verses; series three includes four-line verses.

Even given the constraints the needlework medium and the widely available source texts for sampling imposed on needlework verse, we can still detect some evidence of, if not exactly conscious preference, a perspicacious fortuitousness of selections from Dyche's text that exceeds its function as a literacy primer. For example, no needleworkers appear to have felt compelled to copy from the first single-line series in Dyche's text. Only one passage from the second, two-line series appears: the letter I couplet, 'If all Mankind wou'd live in mutual Love,/This World wou'd much resemble that above'.[47] Perhaps not coincidentally, the vast majority of the one and two-line verses appear to speak to Dyche's male readers about topics like trade, wealth and the perils of debauchery.

The four-line acrostics in series three appear most frequently on the samplers in the V&A's digital collection. These include the following verses for the letters G, L, M, S and T:

Gay dainty Flowers go swiftly to decay,
Poor, wretched, Life's short Portion flies away.
We eat, we drink, we sleep; but lo anon
Old age steals on us never thought upon.[48]

Lord of this lower World frail Man was made,
The Creatures all to him their Homage paid:
But when for Sin God did him once condemn,
He's neither Master of himself nor them.[49]

Make much of precious Time, while in your power;
Be careful well to husband ev'ry Hour:
For Time will come when you shall sore lament
Th' unhappy Minutes that you have mispent [*sic*].[50]

See how the Lilies flourish white and fair.
See how the Ravens fed from Heaven are.
Then ne'r distrust thy God for Cloath and Bread
While Lilly's flourish, and the Raven's Fed.[51]

The Ant against cold Weather wisely hoards
Provision, which the Summer's Wealth affords:
Reading a Silent Lesson to Mankind,
That they in Diligence be not behind.[52]

A logic characterized by associationism premised on proximity or categorical similarity likely governed the selection of these particular verses. It seems intuitive, for example, that poems which mention flowers, birds and insects would be singled out for copying onto samplers that also depict flowers, birds and insects. Likewise, the verse for 'M', 'Make much of precious Time', speaks to the relations between needlework and the labour of literacy as well as stitching. Verses such as these reminded needleworkers that their time spent in learning how to read was also time well spent in learning moral lessons as well as the practical skill of needleworking, producing a pedagogical loop of material and textual literacy. The samplers that take their verses from sources other than Dyche's *Guide to the English Tongue* – as well as those samplers that contain original verse – evince similar logics of selection.

If you separate the verses from the samplers' cheerful flowery, creaturely motifs, the outlook they offer is remarkably bleak, bleaker still when we remember that the samplers were worked by young girls. Pain, sorrow and loneliness lurk just over youth's horizon; time is slipping away; death waits for no one. Black ravens flock around white lilies, well-fed in the present, but 'Gay dainty Flowers go swiftly to decay'; the 'Creatures' of the world will not subject themselves to sinning humans, and the ants know that winter is coming. 'There was no shortage of doom-laden verses for samplers', Parker observes (133).[53] The tendency has been, therefore, to read such verses as both the means by which women were disciplined and as symptomatic of their capitulation – whether reluctant or cheerful – to such disciplining.

Such interpretations have been bolstered by Freud and Breuer's famous aside on needlework: '[N]eedlework and similar occupations render women especially prone' to hysteria.[54] In the full context of Freud and Breuer's essay, however, needlework leads to hysteria because it produces a 'hypnoid state': a state of day-dreaming. The 'ideas which emerge' in this state 'are very intense but are cut off from … the rest of the content of consciousness'.[55] These dissociative states may be perfectly remembered or not, but they remain disconnected from the memories of real experiences: 'traumas', in Freud and Breuer's terminology. Freud and Breuer here defend their use of hypnosis in psychotherapy, the whole point of which is to trace the slippage between the 'waking dream' and the memory of the traumatic event. As Amanda Vickery has suggested, however, the 'hypnoid' states produced by needlework might be simultaneously understood as an unconscious manifestation of the traumas of patriarchal oppression and as attempts to escape them: 'While the seamstress's hands were busy … who was to say that her mind was not flying free?'[56] As a means of inculcating both moral and practical education for young girls, the sampling of moralistic verse could introduce complexity into a Lockean philosophy of mind. On the one hand, the time spent in stitching afforded opportunities for reasoned reflection and for combining simple, sensory impressions into complex ideas. On the other hand, the mind was absorbed in the rote repetition of stitches and rhymes, leaving it free to wander into regions of fancy where uncanny associations delightful and disturbing could be spun.

Mary Linwood's famous exhibition of her needlework paintings, which was one of London's most popular shows between 1798 and her death in 1845, exemplifies how needlework continued to be associated with poetry as well as with poetic states of reverie and dreamlike absorption. Linwood copied, with impressive precision, the paintings of famous artists in needlework. Linwood's gallery of needleworks was 'legendary'.[57] The needleworks were marvels, and the installation was designed to be a spellbinding 'world of wonders'.[58] By the time of Linwood's death, more than sixty needleworks were on display. The exhibition invited visitors to marvel at Linwood's astounding ability to imitate paintings in thread; it also invited them to read poetry as they examined her needleworks. Visitors to Linwood's gallery could purchase a catalogue that paired each work with lines from Shakespeare, Dryden, Milton, Thompson and Cowper, among others.[59] For Lucy Aikin, Linwood's exhibition occasioned a new poem, titled 'On Miss Linwood's Admirable Pictures in Needle-work'. Aikin celebrates Linwood as the heir to an ancient tradition that had been, regrettably displaced by men. As Aikin writes, 'women's labours' had 'crumbled in decay' because 'the manly arts' had 'chas'd the housewife's humbler steel' and 'rent' the 'ag'd tap'stry' 'from the wall'.[60] In Aikin's estimation, Linwood had revived the art of embroidery and taken it to new heights.

With a 'bold air', a specimen of Linwood's handiwork 'fills at once and charms the wondering eye'.[61] And as Aikin's poem itself attests, Linwood's needleworks also had the power to fill a page by charming a poet's pen as well.

Visitors entered Linwood's exhibition by walking up a grand staircase and then travelling down a long hallway with rooms on either side. In each room, needleworks were hung individually or grouped and staged in such a way as to make the rooms seem like they were filled with three-dimensional figures. Linwood's version of a Gainsborough painting that looks in through a cottage window at children warming themselves by a fire was positioned as if it were itself a window looking out of the room. Her copies of two of James Northcote's historical paintings depicting scenes from the reign of Henry VIII were hung in a room that looked like the Tower of London.[62] Linwood's works were celebrated for their trompe-l'oeil qualities as replicas of master paintings, but they were also appreciated as an immersive installation. In Aikin's poem, Linwood embroiders with a 'matchless grace' that makes her needleworks seem to 'live'.[63] One reviewer similarly reported that Linwood's 'figures assumed form and started into life'.[64] The 'glowing' and three-dimensional quality of Linwood's needlework connected her handiwork to poetry's power over painting to more thoroughly depict or conjure motion and change over time; as the reviewer who admired the liveliness of Linwood's figures put it, her needleworks were like those 'splendid wonders that were recorded by Homer'.[65]

The highlights of the exhibition were two cave rooms, one of which offers a tantalizing glimpse of how Linwood's needleworks immersed their viewers in dreamlike states that were akin to the absorbing experience of reading poetry. Linwood's copies of paintings by George Stubbs – one of lions, the other of tigers – hung in the room where bones were reportedly also 'strewn' about.[66] The needleworks in this room were so 'lifelike' that some visitors thought them to be specimens of taxidermy.[67] William Thackeray said that the cave room, 'at the end of a black passage' would have 'frighten[ed] any boy not born in Africa'.[68] Charles Dickens reported that he stood among its 'gloomy sepulcher of needlework dropping to pieces with dust and age and shrouded in twilight at high noon'; Dickens felt 'chilled, frightened, and alone'– responding here to Linwood's needleworks in a way similar to Addison's reader of Virgil.[69] In 1874, Tom Taylor 'still [had] a dim and somewhat awe-stricken remembrance' of the Linwood Gallery, and he recounts feeling 'thankful' after the 'succession of sensations' he experienced in the cave rooms to have at last entered the 'Scriptural Room' where, presumably, he was relieved to consign Linwood to the role of illustrator and subsume her embroideries to their textual referents.[70]

Within the context of seventeenth- and eighteenth-century discourses about poetry – and given the ways in which these discourses used needlework as metaphors for figuring the form as well as the function of verse – the unconscious, hypnotic, day-dreaming associated with embroidery might be recast as not only aesthetically but also politically productive and provocative. Maureen Daly Goggin and Beth Fowkes Tobin describe the combination of 'bodily knowing' and 'cognitive know-how' that needlework entails, and they conclude that we therefore must understand the 'praxis of needlework' as constituting its own 'epistemology'.[71] Embroidered works from the long eighteenth century suggest that needleworkers' material literacy – the various ways of making that characterized the creation of needleworks – functions as a critical component of needleworking's unique ways of knowing which, in turn, could constitute unique modes of resistance.

Needleworking frequently involved a dexterous negotiation of texts alongside needles, thread and fabric; likewise, needleworking brought communities of makers, readers and writers into dialogue with one another, and it established practices as well as vocabularies that proved applicable for the critical assessment of printed, textual matter. Consequently, eighteenth-century embroidered works make the materiality of literacy especially visible – how, in other words, the consumption as well as the production of texts engages not only minds and imaginations but also bodies and an array of three-dimensional objects. Importantly, embroidered works as well as needlework practices in the period did not necessarily subordinate textiles to texts. Although embroiderers frequently copied or alluded to previously printed materials, embroidery itself often prompted the creation of original poetic expressions in thread as well as in print. The intervention in textual literacies that embroidery's material literacies made possible means that historical examples of needlework need not be appreciated solely for the opportunities they present for 'consciousness raising'; nor need they be prioritized for the explicitness of their messages at the expense of their medium or vice versa. Throughout the long eighteenth century, the medium of embroidery itself was capable of operating as a kind of poetry that simultaneously registered the pleasures of creative, artistic making as well as complex critiques of oppressing ideologies and lived conditions. Or, as Gillian Collins stitched on her sampler in 1785: 'My art hath taught my fingers ends how to write without paper ink or pens.'[72]

Notes

1   The project is now known as the *Postal Art Event, Feministo,* or *Portrait of the Artist as a Young Woman or Housewife.* This variance in naming reflects the ways in which the women's collaboration continued to develop and transform in their geographically dispersed networks.

2   See also Chloe Wigston Smith, 'Gender and the Material Turn', in *Women's Writing, 1660–1830,* ed. Jennie Batchelor and Gilian Dow (London: Palgrave Macmillan, 2016), 149–68, 152.

3   See, for example, Elaine Hedges, 'The Needle or the Pen: The Literary Rediscovery of Women's Textile Work', in *Tradition and the Talents of Women,* ed. Florence Hower (Urbana: University of Illinois Press, 1991), 338–64; Laurie Yager Lieb, 'The Works of Women Are Symbolical: Needlework in the Eighteenth Century', *Eighteenth-Century Life* 10, no. 2 (1986): 28–44; Cecilia Macheski, 'Penelope's Daughters: Images of Needlework in Eighteenth-Century Literature', in *Fetter'd or Free? British Women Novelists, 1670–1815,* ed. Mary Anne Schofield and Cecilia Macheski (Athens: Ohio University Press, 1986), 85–100; Heather Pristash, Inez Schaechterle and Sue Carter Wood, 'The Needle as Pen: Intentionality, Needlework, and the Production of Alternate Discourses of Power', in *Women and the Material Culture of Needlework and Textiles, 1750–1950,* ed. Maureen Daly Goggin and Beth Fowkes Tobin (Aldershot: Ashgate, 2009), 13–29. See also Kathryn King, 'Of Needles and Pens

and Women's Work', *Tulsa Studies in Women's Literature* 14, no. 1 (1995): 77–93 and Anna Rosalind Jones and Peter Stallybrass, *Renaissance Clothing and the Materials of Memory* (Cambridge: Cambridge University Press, 2000), both of which I discuss below alongside Susan Frye's work. See Smith's 'Gender and the Material Turn' for another example that complicates the relationships we have imagined to exist between women's needles and their pens.

4   Amy Tobin, '"I'll Show You Mine, if You Should Me Yours": Collaboration, Consciousness-Raising and Feminist-Influenced Art in the 1970s', *Tate Papers* 25 (Spring 2016), 4, https://www.tate.org.uk/research/publications/tate-papers/25/i-show-you-mine-if-you-show-me-yours, accessed 1 July 2018.

5   Alexandra M. Kokoli, 'Undoing "Homeliness" in Feminist Art: *Feministo: Portrait of the Artist as a Housewife* (1975–7)', *N. paradoxa* 13 (2006): 79.

6   A photograph of the work is included in Rozsika Parker, *The Subversive Stitch* (New York: I.B. Tauris, 1984), plate 102. Hereafter cited parenthetically in the text.

7   These activities are documented in Rozsika Parker and Griselda Pollack, ed., *Framing Feminism: Art and the Women's Movement, 1970–1985* (London: Pandora, 1987). See also Janis Jefferies, 'Crocheted Strategies: Women Crafting Their Own Communities', *Textile* 14, no. 1 (2016): 14–35. Parker's *The Subversive Stitch* inspired two exhibitions: one of historical needlework and another of contemporary feminist experiments in the medium. Germaine Greer reviewed both and liked neither. With regard to the historical needleworks, she claimed that 'as soon as textiles are immobilized, framed and hung upon a wall they become only inferior paintings'. With regard to the contemporary experiments with the medium, Greer felt that the 'stitching' was just a 'secondary element in mixed-media statements which appear to reject and resent traditional female occupation'. See Germaine Greer, 'A Stitch in Time', *The Independent* (London, UK), 3 October 1988: 12.

8   King, 'Of Needles and Pens and Women's Work', 84.

9   Jones and Stallybrass, *Renaissance Clothing and the Materials of Memory*, 142 (emphasis added).

10   Susan Frye, *Pens and Needles: Women's Textualities in Early Modern England* (Philadelphia: University of Pennsylvania Press, 2010), 26.

11   Ibid.

12   Ibid.

13   See also Amanda Vickery's query: '[H]ow Far Can We Read a Sex in a Thing?', *Behind Closed Doors: At Home in Georgian England* (London: Yale University Press, 2009), 261.

14   John Locke, *Thoughts Concerning Education* (London, 1693), 76.

15   Ibid.

16   John Locke, *Essay Concerning Human Understanding*, ed. Peter H. Nidditch (Oxford: Oxford University Press, 1979), 223.

17   Ibid., 119.

18   Ibid., 139.

19   On colour and needlework, see Maureen Daly Goggin, 'The Extra-Ordinary Powers of Red in Eighteenth- and Nineteenth-Century Needlework', in *The Materiality of Color: The Production, Circulation, and Application of Dyes and Pigments, 1400–1800*, ed. Andrea Feeser, Maureen Daly Goggin and Beth Fowkes Tobin (Aldershot: Ashgate, 2012), 29–43.

20   Ibid., 224.

21   Ibid., 223.

22 René Descartes, *Principles of Philosophy* (1644), trans. V. R. Miller and R. P. Miller (Dordrecht: Reidel, 1983): 243.

23 Ephraim Chambers, *Cyclopaedia: Or, An Universal Dictionary of Arts and Sciences*, 2 vols. (London, 1728), 1: 375.

24 Benjamin Martin, *The Philosophical Grammar* (London, 1735), 308–9.

25 Bernard Mandeville, *The Fable of the Bees, Part II* (London, 1729), 178, 184.

26 Donald F. Bond, ed. *The Spectator*, 5 vols. (Oxford: Oxford University Press, 1965), 2: 460. Hereafter, abbreviated as *The Spectator* and cited parenthetically in the text.

27 See also Ruth Geuter, 'Embroidered Biblical Narratives and Their Social Contexts', in *English Embroidery from the Metropolitan Museum of Art, 1580–1700*, ed. Melinda Watt and Andrew Morrall (London: Yale University Press, 2008), 57–77 and Vickery, *Behind Closed Doors*, 240–1.

28 John Dryden, *The Dramatick Works of John Dryden*, 6 vols. (London, 1717), 1: 179. Hereafter cited parenthetically in the text.

29 *The Lucubrations of Isaac Bickerstaff*, 4 vols. (London, 1710–1711), 2: 299.

30 Ibid.

31 Isaac Watts, *The Works of the Late Reverend Dr. Watts, Published by Himself* (London, 1753), 5: 380–1.

32 Ibid., 5: 380–1.

33 Blaise Pascal, *Thoughts on Religion and Other Curious Subjects*, trans. Basil Kennet (London, 1727), 244.

34 Joseph Addison, *The Works of the Right Honourable Joseph Addison*, 4 vols. (London, 1721), 1: 250. Hereafter cited parenthetically in the text.

35 David Hume, *Philosophical Essays Concerning Human Understanding*, 2nd ed. (London, 1750), 31.

36 Ibid.

37 Ibid.

38 See Chloe Wigston Smith's chapter on 'Paper Clothes' in *Women, Work, and Clothes in the Eighteenth-Century Novel* (Cambridge: Cambridge University Press, 2013), 47–82.

39 John Taylor, *The Needle's Excellency* (London, 1631).

40 Ibid.

41 Ben Jonson, *The Workes*, 3 vols. (London, 1640), 3: 714.

42 Edmund Burke, *A Philosophical Enquiry into the Origins of Our Ideas of the Sublime and the Beautiful*, 2nd ed. (London, 1759): 332.

43 Samuel Johnson, *Prefaces, Biographical and Critical, to the Works of the English Poets*, 10 vols. (London, 1779–81), 1: 110.

44 For more on the history of samplers, see especially Clare Browne and Jennifer Earden, ed. *Samplers from the V&A Museum* (London: V&A Publishing, 1999); Carol A. Humphrey, *Samplers* (Cambridge: Cambridge University Press, 1997); and more recently, Goggin, 'Stitching a Life in "Pen of Steele and Silken Inke": Elizabeth Parker's circa 1830 Sampler', 17–42.

45 For two recent studies that focus on samplers in the context of the history of education, specifically, see Leena A. Rana, 'Stories behind the Stitches: Schoolgirl Samplers of the Eighteenth and Nineteenth Centuries', *Cloth and Culture* 12, no. 2 (2014): 158–79 and Judith A. Tyner, *Stitching the World: Embroidered Maps and Women's Geographical Education* (Aldershot: Ashgate, 2015), 1–42.

46 Gordon Goodwin, 'Dyche, Thomas (d. 1722x7)', *Oxford Dictionary of National Biography*, accessed 8 May 2018.

47   Thomas Dyche, *A Guide to the English Tongue*, 2nd ed. (London, 1710), 126.

48   See Mary Wakeling's sampler, 1732, Victoria and Albert Museum, London,
      394–1878. Dyche, *A Guide to the English Tongue*, 127.

49   See Charlotte Roy's sampler, 1806, Victoria and Albert Museum, London, 590–1894.
      Dyche, *A Guide to the English Tongue*, 128.

50   See Ann Wragg's sampler, 1764, The Charleston Museum, HT 6422. Dyche, *A Guide
      to the English Tongue*, 128.

51   See Elizabeth Jane Richards's sampler, *c.* 1800, Victoria and Albert Museum, London,
      T.97–1939. Dyche, *A Guide to the English Tongue*, 128.

52   See Mary Shields's sampler, 1827, The National Museum of American History
      TE.T11676. Dyche, *A Guide to the English Tongue*, 129.

53   See also Parker, *The Subversive Stitch*, 13 and 89.

54   Sigmund Freud and Josef Breuer, *Studies on Hysteria, 1883–1885* in *The Complete
      Psychological Works of Sigmund Freud*, 24 vols., ed. James Strachey et al. (London:
      The Hogarth Press, 1953–74), 2: 12.

55   Ibid.

56   Vickery, *Behind Closed Doors*, 238.

57   For a thorough account of Linwood's exhibition, see Heidi A. Strobel, 'Mary
      Linwood, Thomas Gainsborough, and the Art of Installation Embroidery', in
      Materializing Gender in Eighteenth-Century Europe, ed. Jennifer G. Germann and
      Heidi A. Strobel (Wooster: Ashgate, 2016), 173–91. See also Richard Altick, *The
      Shows of London* (Cambridge: Belknap, 1978), 400.

58   *Public Characters of 1799–1800* (London, 1799), 554.

59   *Exhibition of Miss Linwood's Pictures in Needlework at the Hanover-Square Concert
      Rooms* (London, 1810) is a representative catalogue featuring poetry associated with
      each needlework.

60   Lucy Aikin, 'On Miss Linwood's Admirable Pictures in Needle-work', *Monthly
      Magazine*, March 1798, 287.

61   Ibid.

62   Tom Taylor, *Leicester Square: Its Associations and Its Worthies* (London, 1874), 461.

63   Aikin, 287.

64   *Public Characters*, 555.

65   Ibid., 561.

66   Parker, *The Subversive Stitch*, 145.

67   Altick, *The Shows of London*, 401.

68   Quoted in Ibid.

69   Quoted in Ibid.

70   Taylor, *Leicester Square*, 461.

71   Maureen Daly Goggin, 'Introduction: Threading Women', in *Women and the Material
      Culture of Needlework and Textiles, 1750–1950*, ed. Maureen Daly Goggin and Beth
      Fowkes Tobin (Farnham: Ashgate, 2009), 1–12.

72   Gillian Collins's sampler, 1785, Victoria and Albert Museum, London, B.539–2016.

4

Domestic crafts at the *School of Arts*

Chloe Wigston Smith

Hannah Robertson published *The Young Ladies School of Arts* in 1766 in Edinburgh. In contrast to domestic and housekeeping manuals since the late seventeenth century, Robertson's manual reserved significant space for domestic crafts. These 'nicer arts', as she labelled them, included conventional accomplishments like painting, as well as popular techniques such as shell work, filigree, gilding and japanning.[1] Her title page to the first edition accurately positions the range of techniques assembled in the *School of Arts*, pitching the collection's 'Great variety of practical Receipts' and its 'great many curious receipts, both useful and entertaining'. Whether useful or curious, Robertson's manual made craft and artisan knowledge accessible, unmooring it from the traditional apprenticeship and guild systems, which had largely excluded women from formal instruction. Her recipes could support women's management of households and domestic servants, but her manual also promoted the commercial potential of feminized craftwork. Robertson enabled her audience to work as professional designers, to establish the same kind of art schools that she ran and to produce crafts to ornament their homes and those of others. The *School of Arts* affirmed her energetic commitment to placing material tools in women's hands, an agenda she viewed as a financial, cultural and British nationalist project.

Robertson often appears in an anecdotal role in scholarship on women, education and material culture or in Jacobite stories as an illegitimate Stuart descendent.[2] The popularity of the *School of Arts* confirms that young British women desired to purchase a manual that would teach them how to lacquer cabinets and design grottos, alongside other popular decorative practices from China and India.[3] Robertson's manual has thus stood in the service of broader claims about gender and the global politics of design. This chapter, however, positions Robertson's manual as a written production worthy of sustained scrutiny in its own right, one that uncovers the material literacies of home craft production. The *School of Arts* merits a closer look, as a textual production that relates key insights into maker's knowledge and the place of gender in such instruction. I examine Robertson's approach to craft technique via three intersecting parts: the text of craft; the manual's diverse definitions of craft; and finally, the gendered economics of maker's knowledge. First, Robertson touches on the fissures between language and material practice that anticipate Richard Sennett's theorization of making.[4] On the one hand, she discloses the tools and techniques of material self-sufficiency via detailed verbal instruction. On the other hand, Robertson acknowledges that 'practice is the only thing that can teach properly' (xxiii, 1st ed.) in accordance with Sennet's emphasis on manual demonstration. The *School of Arts* unlocks multiple techniques of doing and making through extensive textual explication. Second, the manual reorganizes hierarchies of domestic artefacts. Robertson moves between detailed technical instructions to the aesthetic ideals for landscape painting, creating a range of empirical registers for her manual. Her instructions call for toxic and non-toxic ingredients and chemicals, highlighting the home workshop's scientific potential. Such facility with the discourses of art, craft and chemistry reinforces the manual's pedagogical aims. Robertson offers access to techniques that roam from the haberdasher's door to the chemist's, while simultaneously commercializing craft knowledge. Her instructions, as my third section shows, enable work both inside and outside the home, and throughout emphasize the personal and pecuniary profits of feminized making. Robertson ultimately underscores the feminist potential of craft knowledge, sharing her manual and aesthetic experience with readers and future practitioners.

The text of technique

Robertson devotes the bulk of her manual to recipes for food and cosmetics, and craft techniques. But the brief preface with which she opens reveals her teacherly aims. She strives to make her instructions as clear as possible, ordering her items according to developmental principles and physical ability. For the second edition, she elaborated on her views on the place of manual instruction in the education of children, explaining her care to follow the abilities of children by beginning with 'receipts for making flowers' (viii, 2nd ed). Robertson was making a revolutionary intervention in female education by starting with flower painting instead of stitching. To her mind, traditional instruction introduced needlework at far too young of an age, scaffolding further finicky techniques onto knitting and plain stitching:

> As soon as a girl has learned to read properly, she is commonly put to … the knitting of stockings, which is easily acquired: Then, the next thing is a piece of gauze put into her hands, called a sampler, on which she is to sow letters and different figures, in doing which she finds so much difficulty that she takes a real disgust at school. (vii, 2nd ed.)

Robertson advocates instead for a Dutch approach to education in which children are encouraged to create toys and other objects that reflect age-appropriate interests, as opposed to the programmatic sampler regime (viii, 2nd ed.). If, as Rozsika Parker and Maureen Daly Goggin have shown, and also Crystal B. Lake in this volume, the sampler was a staple of Georgian education that combined sewing instruction with alpha-numeric literacy and other topics, Robertson's description of the sampler as 'too abstruse and intricate for … tender years' (vi, 2nd ed.) demonstrates that its ubiquitous presence in homes and charity schools was not without controversy.[5] For her, the mismatch between the size, strength and abilities of young fingers and the demands of the sampler provokes 'real disgust'. Accordingly, children should be taught sewing skills at the later age of ten or eleven, as their mothers would also find the sampler 'a very irksome task, and cost them much trouble to execute it' (vii, 2nd ed.). Women's crafts must be matched to skill, age and ability to ensure their futurity.

Even as Robertson strives for clarity in her recipes, she acknowledges the familiar difficulties of text's ability to articulate, describe and communicate the tacit magic of making.[6] In her preface she details the care that she has taken to write out instructions for different types of decorative practices with the aim of carefully matching language to manual tasks: 'I have … placed the receipt for making flowers, jappanning, &c. first. They are made as distinct as possible, though practice is the only thing that can teach properly' (xxiii, 1st ed.). Here Robertson prioritizes practice over text. Despite her efforts to remain 'distinct' in her textual accounts, she defers authority to the physical repetition of technique. Such practice might be informed by her instructions, but it ultimately resides outside the pages of books. Her attention to first-hand experience anticipates Sennett's preference for face-to-face demonstration: 'Language struggles with depicting physical action, and nowhere is this struggle more evident than in language that tells us what to do.'[7] In this turn to practice, Robertson echoed dozens of authors, overwhelmingly men, who endeavoured to communicate methods of production and manufacture via encyclopaedias, guides and pamphlets on particular eighteenth-century industries. These authors, as Kate Smith has shown, 'actively asserted that attempting to read their descriptions with little experience of the processes involved was problematic'.[8] The diagram offered one solution to ease the absence of feeling and seeing that adhered to textual explications of making. For John Bender and Michael Marrinan, Denis Diderot's diagrams in the *Encyclopédie* developed new cognitive ways to communicate the knowledge of instructions through a vivid combination of text and image, demanding that audiences correlate verbal, visual and spatial information.[9]

Robertson's manual, however, includes no diagrammatic illustration and moreover makes illustration secondary to hundreds of pages of textual explication. With just a handful of illustrations, the *School of Arts* makes technique subject to text. Various editions include four to five plates – all designs for flowers – and her directions to the binder indicate their intended placement next to specific techniques.[10] The small number of plates, and their straightforward presentation, was likely not Robertson's choice, but the result of economic exigency. Her extant plates include neither the words nor the keys that were central to the diagram's comprehensive linking of text to image. In some instances, they are not mentioned in the text at all. For instance, Robertson's plate of an anemone (Figure 4.1) is one of three flowers that accompanies 'Instructions for painting on Paper and Silk, and colouring Prints, Drawing, &c.' (164–8, 2nd ed.). This section details the proper colours, hues and shading for twelve flowers – the passion flower, anemone and carnation are illustrated – as well as mixing techniques for paint. Although the illustrations lack colour, their shading and varied lines communicate a degree of technical information. Her extensive textual guide to the particulars of each flower modulates her insistence at the start of this section that, 'It is impossible to give directions here for drawing; that must be left to taste and genius of the performer'

Figure 4.1 *Anemone,* Hannah Robertson, *The Young Ladies School of Arts* (Edinburgh, 1767). Courtesy The Lilly Library, Indiana University, Bloomington, Indiana.

(164, 2nd ed.). Robertson may presume the impossibility of expression at the start, but her directions stress the place of textual communication, while reserving space for the performer's individual choices. Robertson thus presents an alternative to Enlightenment culture's commitment to the illustrative diagram as a central vehicle for the production of manual knowledge, as she embraces language to communicate both her techniques and authority.[11] Language here supports instruction for the non-elite production of material literacies, for those without the means or ability to purchase and interpret expensive models of knowing and making. Robertson's illustrations proffer an aesthetic model to imitate; her text works hard and at length to explicate her techniques, illuminating via language the materials and stages of production that generate the whole finished object.

In her preface, Robertson positions herself as a canny practitioner intimately familiar with manual repetition, who has developed her authority over many years of practice:

> The Receipts contained in the following Treatise are founded on many years experience, and are calculated for the improvement or amusement of the young Ladies, as well as for the use and advantage of those who may have the care of childrens education, who, though properly instructed themselves, will find it an advantage to have receipts at hand to put them in mind of what they had formerly learnt. (vi, 2nd ed.)

She moves from tacit knowledge and the empiricism of repeated technique to the importance of text. Moreover, she grounds her practice and transmission of *techne* to her authority as a mother and educator. She envisions her audience as a community of new students and experienced educators, able to rely on language to learn and relearn what they were taught long ago. The text thus functions as an *aide memoire* for the haptic memories of older women; language promises to revive the knowledge that their hands already know.

The curious branches of craft

Such words – their placement, presence and position – varied over the course of the manual's several editions. Robertson named the varied practices assembled in the *School of Arts* 'different curious branches' (ix, 1st ed.). The first edition, appearing in 1766 in Edinburgh, included 374 individual recipes. In her preface, Robertson speaks to readers in 'this northern climate' and notes that she uses 'Scots measure' (xx, 1st ed.). She positions her recipes not for those 'in high life', but for women and servants keen 'to observe oeconomy of a family' (xx, 1st. ed.). At first Robertson placed the recipes for food and drink at the start, leading with 'Rules to be observed by the Cookmaid'; such rules would drop out of later editions as Robertson's attention to cookery contracted and she reoriented her text to emphasize craft over cookery. This shift was buttressed in later prefaces: by 1767 Robertson had replaced her initial discussion of food and ingredients with her expanded views on the importance of craft instruction and the usefulness of handmade goods. The second edition also severed cookery recipes from craft instruction to create a two-volume format 'for the conveniency of the buyer' (v). Robertson explained the rationale behind her two-part structure: 'the first [is] to contain the nice arts for the young Ladies; the second part to contain the recipes for Cookery, &c. As many mistresses of families, house-keepers, and other, may have occasion for the Cookery part that have no occasion to purchase that containing the nicer Arts' (v, 2nd ed.). By the sixth edition in 1784, Robertson had dropped the two-volume format, but continued to lead with instructions for crafts. Her shifting structure reinforces hierarchies of household knowledge: women with the time or incentive to practise 'nicer arts' could so do thanks to the labour of domestic servants.[12]

The *School of Arts* covered an impressive range of topics. In addition to popular techniques such as gilding and japanning (the lacquering of cabinets), Robertson guides her audience through mounting fans; making grottoes and mosaics; creating many different colours of ink (including invisible ink, starting in the second edition); household cleaning tips for laundry; and recipes for cosmetics and perfumes. Some recipes offered insights on how to reuse materials, such as instructions for pulling gold and silver thread from lace (63, 2nd ed.). Other sections decoded the meanings of design. Starting in 1777 in the fourth edition, Robertson explicated, at length, embroidery emblems, such as animals, vegetables and flowers, delving into their mythical and religious valences before moving to the English and Scottish royal orders (27–38, 4th ed.). Her aesthetic advice covered not only the appearance of small emblems, but also enfolded instructions for landscape painting. Robertson ventures into the territory of art instruction, several years before William Gilpin's guides for amateur artists. Techniques for polite art sit alongside more visceral pursuits: as Beth Fowkes Tobin notes in this volume, Robertson includes step-by-step instructions for taxidermy projects, such as dissecting a freshly killed bird with small scissors (35, 2nd ed.).

Over the years, Robertson expanded the number of crafts at the expense of kitchen recipes. Tables 4.1 and 4.2 show how she adapted her table of contents, making good on the promise that new editions included additional recipes.[13] As noted above, the most dramatic changes occurred between the publication of the first and second editions, for which Robertson flipped the order of her instructions to prioritize crafts and slashed the number of cookery recipes. Between 1766 and 1767, the percentage of food recipes lowered from 84 to 53.3. Food would return to dominate the contents in the third edition in 1770, rising to 77.5 per cent of the overall total. Notably both the first and third editions contain the largest number of overall recipes at 374 and 408, respectively. Slimmer editions reduced the number of food recipes while continuing to make space for craft techniques. Indeed, craft instruction rose over the years, growing from thirty-two techniques in 1766 to seventy-three in 1784. The steady increase of craft instruction is further buttressed by Robertson's paratexts, which focus on handicrafts and material literacy.

Similar to other craft books, Robertson's recipes cover natural dyes and paints, but also incorporate chemical ingredients.[14] Her ingredients include many forms of vinegar, copper sulphate, ground white lead, red lead, copper nitrate, aqua fortis, liquid mercury, turpentine and aluminium sulphate.[15] Pamela H. Smith has argued that such approaches generated a new form of vernacular science that blended craft with natural philosophy.[16] More often than not, the new vernacular science, which emerged out of artisan culture in Renaissance Europe, has been linked to the circulation of manual knowledge between male masters and apprentices. As Tobin has noted, Robertson clearly genders her imagined readers as 'ladies', but her techniques overlap with those

Table 4.1 Changes in recipe numbers by type in six editions of the *School of Arts*.

	1766	1767	1770	1777	1784	1806
Total no. recipes	374	208	408	215	209	155
Food	314	113	316	117	112	97
Craft	32	58	55	74	73	42
Cosmetics	17	13	13	11	12	10
Cleaning	9	9	8	8	8	4
Health	2	14	13	2	2	1
Animal breeding	0	1	3	3	2	1

Table 4.2 Changes in recipe ratios in six editions of the *School of Arts*.

Percentages	1766	1767	1770	1777	1784	1806
Food	84.0%	53.3%	77.5%	54.4%	53.6%	62.6%
Craft	8.6%	27.9%	13.5%	34.4%	34.9%	27.1%
Cosmetics	4.5%	6.3%	3.2%	5.1%	5.7%	6.5%
Cleaning	2.4%	4.3%	2.0%	3.7%	3.8%	2.6%
Health	0.5%	6.7%	3.2%	0.9%	1.0%	0.6%
Animal breeding	0.0%	0.5%	0.7%	1.4%	1.0%	0.6%

from comparable books that were aimed at men.[17] Thus Robertson intervenes to re-present her crafts as available and accessible to women practitioners.

Sometimes Robertson includes proper handling methods for her toxic and volatile compounds, but at other times she remains silent. In her instructions for transferring a mezzotint print to glass, for instance, she advises coating a piece of glass with Venice turpentine and then holding the glass to the flame until the liquid evenly coats the surface. Turpentine is flammable. On occasion, she details safe handling methods. In her second edition, Robertson describes how to clean shells with a 'stubborn crust' (14, 2nd ed.) in a bath of aqua fortis: 'In this sort of work, the operator must always have the caution to wear gloves, otherwise the least touch of the aqua-fortis will burn the fingers and turn them yellow; and often if it is not regarded, will eat off the skin and nails' (16, 2nd ed.).[18] Many instructions require access to fire; Robertson repeatedly instructs readers to bring liquids to 'boiling hot' (30, 42, 2nd ed.) or that materials be 'laid on hot' (31, 2nd ed.). These arts and crafts would have needed, then, to be produced in kitchens or in close proximity to fireplaces, suggesting how her manual encouraged the transformation of spaces used for cooking or entertaining into generative settings that might be filled with noxious fumes. Robertson's 'curious branches' of craft thus held the potential to shift and shape the function of the domestic interior, while infusing it with smells, fluids, mess and chaotic productivity. Her manual turns domestic space into the artisan's workshop or artist's studio. Such fusing of the home with the manufacturing of craft sits at odds with changing perceptions of domestic space. As Charles Sumarez Smith has argued, eighteenth-century architecture insisted on the growing distance between the home and manufacturing. Robertson, by contrast, would have women pursue production, whose commercial potential she underlines, within the home.[19] In this way, her guide returns to urgent questions about the nationalist promotion of domestic manufacturing, as described by Ruth Mack as a debate over whether design fell under 'the domain of craft or of high art'.[20] The majority of Robertson's techniques point to craft and maker's knowledge, but her title insists on 'arts' and her preface highlights the 'nicer arts' that appear alongside the production of pickles, best practices for laundry and the occasional remedy for cramps or gout.

For Robertson, all of these branches of knowledge belong to British women, no matter how refined, caustic or scientific the ingredients or methods are. She links her maker's knowledge to the right of British women to domesticate imported goods: 'I should esteem it the greatest happiness, if I should prove the instrument of prompting my neat-handed country-women, and even those in a higher sphere of life, to employ their talents in these useful and ornamental arts, which I look upon as their proper province and profession' (xxiv, 1st ed.).[21] In 1777, Robertson claimed that women's domestically produced artefacts might rival the quality of foreign imports: 'As most of the curious articles recommended in this Book, and which derive their beauty and perfections from the invention and ingenuity of our sex, are generally brought from abroad at a great expence … they appear, when properly done as little inferior to those brought from abroad' (ix–x, 4th ed.). To that end, Robertson includes instruction for making 'glass jars, and other pieces of ornament, to look like china' (21, 4th ed.). Such techniques encouraged readers to imitate imported goods at home, with their own hands. Robertson also attempted to appeal to the domestic economy of her readers: 'Would the British Ladies for a moment reflect what sums of money are laid out for those trifles' (x, 4th ed.). Robertson's position overlaps with the widespread support of domestically produced goods in the eighteenth century, but her insistence that such activities fall under the 'proper province and profession' of British women takes this agenda a step further by positioning craft, creativity and art as a feminist, nationalist project.

The commerce of domestic craft

The *School of Arts* provided the means for women to establish themselves as either leisured amateur artists or professional designers, illuminating Amanda Vickery's claims of how craft cut across class divides: 'For every celebrated female amateur there was an army of women for whom craft skills furnished a livelihood, as teachers, tutors, educational authors, and occasionally exhibitors.'[22] Robertson pitches her recipes as tools that will support labouring women, earning them higher wages than needlework, which was seriously undervalued in the marketplace:

> I would in particular recommend the practice and knowledge of those nice arts, to young women that have no fortunes, or may be left in low circumstances: It is too well known how small a value is set on womens work, so that the cleverest at the needle can scarcely earn subsistence; but a knowledge of the different curious branches mentioned in this Treatise, will greatly make for their advantage. Any girl capable of painting, japanning, gum-flowers, pongs, &c. will always find employment among fashionable people, and especially in towns of trade and commerce, by which they may earn a tolerable living. (xxiii-xiv, 1st ed.)[23]

Such advice foregrounds the economic advantages of home decoration and how Robertson sees her maker's knowledge as a hedge against the undervaluing of women's work in the eighteenth century.

It was a story all too familiar to Robertson herself. The *School of Arts* was Robertson's first publication; her second, an autobiography, recounted episode after episode of how the condition of womanhood in the eighteenth century – its attendant losses, particularly of adult children – conspired against both happiness and economic gain. The title page of the *Life of Mrs. Robertson* (1791) identifies its author as the granddaughter of Charles II, painting a sorrowful portrait of a woman who 'has been reduced, by a variety of uncommon events, from splendid affluence to the greatest poverty, and after having buried nine children, is obliged, at the age of sixty-seven, to earn a scanty maintenance for herself and two orphan grandchildren, by teaching embroidery, filligree, and the art of making artificial flowers'.[24] Robertson's forty-seven-page autobiography tacks between her Stuart connections, numerous personal trials and her efforts to support herself and her family through craft. If the *School of Arts* captured Robertson as she was putting into place the craft business and manual that would propel her towards some economic stability, her subsequent memoir lifted the curtain to reveal the relentless personal and financial loss that followed.

Throughout her memoir, Robertson returns to material literacy as formative to her identity. She describes herself as retreating to craft as a six-year-old child, following the remarriage of her mother. She refused to attend the local school and instead: 'retired unseen to a closet, which I called my own; laid out the little money I could get in paint and paper; and thus early, and without any mistress but kind nature, began to practice myself in embroidery, drawing, making flowers, and various other elegant works of fancy' (5). Such early glimpses of her skill and creativity seep into adulthood. After marriage and motherhood, Robertson notes her efforts to make time for creative handicrafts:

> The short intervals of leisure, which my duties allowed, were chiefly devoted to those favourite arts which I ever had cultivated with delight; I studied nature with a view to imitate her most elegant productions, – a new creation rose beneath my hands, – I formed flowers of art – I painted, and I embroidered; so that (like

Penelope of old) I charmed away with works of fancy, the tedious hours, during the absence of my lord. (17)

In these examples, Robertson links craft to a fanciful escape from the dual constraints of education and home. Craft allows the expression of feminine imagination – and powerful links to classical ideals – that institutions, such as schools and marriage, seek to diminish through obedience and duty.

Robertson highlights craft's commercial potential, as opposed to its creative rewards, with increasing urgency over the course of her autobiography. Her marriage plunged her into deep financial insecurity, from which she clawed her way out through craft: '[I] was able by my talents to provide for my family' (36). As a married woman, Robertson established what she called 'her little business' (24), a shop in Edinburgh, where she set about 'practising and teaching the various arts I understood' and soon afterwards published the *School of Arts*.[25] As an author whose claims to gentility were compromised by her debtor husband, Robertson used her manual to burnish her social connections: the first edition claimed the patronage of the Countess of Fife and a two-page list of subscribers. Her insolvent husband died in a snowstorm shortly after the publication of the *School of Arts*, at which point her crafts, lessons and text became her main sources of income: 'I collected the small ruins of my fortune, and advertised my talents: the skill I had acquired in my favorite arts, might now, I thought, be rendered serviceable to my family' (31).

In early editions of the *School of Arts*, Robertson papered over the pressures of providing for her children during and after her marriage, even as she underlined the potential of her techniques to support women when work like sewing had failed them. Her promise to 'young women that have no fortunes, or may be left in low circumstances' was, after all, based on first-hand experience, to which she more openly refers in the 1777 preface to the fourth edition:

> As I myself have been one of those children of affliction, being in the early part of my life left with a number of infant children, whose daily subsistence I found most depended on the works of my hands, and who formerly was no stranger to the smiles of good Dame Fortune, I therefore can by my own experience, advise those of my own sex, who may be left in the same unhappy situation to exert their talents to procure bread for themselves and infant family. (xi, 4th ed.)

Robertson offers herself as a reliable model of craft fortune in a vignette in which the father, stepfather and husband who each failed her make no appearance. Instead she positions herself as the maternal provider who triumphs through the 'works of my own hands'.

Despite this turn to the personal, Robertson's various prefaces to the *School of Arts* largely conceal her direct knowledge of craft's support of women in an 'unhappy situation' or 'low circumstances'. Shortly after publishing the *School of Arts*, Robertson abruptly closed her Edinburgh shop and moved her three youngest children to York, where her eldest son was apprenticed to the eminent architect John Carr. Her move was provoked by the 'misery of shame' (35) prompted by her eldest daughter's decision to abandon her unhappy marriage, under the influence of an 'insinuating lover' (35). Although she feels 'ashamed of appearing in the face of day' (36), she also mourns 'her lost Anna' (35), sharing a compelling mixture of emotion and embarrassment. At this point in the narrative, the pace quickens as Robertson passes more quickly through the ensuing years of poverty, illness and deaths of her children. She does, however, make space for the craft work and instruction she pursued. In York, for instance, she expanded the reach of her lessons: 'I found employment in the boarding schools, particularly, in the line of filigree; which from having been long neglected, appeared like a new art' (37).[26] She combined craft instruction with commissions for elite families, relying on her children, especially a talented young son, during bouts of illness. As Robertson recounts:

> A lady's Barcelona handkerchief was brought for me to ornament with a painted border – I was incapable of undertaking it, but my little son (altho' wholly unpractised in these arts) took the handkerchief, and painted it. The public approved his taste, and other fashionable ladies in York sent handkerchiefs, shawls, &c. to be painted: and thus untaught, except by nature, my youngest son with his little hands maintained me. (37)

Her son completes her work, his natural talent mirroring her own childhood abilities. Notably he practises painting (not needlework) and thus follows Robertson's own pedagogical commitment to what small hands might accomplish.

If Robertson's lean years in York were prompted by the closure of her Edinburgh shop, then her years following York rewrote this narrative of personal shame into one of triumphant feminine community. Robertson left York after tracking down her eldest daughter, who had found employment as a domestic servant in Dublin. After seven years apart, Anna took up her mother's trade 'practising the elegant arts she had been taught' (39), first in Hull and then in Manchester and Liverpool. By 1782, Robertson had followed Anna to London, where she opened a successful shop in the fashionable Strand. She commissioned a trade card to publicize the products and services of the 'Authoress of the Young Ladies School of Arts' (Figure 4.2). If her business card underscores her elite connections to the 'First Nobility', her autobiography offers firm evidence of the very real rewards of craft for women. Craft would rescue both her adult daughters from unhappy marriages and the eldest from the consequences of leaving a husband for a lover.[27] Robertson embeds the description of the shop's success within its origins as a mother-daughter venture: 'I therefore took a house, and with my daughters, Minia opened a shop (the first of the kind in London) for various works of fancy: and here … I might have made a second fortune: our shop was crowded with nobility, and we were also employed in teaching many of the first families' (40). As Tobin notes in this volume, Robertson's shop combined an innovative blend of craft materials and instruction. Its success, however, was cut short by the deaths of Robertson's adult daughters, which propelled her from London to Northampton to seek other work and finally to Birmingham, where she composed her autobiography in a 'dismal apartment in an upper story' (45), far removed from the once-thriving shop in the Strand.

Mrs. HANNAH ROBERTSON,
Artificial Flower Maker & Fancy Trimmer,
AUTHORESS of The YOUNG LADIES SCHOOL of ARTS,

At No. 331, *oppofite* SOMERSET HOUSE, *Strand.*

Has had the Honour of Teaching feveral of the Firft Nobility in
England and Scotland, likewife the Governeffes of the Boarding
Schools in Edinburgh, the following Branches:

Fillagree in all its Branches; Flowers of all Kinds in ftained Cam-
brick, Paper, Tiffany, Combed Flofk Silk, or Ifinglafs.
Gold and Silver Flowers fimilar to thofe from Madeira.
Birds done in Silk and Paper fo as to imitate in their beautiful and
variegated Colours, Nature.
Embroidery of all Kinds taught.
Fillagree Paper, &c. for making any of the above Articles, furnifhed at
the fhorteft Notice.
Painted Trimmings of all Kinds for Gowns, &c.
Grown up Ladies waited on at their own Apartments.

Figure 4.2 Trade card for Hannah Robertson, *c.* 1783, 2002.24 © Museum of London.

Robertson's defence of women's labour with material artefacts has received minimal attention in the limited scholarship on her life writing. In her autobiography, Robertson credits her manual skills with her survival, calling them arts 'to which I am indebted for an unfailing resource, during a long, and painful series of adverse fortune' (5). In the *School of Arts*, she positions craft as a safety net against the uncertain and the unknown: 'Such Ladies as have time to spare, may do a great many pretty things at a small expence, for their own pleasure, or the ornament of their houses; and in case of misfortunes happening in life, which is not uncommon their knowledge in such things will be of real use to them' (xxiii, 1st ed.). This loyalty to craft makes the conclusion of her own story even bleaker, as she struggled to provide for two grandchildren, following the deaths of most of her children (including her cherished artist son).[28] At the age of sixty-seven, Robertson feels that her years 'disqualify me for a laborious attention even to my favourite arts' (46). Poverty means her imagination has 'insensibly become less capable of new creations, even in my former delightful works of fancy' (46). Together age and indigence diminish Robertson's access to crafts, robbing her of both the time and creativity to exercise her material literacy.

The publication of a second edition in Edinburgh in 1792 confirms that Birmingham was not Robertson's final destination. There she added an opening dedication to her orphaned grandchildren, in which she anticipated her imminent death from 'a broken heart'.[29] Twelve years later, however, Robertson would publish a tenth edition of the *School of Arts* in Edinburgh, that dropped 'young' from the original title in favour of the *Ladies School of Arts*, and included her autobiography.[30] Here she gathered together her two texts, merging her roles as a professional instructor of craft with the personal experience of a woman who relied on craft to express and support herself. Robertson placed her autobiography first, cutting the preface on education that had opened earlier stand-alone editions of *School of Arts*. A new preface introduced the combined parts. Its first three sentences find Robertson addressing her readers in first person, after which the prose shifts, without explanation, to third person. The ensuing pages underscore the 'many disastrous events of her life'.[31] Its conclusion details Robertson's wish that her instructions outlast her: 'She hopes this work will be read with some advantage when she is no more, and supply in some measure those instructions which she delighted to impart while youth, health, and vigour continued to support her' (x). No longer able to lead lessons herself, Robertson leaves her words to inspire future craftswomen. A final italicized passage promises that a second volume would be '*finished with all convenient speed*' and would contain '*all the valuable and useful recipes, commonly followed on the modern art of cookery*' (x).[32] Numbering 155 techniques, the 1806 volume remains Robertson's slimmest edition, yet as Tables 4.1 and 4.2 above show, cookery recipes actually outnumber craft instructions, an imbalance that remains unacknowledged in her promise to publish a second volume of recipes.

No second volume on the '*modern art of cookery*' appeared. The end of her autobiography in the 1806 edition indicates that Robertson likely died before penning the volume; there, a brief, appended note shifts from the first person of Robertson's tale to third person, filling readers in on her return to Edinburgh where 'she has laboured under a severe asthmatic complaint; during the intervals of her trouble she has continued to teach the arts treated of in the annexed publication' (82). Despite her poor health and many years of loss, Robertson continued to rely on craft instruction for her financial survival. At the age of eighty-two, 'those labours which she formerly delighted in are now become a burden to her, and therefore she relies on the sale of the present publication as some means of affording her assistance in her present weakly situation; and which she hopes, the sale of in some measure will have a tendency to alleviate' (82). By this time in her life, the creation of crafts lay beyond the edges of Robertson's failing body, but the verbal expression of technique still held the promise of essential income.

Maker's knowledge, for Robertson, offered a path towards redemption and solvency, towards an artistic life that reimagined how women might choose to use their hands and express their creative energies, what they might make in their homes and the homes of others, without the help or hindrance of men. In her autobiography, Robertson recalls with fondness, the material literacy she acquired as a precocious child; such 'arts which ever after continued the amusements of my leisure in prosperity, and to which I am indebted for an unfailing resource, during a long, and painful series of adverse fortune' (5, 1st ed.). In the *School of Arts*, Robertson's 'operator' (16, 2nd ed.) – her chosen name for the woman who performs her craft instruction – inscribes the professional vision of craft at stake in her manual. Samuel Johnson understood 'operator' as 'one that performs any act of the hand'.[33] It is a label that returns us to the haptic agency of craftwork that underscores the professional potential of the 'nicer arts'; Robertson's operator is the agent of her manual destiny.[34] The operator uses Robertson's verbal tools to make new objects that combine the technical with the artistic, the scientific with the domestic, tuned to the ways that the language of craft contributes to the manufacture of new artefacts and goods.

Notes

1   Hannah Robertson, *The Young Ladies School of Arts*, 2nd ed. (Edinburgh, 1767). Hereafter cited parenthetically in the text. Robertson published many editions of her manual with many variants across the years, as I discuss in the second section of this essay. Each quotation sources the first appearance of the evidence under discussion.

2   For brief references to Robertson, which note her contributions to material culture, Orientalism, life writing and parenthood, see Carrie Rebora Barratt, *John Singleton Copley in America* (New York: Metropolitan Museum of Art, 1995), 63; Jennine Hurl-Eamon, *Marriage and the British Army in the Long Eighteenth Century: The Girl I Left behind Me* (Oxford: Oxford University Press, 2014), 160; Eugenia Zuroski Jenkins, *A Taste for China: English Subjectivity and the Prehistory of Orientalism* (Oxford: Oxford University Press, 2013), 127; Susan M. Stabile, *Memory's Daughters: The Material Culture of Remembrance in Eighteenth-Century America* (Ithaca: Cornell University Press, 2004), 19, 105; Cheryl Turner, *Living by the Pen: Women Writers in the Eighteenth Century* (New York: Routledge, 1992), 121; Maria Zytaruk, 'Mary Delany: Epistolary Utterances, Cabinet Spaces & Natural History', in *Mrs. Delany & Her Circle*, ed. Mark Laird and Alicia Weisberg-Roberts (London: Yale University Press, 2009), 131–49, 140–1.

3   In the late seventeenth century, Hannah Woolley covered a small number of decorative techniques (including the lacquering of picture frames and making paint colours) in *A Supplement to the Queen-like Closet* (London, 1674).

4   Sennett, *The Craftsman*.

5   On embroidery and the sampler in the eighteenth century, see Parker, *The Subversive Stitch*, 113–39; Goggin, 'The Extra-Ordinary Powers of Red in Eighteenth- and Nineteenth-Century Needlework', 29–43.

6   In her study of Renaissance artisans, Pamela H. Smith emphasizes the potential of the nontextual and nonverbal literacy of 'artisanal literacy'. See Pamela H. Smith, *The Body of the Artisan: Art and Experience in the Scientific Revolution* (Chicago: University of Chicago Press, 2004), 8.

7   Sennett, *The Craftsman*, 179.

8   Smith, *Material Goods, Moving Hands*, 43. On the circulation of useful knowledge among artisans and to the public, and related questions of *episteme* versus *techne* in the production of maker's knowledge in the eighteenth century, see Maxine Berg, 'The Genesis of "Useful Knowledge"', *History of Science* 45, no. 2 (2007): 123–33 and Liliane Hilaire-Pérez, 'Technology as a Public Culture in the Eighteenth Century: The Artisans' Legacy', *History of Science* 45, no. 2 (2007): 135–53.

9   John Bender and Michael Marrinan, *The Culture of Diagram* (Stanford: Stanford University Press, 2010), 8; see also 60 on the assemblage of data in the diagram.

10 The first edition has no plates. The 1767 second edition includes five plates and the 1770 third edition four plates. These illustrations drop out of the sixth edition in 1784. The copy of the 1767 edition at the British Library includes three of the five plates; each inserted plate is attached to a folded sheet of paper. The illustrations were likely drawn from other print sources, as indicated by the months that appear in the corner for each flower (July for the carnation; September for the passion flower; May for the anemone).

11 Robertson's cookery recipes likewise insist on the precision of language and accessible words. She critiques impractical recipes with obscure, expensive

ingredients, by noting that the substitution of 'slice' of bacon for 'bard' immediately cues cooks to the correct ingredient ('bard' was the French word for slice); see *School of Arts*, 1st ed., 1766, xxi.

12 Robertson indicated the aspirations she held for her volume by dedicating it to the Countess of Fife. For the first edition, Robertson penned a dedication that ran to three pages (1766, xvii–xix). Passages from the dedication were repurposed in subsequent prefaces.

13 My thanks to Shane Hamilton for his guidance in creating these tables.

14 See Beth Fowkes Tobin on this correspondence in 'Women, Decorative Arts, and Taxidermy', in *Women and the Material Culture of Death*, ed. Maureen Daly Goggin and Beth Fowkes Tobin (Aldershot: Ashgate, 2013), 311–30, 319. For an overview of technique and aesthetic sources for crafts, see Ariane Fennetaux, 'Female Crafts: Women and Bricolage in Late Georgian Britain, 1750–1820' in Maureen Daly Goggin and Beth Fowkes Tobin, ed. *Women and Things 1750–1950*, 91–108 (Aldershot: Ashgate, 2009), 91–108.

15 The list includes: gum Arabica; gum tragacanth; gum-gatta; several forms of vinegar; verdegrease (either copper sulphate or copper chloride); ground white lead; red lead verditer (copper nitrate); aqua fortis; quicksilver (liquid mercury); Venice turpentine; powdered and burnt alum (likely aluminium sulphate); Venice varnish; oil of turpentine; and white rosin (the non-volatile part of turpentine). My thanks to Glenn Hurst for his help in understanding the toxicity of many ingredients and how they would be treated in the chemistry lab today. In his words, several are 'flammable', 'very toxic', 'harmful', 'very dangerous' and 'exceedingly dangerous' (email correspondence received on 12 March 2018).

16 Pamela H. Smith, 'Making as Knowing: Craft as Natural Philosophy', in *Ways of Making and Knowing: The Material Culture of Empirical Knowledge*, ed. Pamela H. Smith, Amy R. W. Meyers and Harold J. Cook (Ann Arbor: University of Michigan Press and Bard Graduate Centre, 2014), 17–47, 19.

17 Beth Fowkes Tobin, *The Duchess's Shells* (London: Yale University Press, 2014), 89.

18 As Tobin notes in her contribution to this volume, Robertson conveys her extensive knowledge of shells and the chemical reactions prompted by her techniques for treating them.

19 See Ann Bermingham's *Learning to Draw: Studies in the Cultural History of a Polite and Useful Art* on the commercialization of amateur art in early nineteenth-century Britain, as well as the dismissal of women's handicrafts at the onset of industrial capitalism (London: Yale University Press, 2000), 127, 160.

20 Mack, 'Hogarth's Practical Aesthetics', 21–46, 22.

21 The second edition, in 1767, swapped 'neat-handed' for 'pretty' (x).

22 Vickery, 'The Theory and Practice of Female Accomplishment', 94–109, 96.

23 A pong was a flower sprig. The sixth edition (1784) altered the wording to elevate her audience to 'young women in a middling station of life; and in case of misfortunes happening, which is not uncommon' (ix).

24 Hannah Robertson, *The Life of Mrs. Robertson* (Derby, 1791), title page. Hereafter cited in the text. A second edition was published in 1792 in Edinburgh. According to Joanne Bailey, Robertson's 'legacy as the illegitimate granddaughter of Charles II was inner nobility in the face of appalling loss'. See *Parenting in England 1760–1830* (Oxford: Oxford University Press, 2012), 133.

25 Robertson relied on her material literacy as her husband's finances fell into increasing disarray. First in Perth, Robertson established a school (which quickly failed), where

she then attempted 'to sell a few millinery articles' (23). She was far more successful in her enterprise in Edinburgh.

26 There were two official schools in York that could have incorporated such instruction: the convent-run school for girls, known as Bar Convent, established in 1686 as the first Catholic school for girls in England; and the Grey Coat School, founded in 1705, a charity school for orphans and children of the poor, who were taught sewing, spinning and other tasks to prepare them for domestic service. In the mid-1780s, two further charity schools, a spinning school and a knitting school, were established, but Robertson had moved to London by then. There was also Ackworth School (1757–73), near Pontrefract, an outpost of the London Foundling Hospital, established by Thomas Coram. See W. B. Taylor, *The Blue & Grey Coat Schools and St Stephen's Home of York 1705–1983* (York: Sessions Book Trust, 1997). In York, Robertson published the sixth edition of *School of Arts*.

27 Her adult daughter, Minia, depended on the success of the shop as her husband, a midshipman 'could not maintain her' and their children (*The Life of Mrs. Robertson*, 40). On how craft played a role in elite circles, see Elizabeth Eger, 'Paper Trails and Eloquent Objects: Bluestocking Friendship and Material Culture', *Parergon* 26, no. 2 (2009): 109–38.

28 By this time, only Robertson's oldest son was alive, but he had absconded to France, leaving his child, likely illegitimate, to the care of his impoverished mother.

29 Hannah Robertson, *The Life of Mrs. Robertson, Grand-Daughter of Charles II* (Edinburgh, 1792), n.p.

30 After her death her tale would find another outlet in the compilation *The Weekly Entertainer; or Agreeable and Instructive Repository*, Vol. 52 (Sherborne, 1812).

31 Hannah Robertson, *The Ladies School of Arts*, 10th ed., Vol. 1 (Edinburgh, 1806). Hereafter cited in the text.

32 Potential subscribers could seek out the author-teacher herself at Lochends Close, Canongate in Edinburgh.

33 Samuel Johnson, *A Dictionary of the English Language* (London, 1755–6), 'Operator' s.v.

34 See *OED*, 'Operator' s.v. definition 1a 'A person (professionally) engaged in performing the practical or mechanical operations of a process, business, etc.' and 3 'A person who does or effects something; a worker, an agent; †a maker, creator (*obs.*).'

5

'To embroider what is wanting': Making, consuming and mending textiles in the lives of the Bluestockings

Nicole Pohl

I have to go to market and provide for the family, to look after the servants, to help in taking care of you children, and in teaching you, to see that your clothes are in proper condition, and assist in making and mending for myself, and you, and your papa. All this is my necessary duty; and besides this, I must go out a visiting to keep up our acquaintance; this I call partly business, and partly amusement … Now a great many of these employments do not belong to Lady Wealthy, or Mrs. Rich, who keep housekeepers and governesses, and servants of all kinds, to do every thing for them. It is very proper, therefore, for them to pay more attention to music, drawing, ornamental work, and any other elegant manner of passing their time, and making themselves agreeable.[1]

This extract from John Aikin and Anna Barbauld's 'Dialogue, on Things to Be Learned' (1792) highlights two crucial aspects of women's work in the eighteenth century.[2] Mamma explains to her daughter Kitty both the gendered division of labour for the middling classes and the dichotomies of textile work for the middling and upper classes. She indicates that whilst the 'work' of making and preserving clothes was a necessity for the middling class woman (and that of the labouring class), textile 'work' was both ornamental and decorative for the upper-class woman, forming part of a wider skills set appropriate to polite society. Mending, as Mamma highlights, was a standard element of making clothes, ensuring that they remained in 'in proper condition' and requiring the same accomplished skill sets as ornamental textile work.[3] This 'stewardship of objects' was part of 'a *modus operandi* that ran through the whole fabric of eighteenth-century society and characterised several of its key developments'.[4] As Ariane Fennetaux, Amélie Junqua and Sophie Vasset have argued, the different incarnations of this stewardship, 'recirculation, reuse, repair, refashioning, transformation', 'affected the whole of society'.[5]

However, following on from Mamma's advice to her daughter, there were subtleties in these material stewardship practices that articulated both class and social mobility, and that played out in the material variants of textile 'work'. My case study of the

Bluestocking Circle, in particular Elizabeth Montagu (1718–1800) and Montagu's younger sibling, Sarah Scott (1721–95), will explore the complex relationship the two women had with textiles: one that was driven by fashion consciousness, domestic economy and most importantly by the idea that textiles like fashion were 'an emblem of material self-advancement, [and] … a badge of moral worth'.[6] The contrasting lives of Elizabeth Montagu, coal magnate and the 'Queen of the Blues', and her sister, the writer and reformer Sarah Scott, offer a glimpse of how textiles functioned as subtle markers of social mobility both within a social circle (the Bluestockings) and a family. As the sisters' mostly unpublished letters demonstrate, even within familial networks material culture and practices could underscore disparities in social status and wealth. Whilst we can assume that both sisters were brought up and taught the same textile skills, their skills, textile experience and technical knowledge of materials diverged.[7]

Elizabeth Montagu designed and commissioned unique decorative fibre arts and interiors such as her famous feather work for public display. She profited from her brother working as a captain for the East India Company who was able to bring her (and her sister) fine art objects and exotic fabrics, but Montagu was of course able to afford luxury items and clothes herself. Her sister, Sarah Scott, forced by diminished social and economic circumstances, became well versed in practical dress-making, mending and alteration, and 'upcycled' home decoration. Sarah Scott also functioned as a 'project manager' and 'proxy shopper' for Montagu, sourcing and managing luxury textiles for clothes and interiors.[8] While the sisters shared an extensive material literacy of technique and handicraft, their letters demonstrate very different textile skill sets and attitudes towards the making, consumption and recycling of textiles which were shaped by costs and circumstances.

The Bluestockings and female accomplishments

Alas, it is plain mankind look upon thought as the greater evil for there is no disease for which many cures have been found out, those who have many ways of killing time are always term'd ingenious, amongst the Diverse Instruments for destroying time how pretty are knitting needles, knotting shuttles, & totums & Cards & Counters, I begin to think no Woman has a chance to be Reasonable who is born with more than one hand & one Eye, for if she can be ingenious with her hands she has no chance to be so with her head.[9]

Elizabeth Montagu's early insight into the seemingly mutually exclusive occupations of reading and female accomplishments was the product of her own education.[10] Montagu and her sister came from a respectable Yorkshire family, the Robinsons. Matthew Robinson's direct relatives were heirs of the estates of West Layton and Kirby Hall, North Yorkshire, and their mother, Elizabeth Drake, was the daughter of Councillor Robert Drake of Cambridge. Elizabeth Drake enjoyed a thorough education by the reformer and scholar Bathsua Makin (1600–75).[11] In her *Essay to Revive the Antient Education of Gentlewomen, in Religion, Manners, Arts & Tongues: With an Answer to the Objections against This Way of Education* (1673), Makin emphasized the tension between female accomplishments and 'higher', intellectual pursuits:

> I do not deny but Women ought to be brought up to a comely and decent carriage, to their Needle, to Neatness, to understand all those things that do particularly belong to their Sex. But when these things are competently cared for, and where there are Endowments of Nature and leisure, then higher things ought to be endeavoured after. Meerly to teach Gentlewomen to Frisk and Dance, to paint their Faces, to curl their Hair, to put on a Whisk, to wear gay Clothes, is not truly to adorn, but to adulterate their Bodies; yea, (what is worse) to defile their Souls.[12]

Makin was not alone in raising this point. Whilst conservative authors of conduct books such as James Fordyce and Erasmus Darwin supported the teaching of needlework and other crafts for young girls and women during the eighteenth century as part of female accomplishments, other authors such as Makin, but also Hannah More and Mary Wollstonecraft, found little value in these skills and saw them as distractions from more worthy (intellectual) pursuits, a position that the Bluestockings also took.[13]

However, Hester Chapone (1727–1801), another prominent Bluestocking and conduct book writer, identified the occupation of needlework and other crafts less as markers of virtue and femininity than of class and therefore necessities. In her *Letters on the Improvement of the Mind, Addressed to a Young Lady* (1773), Chapone echoed the insights of Aikin and Barbauld's Mamma:

> Ladies, who are fond of needlework, generally choose to consider that as a principal part of good housewifery: and, though I cannot look upon it as of equal importance with the due regulation of a family, yet, in a middling rank, and with a moderate fortune, it is a necessary part of a woman's duty, and a considerable article in expence is saved by it … But, as I do not wish you to impose on the world by your appearance, I should be contented to see that you worse dressed, rather than see your whole time employed in preparations for it, or any of those hours given to it, which are needful to make your body strong and active by exercise, or your mind rational by reading.[14]

Indeed, some middling women such as Mary Wollstonecraft or impoverished ladies were forced to earn money through needlework. In this context, the 'work' of needlework was valued as economic necessity or as part of middling class 'good housewifery', but any employment of these skills for personal embellishment and vanity was rejected as work and classified as 'fancy', sprung from the idleness of leisure.[15]

This opposition was echoed in the rhetoric of needlework ('fancy work' and 'plain sewing') and reflects the thorny eighteenth-century relationship between rank and virtue, birth and worth, excess and moderation.[16] This is perhaps why Elizabeth Montagu's friend, the Duchess of Portland, was drawn to 'amusements … of the Rural Kind, working, Spinning, Knotting, Drawing, Reading, writing, walking & picking Herbs to put into an Herbal'.[17] In a letter to Mrs Port, Mary Delany reported:

She [the Duchess of Portland] desires her kind compliments to Mr. Granville and her spinning mistress, and bids me enclose the remains of her lock of wool, to show you how near she spins it off, and makes *no waste of ends*, all which she hopes you will approve of. In the midst of her philosophical studies she used to start up and go to her wheel for a quarter of an hour's relaxation, and intends that spinning shall be one of her employments, and chief amusements when she goes to town; her last wheel and reel stand in the anti-chamber of her great dressing room.[18]

The Duchess of Portland clearly staged herself as a frugal and accomplished housewife and underscored her simple 'work' as civic virtue. However, in opposition to the middling and labouring poor, the tools and equipment were high-end. In her letter to Frances Hamilton on 10 October 1783, Delany described the luxurious materials and tools that Queen Charlotte used:

I found the Queen very busy in showing a very elegant machine to the Duchess of Portland, which was a frame for weaving of fringe, a new and most delicate structure, and would take up as much paper as has already been written upon to describe it minutely, yet it is of such simplicity as to be very useful ... The King, at the same time, said he must contribute something to my work, and presented me with a gold knotting shuttle, of most exquisite workmanship and taste; and I am at this time, while I am dictating the letter, knotting white silk, to fringe the bag which is to contain it.[19]

In 1770, Queen Charlotte visited the Duchess of Portland at Bulstrode and was taken with a new treadle wheel, a 'little' or 'Saxony' treadle wheel that allowed the spinner to sit down. In the same year, the Duchess of Northumberland marvelled at the display of spinning wheels in Paris when Lady Berkeley 'had 100 spinning Wheels brought into Coach to chuse of'.[20]

But the material practice of this work for the actual labouring poor was very different. The skills taught to the labouring poor or financially compromised middling classes were summarized under the umbrella term 'work': spinning, knitting, plain sewing, mending and sampler making to teach also literacy, numeracy and geography. In Wollstonecraft's *The Wrongs of Woman* (1798), Jemima regrets that 'not having been taught early, and my hands being rendered clumsy by hard work, I did not sufficiently excel to be employed in the ready-made linen shops'.[21] The Victoria and Albert Museum Textile Collection in London indicates that samplers made in Quaker Schools or Charity schools used coarser materials to practise useful stitches such as darning stitches and Hollie point which could be used in the production and refashioning of clothes.[22] Samplers exercised various alphabets in reversible stitches, pious verses or religious symbols, and taught geography in the form of embroidered maps, or mathematics in the form of cross stitch multiplication tables. As John Styles has shown, the textiles used as identifying tokens by mothers who had to leave their babies in the care of the Foundling Hospital were of different quality and provenance, ranging from cheap to mid-priced textiles, woven and printed.[23] Styles also suggests that labouring class girls, particularly in the country side, were taught spinning and knitting rather than sewing or embroidery as the former skills guaranteed employment; thus, bought and manufactured textiles for everyday use were not uncommon for the labouring poor.[24]

Basic craft equipment such as needles, bobbins and brass thimbles were inexpensive and not reliably mentioned in inventories but more elaborate tools and equipment made out of expensive materials such as horn and silver were dear.[25] Whilst genteel and aristocratic women such as the Bluestockings used fibre arts to perform (rustic) simplicity and frugality as emblems of civic virtue and femininity, nevertheless using expensive equipment, middling-class women and the labouring poor practised these arts as part of their everyday household duties, and as work.[26] Below I focus on unpublished epistolary exchanges between Montagu and Scott that reflect these tensions between leisure and work, social mobility and economic standing.

Feathers and artichokes: Elizabeth Montagu

Montagu, the 'Queen of the Blues', was infamous for her 'laboured' finery and display of ostentatious wealth in her residences and personal attire. When at one of the Bluestocking assemblies at Hill Street, Delany complained, 'Was dazzled with the brilliance of her assembly. It was a moderate one, they said, but infinitely *too numerous* for *my senses*'.[27] Montagu strived for the exemplary display of 'Virtue, prudence and Temperance, [that] should sometimes keep open House, and shew there is a golden mean between churlish severity of manners and lean and sallow abstinence in diet; and indecent gayety of behaviour, and that swinish gluttony which ne'er looks to Heav'n midst its gorgeous feast but crams and blasphemes its feeder'.[28] However, temperance in display, as we will see, was not always Montagu's forte.

Montagu met the fellow Bluestocking Delany (then still Pendarves) in 1735 through their mutual acquaintance the Duchess of Portland.[29] Their correspondence contained items such as feathers and shells, flowers and fabrics that were gathered and collected from many sources. Montagu went so far as to instruct her naval brother Robert to bring back shells and feathers from his journeys, asked her sister Sarah Scott to obtain feathers and ordered 'people upon all our Coasts to seek for shells, but have not yet got any pretty ones'.[30] She even asked her infamous cousin Sir Thomas Robinson, governor of Barbados, to send some shells to the Duchess:

> He shall get some shells for your Grace. He should pay you the homage of old when the conquered Nation sent some of their Earth and water to their Conquerers; he ought to do your Grace homage in every element where he has any command, and if you want either fish, beast, or bird, give him your orders, and with more than the power, take the style, of a Queen.[31]

Objects in the National Portrait Gallery's 2008 exhibition *Brilliant Women: Eighteenth Century Bluestockings* document these tokens of friendship and mutual intellectual interests in the shape of 'natural curiosities', friendship boxes, snuff boxes, poems and manuscripts.[32] Luxurious objects such as these commemorated the networks of Bluestocking friendship as distinct, personal and, most importantly, exclusive.

When Montagu purchased her house in Portman square in 1775, she not only employed renowned artists and architects such as James Stuart, possibly Angelica Kauffman, Giovanni Battista Cipriani and Matthew Boulton but also added her own designs and ideas.[33] The refurbishment and decoration of the house took ten years and resulted in, as James Harris praised, 'an Edifice which for the time made me imagine I was at Athens in a House of Pericles, built by Phidias'.[34] The pinnacle of taste and ornament was the 'Feather Room', decorated with Montagu's original designs made of feathers and later immortalized by William Cowper in his poem, 'On the Beautiful Feather-Hangings, Designed for Mrs Montagu'.[35] The screens were assembled and mounted on canvas in Sandleford, Berkshire, by Montagu's chief seamstress Betty Tull and her assistants Miss Pocklington and Mrs Fry, and managed by her sister Sarah Scott.

Montagu had some experience with feather work, notably with her friends the Duchess of Portland and Mary Delany. Scott had also tried her hand at feather painting in the 1750s and Mary Anstey, Christopher Anstey's daughter and frequent visitor to Montagu's assemblies, also dabbled in the art. Anstey however deplored that her feather work was not as sophisticated:

My feather screen makes but a poor figure yet. Could I carry it with me into company, as one does a piece of knotting it would be soon finished. But as I can only imploy that time upon it wch I have to myself it goes on but slowly. For when one can retire onto ones chamber & be still the houres may be better spent in the sorting of feathers.[36]

Montagu was blessed that Scott supported her by collecting feathers and co-ordinating some of the work. She wrote to her enthusiastically:

I have orderd some whole feathers to be sent out of my stock for present use. I will get some fine goose feathers for them as soon as Gees are slain. In the mean time I am collecting some white feathers which I will send by some opportunity from hence. Betty Tull served me a sad trick in leaving yᵉ feathers, to be moth eaten. I have a great proposal in my head in yᵉ feather way.[37]

In a later letter of 14 October, Montagu wrote to her sister: 'I have sent Miss Pocklington some Goose feathers, let me know if they want more. I have done or rather they have done for me, part of a feather trimming for a Sack. Betty & Mʳˢ Fry soon dispatch a trimming in yᵉ mosaick way'.[38] A 1786 letter to Elizabeth Carter gives us some indication of the seamstresses who contributed to Montagu's ambitious design:

My Feather work, tho of a tedious nature, had made a great progress since I left it; the ingenious Betty Tull, yᵉ clever little girl her elève, an elderly Virgin, & two old Widows having been constantly employ'd at it, besides casual assistance. Betty & yᵉ little Girl are yᵉ only Persons who can do ye fine parts but ye inferior artists do ye ground, & ye mosaic, one Widow Gentlewoman has been employd for above 4 months in stripping ye feathers of ye downy part, & preparing them for use, and, an expence I did not regret, as I had ye pleasure of observing that in the time she moulted her own threadbare garments, & acquired new & warm ones. She is ye Widow of a Farmer reduced to parish Allowance, not by his or her late Husbands fault, but various misfortunes.[39]

Whilst Montagu was instrumental in the design of her interior décor, she clearly relied on the expertise and skills of her staff which were acknowledged in private letters between Scott and Montagu, for instance, but not publicly.

But Montagu was not completely removed from processes of making as she clearly possessed the relevant skills. In a letter of October 1755, Scott wrote to Montagu, 'I much doubt whether Mrs Tull may not understand the quilting of a petticoat almost as well as you.'[40] Although keen to display her wealth in extravagant interior designs, Montagu was mindful of waste and did not want to squander away valuable resources – tapping into her public display of 'Virtue, prudence and Temperance'. In her letter of 1785, Montagu asked Scott again, 'Pray when you eat Artichokes don't be wastefull & throw away ye little leaves round ye choak we use them in our Mosaic work.'[41] This thrift fitted in with Montagu's charitable work. She was conscious of the responsibilities and duties her immense wealth brought and exercised paternalistic charity and benevolence to her colliery workers in Northumberland and the chimney sweeps in London, all displays of charity that, as Eger has rightly argued, were very public.[42] The employment of impoverished gentlewomen and the financial support of Scott and fellow writers such as Sarah Fielding were less ostentatious and were mentioned in private letters between Montagu and her close friends or her sister. Montagu's letter to Carter also indicates the scale of the project and the mixed abilities that the seamstresses and impoverished gentlewomen had. Montagu certainly was a hard taskmaster and Betty Tull's health suffered greatly during the project. When Tull was ill in 1788, Montagu eulogized:

> Poor Betty Tull is I fear going to take her flight to another World. As a virgin she might claim ye white plumes of the Ostrich for her Hearse, but her triumphs over the whole feather'd Race may give her pretensions to evry feather, of every bird, from the Eagle to ye Wren; from the Croaking Raven to the chattering Parrot.[43]

The room however was a legendary success, as affirmed by the *St James Chronicle*'s glowing praise: 'Wholly covered with feathers, artfully sewed together, and forming beautiful festoons of flowers and other fanciful decoration. The most brilliant colours, the produce of all climates, have wonderful effects on a feather ground of dazzling whiteness.'[44]

Montagu marked her social advancement from a modest companion and friend to the Duchess of Portland to the 'Queen of the Blues' and a wealthy coal magnate by producing and commissioning decorative work that moved from the category of private 'fancy work' to public art that was to celebrate 'the warmest approbation of the taste and magnificence of Mrs. MONTAGU'.[45] As Ruth Scobie has argued, the feather screen was an emblem of Montagu's social rise in polite society; 'like the collection of feathers, such a collection of exotic guests within one English house proved social and economic status, but also privileged access to British colonial and commercial networks around the world'.[46] The material and social properties of her 'objectscape' underpinned the idea of a cosmopolitan 'cultured feminine community' which the Bluestockings represented.[47]

'To embroider what is wanting': Sarah Scott

Scott did not rise in the world in an equal fashion to her sister. She married, against the will of her family, the sub-preceptor to the Prince of Wales and mathematician George Lewis Scott on 15 June 1751. Her father, Matthew Robinson, reluctantly provided Scott with a dowry. There are still speculations about the hastened separation that was clearly manoeuvred by Scott's family. In April 1752 Matthew Robinson and his sons removed Scott from her marital home in London. George Lewis Scott refunded half of the dowry to her father but continued to support Scott with £150 per annum.[48] In the ensuing years, the correspondence frequently refers to Scott's general financial worries. George Lewis Scott was not always diligent in his payments and thus compromised his wife's financial security. Thus, Scott returned to Bath as a neither single nor married woman to set up a household with her life companion Lady Barbara Montagu.

Scott belonged to a different social network than her sister.[49] This is marked not only by the gaps between their finances, homes and locales, but also by the nature of the objects they exchanged. Montagu supported Scott with food, materials and money during her financially lean years. She advised her on the latest fashion trends from London, so that Scott could refashion her clothing into a respectable and current state. Both sisters profited from their brother Robert Robinson's frequent journeys to China, but his gifts could also present new challenges. As Scott wrote to Montagu in 1752:

> I am afraid the Captain has been tardy in sending the blue lutestring. the largeness of the flowers on your sleeves & part of the body will be a fault. Upon examination I find my gown is one third plain only, has no sleeves & but one side piece instead of a whole body; I am in some hopes the Captain may find another piece with these little parts on it, but if not I shall be extremely obligd to You for any indian white silk, for the only way will be to embroider what is wanting.[50]

Scott has clearly mastered the skills of embroidery and appliqué but does not mention darning, mending or quilting *per se* in any of her extant letters.[51] It is possible to assume however that those skills were essential for maintaining her wardrobe to a proper standard. Scott frequently refers to refashioning clothes and buying and maintaining different kinds of fabrics, ranging from flannel, poplin, cottons and wool to silk.[52] Ribbons and trimmings became essential items to refresh or remake her wardrobe.

Given her sister's more prominent status in polite society, Scott often asked for advice. In a letter to her sister in February 1740/41, Scott worried 'if there is much alteration in the gown sleeves this year' as she considered how to update her wardrobe.[53] In October 1743, whilst in Bath, Scott indicates that updating her gown was part of her routine to appear respectable in polite society. She 'endur'd the pain of curling to make my countenance appear proper to be admitted to the rooms, & sow'd my trimming on to my buff gown'.[54] Scott even relied on her sister's advice to guide the shape and size of the hoops which provided the foundations for her garments: 'I shou'd be obliged to you if you wou'd in your next letter send me word what sized hoops moderate people; who are neither over lavish or covetous of whalebone, wear; because I intend to write to my hoopmaker to have one ready for me against I come to Town.' Unlike her sister, Scott remained sceptical about the waves of ostentatious fashion:

> I hope our hoops will not increase much, for we are already almost as unreasonable as Queen Dido, & don't encircle much less with our whalebones, than she did with her bulls hide, & I am afraid we are not so excuseable for her ground was to build a Town, whereas what we gain is only for a sort of wall, which in some measure hinders the trade & use of the Citty.[55]

In addition to updating or altering clothes, appearances also required that clothes would be clean which was neither straightforward nor cheap. When Scott wrote to her sister about the state of her clothes, she expressed embarrassment and shame, but recognized the representational value of more expensive fabric: 'My flower'd gown & petticoat is very dirty I shou'd be oblig'd to you if you wou'd tell me whether you think it will be necessary for me to buy a lutestring gown & petticoat as the flower'd is so dirty & my night gowns are shabby enough; but answer this in private.'[56]

In the early years of their correspondence, Montagu and Scott also exchanged embroidery and appliqué patterns for the fashionable aprons and shared in detail the progress of their fancy work:

> I return you my thanks for the leaves, & desire you will not abuse them. they are as distinct as need be, & if I can but make them big enough will be of great service to me; I am afraid the drawing them may have hurt your eyes, & if I had not been just then a little forgetfull of their weakness I wou'd not have set them so hard a task.[57]

Whilst Montagu was able to furnish her homes with silk, feathers, rare shells and lacquer work, Scott decorated her homes with found and natural materials – an early form of upcycling:

> We gilded cones corn acorns poppy heads & various evergreens with flowers & leaves in lead & some fruit in pipe makers clay, with these she [Mrs Isted] made a frame to the glass, & continued the work in a light pattern with small bracketts for eleven pieces of small china, from the top of the room to the chimney spreading over the whole pannel; it is really the lightest & prettiest thing I ever saw, & suits the rest of the room.[58]

Aware of the greater plight of impoverished gentlewomen and the labouring poor, Lady Barbara Montagu and Scott took impoverished servant girls into their houses in Bath and Batheaston, employing them to produce silk flowers and other crafts to teach them skills and economic independence. Scott and her helpers supplied her sister with silk flowers which served as ornaments to Montagu's wardrobe:

> You have not answered me about your Tissue silk, if it is for a Gown I woud advise all the Roses shoud be red, which will be excessively pretty mixed with the green leaves, & I know you have no dislike to Rosecolour. Till I know its destination, I make them do only such a number as at all Events will be red, & the remainder will be done according to your order. If You determine to have this all red Roses, those in that I sent You that are not so, I think we can so far take the colour out of & have them done over with red that it will not be perceived when put least in sight that they have ever been otherwise.[59]

Scott's charitable enterprise echoed the principles of charity and empowerment in her successful reformist novel, *Millenium Hall* (1762). *Millenium Hall* is based on the principles of vertical friendship and self-help that unite the household, tenants, the wider family and villages in the manner of a country estate. The charity work described in the novel does not only keep every member of the estate in their place but also, more importantly, help the poor and disadvantaged to provide for themselves. The women in the alms-houses sew, spin and cook for the benefit of the whole community with the understanding that everyone contributes as best as they can. The community's carpet and rug manufactory functions as a social enterprise, where the profits are invested in a 'fund for the sick and disabled'.[60] Scott was a social reformer; her political convictions and charity projects were informed by principles of Practical Christianity and self-help. Her own charity projects and the fictional *Millenium Hall* community slot into contemporary practices of textiles as basic economic work for the labouring poor or impoverished gentlewomen. The work promoted in *Millenium Hall* is plain sewing, embroidery and rug making, skills crucial to the flourishing mercantile economy.[61]

Fennetaux, Junqua and Vasset have suggested that '[p]oised between the early modern economic model of scarcity and want and the modern world of consumerism and waste, the eighteenth century was marked by a special relationship to the material where objects went through not one but several lifecycles'.[62] This is apparent in the correspondence between Montagu and Scott. Whilst Montagu had no qualms in displaying luxury and wealth, she claimed unimpeachable credentials in the pursuit of 'a golden mean between churlish severity of manners and lean and sallow abstinence in diet' which also included the careful consumption of material objects. Her social ascent to the 'Queen of the Blues', wealthy coal magnate and business woman was counterbalanced by her self-fashioning of herself and her salon as a civilizing force. Her display of quasi-public art in her house at Portland Square marked the boundary between polite and enlightened society and the vulgar.[63] Scott was in many ways Montagu's moral monitor. Her implementation of principles of Practical Christianity in her local community countered Montagu's ostentatious wealth.

Ellen Kennedy Johnson has shown that the 'type of needlework a woman performed was often determined by her social standing or class position. Notions of propriety that were linked to needlework were also affected by class or social rank.'[64] Needlework, textile arts, and the basic production and maintenance of textiles and clothes were not only markers of gender and rank, but also recorded very sensitively how women within the Bluestocking circle negotiated ideologies of femininity and domesticity. But, as I have argued, social aspirations or circumstances were also mirrored in the materiality and maintenance of textiles. Whilst we have to acknowledge that the correspondence between Scott and Montagu is not complete, the indication is that Scott was much more pre-occupied by a 'make-do and mend' approach to textiles and clothes and had the appropriate skills to do so, particularly as she managed her sister's fibre art and interior design projects.[65] Scott was dependant on her sister Elizabeth and her brother Robert, an East India captain, to acquire precious fabrics for her outfits which were, one could argue, 'above her station'. However, in opposition to Montagu, Scott took, and had to take particular care to maintain and recycle these fabrics and garments to maintain a level of respectability in eighteenth-century polite society. Reusing, repairing, refashioning and transforming were necessary socio-economic practices to uphold respectability and propriety for women such as Scott who found themselves in economically compromised situations but who mixed in genteel circles. Scott's material literacy thus straddled the boundaries of social rank by navigating the luxury textile market as a 'proxy shopper' and textile producer for her sister with materials she could not afford but had expert knowledge in. At the same time, in order to meet the social requirement to dress well, she altered, reworked and transformed her clothes, not to be fashionable but to be respectable. Scott's epistolary exchanges with her sister offer us glimpses into her own lifecycle and the lifecycle of her textiles.[66]

Notes

1   John Aikin and Anna Letitia Barbauld, 'Dialogue, on Things to be Learned', *Evenings at Home; or, the Juvenile Budget Opened* (London, 1792–96), Part I, 84–98, 87–8.

2   On women's work in the eighteenth century, see Bridget Hill, *Women, Work and Sexual Politics in Eighteenth-Century England* (Montreal: McGill Queens University Press, 1993); Jennie Batchelor, *Women's Work: Labour, Gender and Authorship, 1750–1830* (Manchester: Manchester University Press, 2010); Christine Hivet, 'Needlework and the Rights of Women in England at the end of the Eighteenth Century', in *The Invisible Woman: Aspects of Women's Work in Eighteenth-Century Britain*, ed. Isabelle Baudino, Jacques Carré and Cécile Révauger (Aldershot: Ashgate, 2004), 37–46; Smith, *Women, Work, and Clothes in the Eighteenth-Century Novel*; Daryl Hafter, ed., *European Women and Preindustrial Craft* (Bloomington: Indiana University Press, 1995); Pamela Sharpe, ed., *Women's Work: The English Experience 1600–1914* (Oxford: Oxford University Press, 1998); Susan Cahn, *Industry of Devotion: The Transformation of Women's Work in England 1500–1660* (New York: Columbia University Press, 1987).

3   Mending textiles encompassed a range of skills and techniques: (a) darning which, according to the *OED*, is a specific way of 'mending (clothes, etc., esp. stockings) by filling-in a hole or rent with yarn or thread interwoven so as to form a kind of texture … To ornament or embroider with darning-stitch'; (b) patching, 'in order to repair, strengthen, protect, or decorate it.' Both techniques included a practical, necessary and ornamental element; see 'darn,' and 'patch', *OED Online*. July 2018. Oxford University Press, accessed 14 September 2018. The stitches for mending and patching were part of traditional sewing and embroidery techniques. See Gail Marsh, *18th Century Embroidery Techniques* (Lewes: Guild of Master Craftsmen Publications, 2006) and Kathleen Kannik, ed., *The Lady's Guide to Plain Sewing, by a Lady*, Books I and II (Springfield: Kannik's Corner, 1993).

4   Susan Strasser, *Waste and Want: A Social History of Trash* (New York: Henry Holt, 1999); Ariane Fennetaux, Amélie Junqua and Sophie Vasset, 'Introduction', in *The Afterlife of Used Things: Recycling in the Long Eighteenth Century*, ed. Fennetaux, Junqua and Vasset (London: Routledge, 2015), 1–12.

5   Ibid., 2–3.

6   John Styles, *The Dress of the People: Everyday Fashion in Eighteenth-Century England* (London: Yale University Press, 2007), 60.

7   There are no direct references in the Montagu–Scott correspondence to how the sisters first acquired their textile skills and material literacy. As this chapter will show, both Scott and Montagu were versed in material terminology and knowledge as consumers and producers.

8   Claire Walsh, 'Shops, Shopping, and the Art of Decision Making', in *Gender, Taste and Material Culture in Britain and North* America, *1700–1830*, ed. John Styles and Amanda Vickery (London: Yale University Press, 2006), 151–87. See also Goggin and Tobin, ed., *Women and the Material Culture of Needlework and Textiles, 1750–1950*.

9   Letter from Elizabeth Montagu to Grace Robinson Freind, 18 May 1742. Montagu Collection, Huntington Library, San Marino, MO 965.

10   Montagu was only twenty-four years old at the time of the letter. Her correspondent was her cousin, Grace Freind (1718–76), née Robinson, the youngest daughter of William Robinson (1675–1720) of Rokeby, Yorkshire, and Anne (?–1730), daughter of Robert Walters, of Cundall, Yorkshire. Mr Freind, her husband, was Rev. William Freind (1715–66), Dean of Canterbury.

11   The education of Drake's own children was paramount and included the services of her stepfather, Dr Conyers Middleton (1683–1750), the famous Cambridge scholar and clergyman, whom the family visited several times per year.

12   Bathsua Makin, *Essay to Revive the Antient Education of Gentlewomen, in Religion, Manners, Arts & Tongues: With an Answer to the Objections against This Way of Education* (London, 1673), quoted in Frances N. Teague, *Bathsua Makin, Woman of Learning* (London: Bucknell University Press, 1998), 109–50, 128.

13   James Fordyce, *Sermons to Young Women* (London, 1766); Erasmus Darwin, *A Plan for the Conduct of Female Education in Boarding Schools* (London, 1797); Hannah More, *Strictures on the Modern System of Female Education* (London, 1799). Wollstonecraft wrote for instance, 'I have already inveighed against the custom of confining girls to their needle, and shutting them out from all political and civil employments; for by thus narrowing their minds they are rendered unfit to fulfil the peculiar duties which nature has assigned them' in *A Vindication of the Rights of Woman: With Strictures on Political and Moral Subjects* (1792; London, 1796), 391.

14   Hester Chapone, *Letters on the Improvement of the Mind, Addressed to a Young Lady* (1773; London, 1774), Vol. II, 62–4.

15   Fancy work included knotting, tambour work, appliqué and tatting. Tatting was fittingly called *frivolité* in French. See Thérèse de Dillmont, 'Tatting' in *The Encyclopedia of Needlework* (1884): http://encyclopediaofneedlework.com, accessed 22 May 2020. See also Vickery, 'The Theory and Practice of Female Accomplishment', 94–109. On the more subversive potential of needlework, see Fennetaux, 'Female Crafts: Women and Bricolage in Late Georgian Britain 1750–1820', 91–108; and also Crystal B. Lake in this volume.

16   In French literature of the time, the opposition is set up between 'tisser' and 'broder' and is on the same ideological continuum as the binary juxtaposition of 'text' and 'textile'. See 'Ouvrages de dame? Ouvrages d'une dame? Présentation du thème', *Cahiers Isabelle de Charrière/Belle de Zuylen Papers: 'Women's work: pens and needles of Belle de Zuylen'*, 1 (2006): 9–17. On the opposition between needle and pen, see Carol Shiner Wilson, 'Lost Needles, Tangled Threads: Stitchery, Domesticity, and the Artistic Enterprise in Barbauld, Edgeworth, Taylor, and Lamb', in *Re-Visioning Romanticism: British Women Writers, 1776–1837*, ed. Carol Shiner Wilson and Joel Haefner (Philadelphia: The University of Pennsylvania Press, 1994), 167–90; King, 'Of Needles and Pens and Women's Work', 77–93. See also Crystal B. Lake in this volume.

17   Letter of Margaret Cavendish Bentinck, Duchess of Portland, to Elizabeth Montagu, 30 June 1738. Elizabeth Montagu Collection, Huntington Library, San Marino, MO 176.

18   Mary Delany, 'Letter of Mary Delany to Mrs Port, 19 November 1771' in *The Autobiography and Correspondence of Mary Granville, Mrs. Delany*, ed. Augusta Waddington Hall, 3 vols. (London: Bentley, 1861), II, 370.

19   Mary Delany, *Letters from Mrs. Delany (Widow of Doctor Patrick Delany) to Mrs. Frances Hamilton, from the Year 1779, to the Year 1788: Comprising Many Unpublished and Interesting Anecdotes of Their Late Majesties and the Royal Family* (London: Longman, Hurst, Rees, Orme and Brown, 1821), 33.

20   Sylvia Groves, *The History of Needlework Tools and Accessories* (Newton Abbot: David & Charles, 1973), 30.

21   Mary Wollstonecraft, *Mary and the Wrongs of Woman*, ed. Gary Kelly (Oxford: Oxford University Press, 1980), 113.

22   Browne and Wearden, *Samplers*. However, the archives of the London Foundling Hospital qualify the idea that the labouring poor used solely coarse and cheap materials on the one hand and that the labouring poor were all proficient in 'plain sewing' on the other hand.

23   See also Ariane Fennetaux who indicated that re-using fabrics and clothes was motivated by thrift and sentimental reasons, 'Sentimental Economics: Recycling Textiles in Eighteenth-Century Britain', 122–42.

24   John Styles, *Threads of Feeling: The London Foundling Hospital's Textile Tokens, 1740–1770*, 58–61.

25   See, for instance, the trial of Sarah English in 1744 where she steals aprons and handkerchiefs in order to pawn the goods for a spinning wheel which would secure her a future income: *Old Bailey Proceedings Online* (www.oldbaileyonline.org, version 7.0, 25 October 2012), April 1744, trial of Sarah English (t17440404-4).

26   The Bluestockings had to carefully negotiate gendered discourses on feminine virtues and intellectual accomplishments. See Nicole Pohl and Betty A. Schellenberg, 'Introduction: A Bluestocking Historiography', *Huntington Library Quarterly* 65, no. 1 (2002): 1–19 and Harriet Guest, *Small Change: Women, Learning, Patriotism, 1750–1810* (Chicago: University of Chicago Press, 2000).

27 Delany, *Autobiography,* II, 97.

28 Letter of Elizabeth Montagu to William Pepys, 14 August 1781, Montagu Collection, Huntington Library, San Marino, MO 4069.

29 Emily J. Climenson, *Elizabeth Montagu: The Queen of the Bluestockings,* 2 vols. (New York: E. P. Dutton, 1906), 1: 18.

30 Ibid., 1: 18. Letter of Elizabeth Montagu to Margaret Harley Cavendish Bentinck, Duchess of Portland, 5 May 1741, Montagu Collection, Huntington Library, San Marino, MO 297.

31 Letter of Elizabeth Montagu to Margaret Cavendish Bentinck, Duchess of Portland, January 1741/42, Montagu Collection, Huntington Library, San Marino, MO 317. On the exotic materials for the feather screen, see Ruth Scobie, "'To Dress a Room for Montagu": Pacific Cosmopolitanism and Elizabeth Montagu's Feather Hangings', *Lumen* 33 (2014): 123–37.

32 Elizabeth Eger and Lucy Peltz, *Brilliant Women: Eighteenth-Century Bluestockings* (London: National Portrait Gallery, 2008), 37–9.

33 See Rosemary Baird, *Mistress of the House: Great Ladies and Grand Houses 1670–1830* (London: Weidenfeld & Nicholson, 2003), 284 (n. 129), on Kauffman's involvement.

34 Letter of James Harris to Elizabeth Montagu, 4 November 1780, Montagu Collection, Huntington Library, San Marino, MO 1133.

35 *Gentleman's Magazine* 58 (June 1788), 542.

36 Letter of Mary Anstey to Elizabeth Montagu, 1752, Montagu Collection, Huntington Library, San Marino, MO 108.

37 Letter of Elizabeth Montagu to Sarah Scott, 28 August [1774], Montagu Collection, Huntington Library, San Marino, MO 5959.

38 Letter of Elizabeth Montagu to Sarah Scott, 1774, Montagu Collection, Huntington Library, San Marino, MO 5964.

39 Letter of Elizabeth Montagu to Elizabeth Carter, 25 September 1781, Montagu Collection Huntington Library, San Marino, MO 3517.

40 Letter of Sarah Scott to Elizabeth Montagu, October 1755, Montagu Collection, Huntington Library, San Marino, MO 5250.

41 Letter of Elizabeth Montagu to Sarah Scott, [26 August 1785], Montagu Collection, Huntington Library, MO 6113.

42 Elizabeth Eger, *Bluestockings: Women of Reason from Enlightenment to Romanticism* (London: Palgrave Macmillan, 2010), 78.

43 Letter of Elizabeth Montagu to Elizabeth Charlton Montagu, 17 December 1788, Montagu Collection, Huntington Library, San Marino, MO 2975.

44 *St. James's Chronicle or the British Evening Post* (London, 11–14 June 1791).

45 Ibid.

46 Scobie, "'To dress a room for Montagu'", 126.

47 Jo Dahn, 'Mrs. Delany and Ceramics in the Objectscape', *Interpreting Ceramics,* 1 (2000).

48 Climenson, *Elizabeth Montagu,* II: 7.

49 On the different social networks that Elizabeth Montagu belonged to, see Anni Sairio, *Language and Letters of the Bluestocking Network: Sociolinguistic Issues in Eighteenth-Century Epistolary English* (Helsinki: Société Néophilologique, 2009).

50 Letter of Sarah Scott to Elizabeth Montagu [1752], Montagu Collection, Huntington Library, San Marino, MO 5228.

51   Scott likely acquired mending and darning skills as part of her education in needlework and embroidery. Print sources suggest that such mending knowledge was assumed: *The Lady's Magazine; or Entertaining Companion for the Fair Sex, Appropriated Solely to Their Use and Amusement* (1770–1847) does not give specific guidance on mending or darning in the period between 1770 and 1818. I thank Jennie Batchelor for her help with this question.

52   Flannel was fine woven woollen fabric made from carded wool or worsted (*OED*); Poplin was strong plain-woven fabric made from various fibres (*OED*).

53   Letter of Sarah Scott to Elizabeth Montagu, 4 February [1740/41], Montagu Collection, Huntington Library, San Marino, MO 5164.

54   Letter of Sarah Scott to Elizabeth Montagu, [October 1743], Walpole Collection, Lewis Walpole Library.

55   Letter of Sarah Scott to Elizabeth Montagu, 25 February [1740/41], Montagu Collection, Huntington Library, San Marino, MO 5167.

56   Letter of Sarah Scott to Elizabeth Montagu, [29 June 1744], Montagu Collection, Huntington Library, San Marino, MO 5189. On the importance of cleanliness, see Styles, *The Dress of the People*, 78.

57   Letter of Sarah Scott to Elizabeth Montagu, 11 [November 1741], Montagu Collection, Huntington Library, San Marino, MO 5170. Leaves were used for embroidery work (*OED*). Whilst embroidered aprons were fashionable, they were not always acceptable. The Duchess of Queensbury was publicly criticized by Richard Nash, then Master of Ceremonies at Bath, for wearing an apron to the Assembly Rooms.

58   Letter of Sarah Scott to Elizabeth Montagu, [1752], Montagu Collection, Huntington Library, San Marino, MO 5223.

59   Letter of Sarah Scott to Elizabeth Montagu, [1763], Montagu Collection, Huntington Library, San Marino, MO 5308. Silk tissue is an organza type of fabric from India. Scott used it for the silk roses as it can be perfectly transformed by crumpling a piece in water.

60   Sarah Scott, *A Description of Millenium Hall*, ed. Gary Kelly (1762; Peterborough: Broadview Press, 1995), 247.

61   On material practices in the novel, see Smith, 'Gender and the Material Turn', 159–78.

62   Fennetaux, Junqua and Vasset, 'Introduction', 2.

63   See Carol Duncan, *Civilizing Rituals: Inside Public Art Museums* (London: Routledge, 1995). See also Emma Major, 'The Politics of Sociability: Public Dimensions of the Bluestocking Millennium', *Huntington Library Quarterly* 65, no. 1 (2002): 175–92.

64   Ellen Kennedy Johnson, 'Alterations: Gender and Needlework in the Late Georgian Arts and Letters' (PhD Dissertation, Arizona State University, 2009), 5.

65   For a printed version of Scott's letters, see Nicole Pohl, ed., *The Letters of Sarah Scott*, 2 vols. (London: Pickering and Chatto, 2014). *The Elizabeth Montagu Collection Online* (*EMCO*) will publish the remaining letters by Montagu to Scott and indeed all extant letters by Elizabeth Montagu. See http://www.elizabethmontaguletters.co.uk/emco, accessed 22 May 2020.

66   On the lifecycle methodology, see Pernilla Rasmussen, 'Recycling a Fashionable Wardrobe in the Long Eighteenth Century in Sweden', *History of Retailing and Consumption* 2, no. 3 (2016): 193–222. See also *The Social Life of Things: Commodities in Cultural Perspective*, ed. Arjun Appadurai (Cambridge: Cambridge University Press, 1986).

6

Material literacies of home comfort in Georgian England

Jon Stobart

Comfort has been seen as a key imperative in shaping the material culture of the eighteenth-century home. An array of technical innovations in heating, lighting, plumbing and furniture design was coupled with changes in the way that houses were conceived and organized to produce places that were more comfortable and convenient (ideas that frequently went together).[1] The emphasis in the scholarship is firmly on the material. Homes were purposely designed and furnished for physical comfort; they were warmer, lighter and better ventilated, and filled with furniture designed for bodily ease and convenience of use. In different ways and different contexts, John Crowley and Joan DeJean argue that these material changes were increasingly driven by a shift in the mentality of homeowners who came to prioritize comfort and convenience over display and took pleasure in the luxuriant comfort and ease of their 'modern' homes.[2] In this way, comfort was seen as a measure of material and societal progress, a notion encapsulated in William Cowper's 1785 poem *The Sofa*: 'Thus first necessity invented stools,/Convenience next suggested elbow chairs,/And luxury th' accomplished Sofa last.'[3] We move from necessity through convenience to luxury as the physical comfort of the sitter increases over time. More generally, comfort was seen as a way in which luxury could be critiqued from a morally secure stand-point. Marie-Odile Bernez argues that it allowed English contemporaries to critique the waste, inequality and immorality of foreign and especially French luxury with virtuous English qualities. In England, the argument went, comfort and convenience were available to all; material comfort was inclusive, a reflection of English democracy as well as the country's social and economic development.[4] In this way, it was aligned with what Jan de Vries has termed the 'new luxury', defined by both its inclusivity and its association with polite sociability.[5]

Unlike taste, which had to be learnt and constantly honed, most obviously through interaction with the right people and places, some level of comfort could be achieved by anyone with the means to buy the requisite material objects. Jobbing architects such as William Halfpenny could offer off-the-peg designs for 'convenient and decorated' houses, putting into practice the grander conceptions of men like Jacques-François Blondel, whose designs paired specialized rooms and apartments with *dégagements* offering convenient access between different parts of the house.[6] Equally important, but less well understood, are the ways in which furniture makers valorized and communicated ideas of comfort to their customers. Akiko Shimbo notes the growing interest in comfort in the promotional material produced by some furniture makers and the often celebratory descriptions they included in publications such as Rudolph Ackermann's *Repository*. Furniture makers were eager to highlight the technical ingenuity that increased the comfort and convenience of their products: chairs with complex reclining mechanisms, desks with hidden drawers and fold-out reading tables, and the like.[7] These publications merit closer inspection in terms of what they can tell us about how comfort and convenience were conceived and valorized by producers and how they were understood and valued in material terms by consumers. How did producers promote the idea and ideal of comfort, and what aspects of comfort did consumers seek out and value? More specifically, in what ways did these link to material literacy and corporeal experience?

This chapter attempts to address these questions through the promotional material produced by furniture makers, the bills and accounts of customers, and the diaries and correspondence of genteel and aristocratic householders. It starts by assessing how the discourse of pattern books and directors, and invariably upbeat descriptions in publications like Ackerman's *Repository*, linked comfort on the page to experiential comfort for buyers and owners. It traces this link further through analysis of the furniture and furnishings promoted through trade cards and auction catalogues to explore how comfort and convenience were described and marketed in overtly commercial publications. Finally, it considers how consumers conceived and appreciated comfort in terms of the materiality of their homes: what evidence is there that they internalized the rhetoric of pattern books and promotional sources? Overall, my chapter connects comfort in the home and of the body to processes of commercial exchange through discourses and understandings of material objects: language and illustrations were used to reach into the homes of consumers by focussing on the physical experience of furniture.

Constructing comfort: The discourse of pattern books and puffs

Following Thomas Chippendale's 1754 *Gentleman and Cabinet Makers' Director*, growing numbers of English furniture makers issued pattern books that promoted both their skills and taste, and highlighted their flexibility in terms of furniture design. In his pattern books, Chippendale showcased the wide range of different styles for his chairs and other pieces.[8] This virtuosity was framed in terms of taste and utility, rather than comfort; the subtitle of his *Director* is a 'large collection of the most elegant and useful designs', reflecting the broader importance of utility and elegance in furniture advertisements. No mention is made of comfort or its synonyms, convenience and ease. Much the same was true of other pattern books published in the 1760s to the 1780s. Robert Manwaring's *The Chairmaker's Guide* (1766) and John Crunden's *Joyner and Cabinet Maker's Darling or Pocket Director* (1770), for instance, both place similar emphasis on elegant taste and beauty, Manwaring's book being subtitled 'upwards of two hundred new and genteel designs'.

It is not until the 1780s that the material qualities of furniture were explicitly promoted in ways that align with the ease they might afford their owners. A. Hepplewhite's *Cabinet Maker's and Upholsterer's Guide* (1788) makes no mention of comfort *per se*, but it does highlight the way in which a gentleman's stool 'by being so easily raised or lowered at either end, is particularly useful to the afflicted'.[9] Such promises drew on the association often made between comfort and the ill or invalid: their physical ailments rendered them more in need of bodily ease.[10] Thomas Sheraton's 1803 discussion of beds brings this into sharper focus. He argues that the material qualities of beds 'should be regulated by the nature of the constitution of those who are to sleep on them'. Firm beds are necessary for 'those who are of a delicate frame, to whom scarcely anything can be more hurtful than to sink into soft down. These sorts of beds are better adapted to the robust and healthy, who can with propriety sustain the indulgence'.[11] In this way, Sheraton directly equates the material qualities of the soft feather bed with the level of comfort it would provide to different people, but his *Dictionary* more often links physical ease with the use of furniture. When discussing dining tables, for instance, he suggests that the size 'may easily be calculated, by allowing 2 feet to each person' sitting at table; less than this cannot with comfort be dispensed with'.[12] Chairs were differentiated by use, their names 'answering to their appropriate design, whether for furnishing particular rooms, or for our accommodation in cases of ease and convenience'. Hence we have a variety of arm chairs 'for ease … and for particular rooms'.[13] However, whilst design was linked to comfort, Sheraton does not identify the precise material qualities that make the arm chair easy or comfortable. He does note that cushions from the back of a sofa might 'serve at time for the bolsters, being placed against the arms to lol against'.[14] Writing some twenty years later, Richard Brown is more specific: 'Many chairs are very uncomfortable to sit upon, in consequence of the raised carved work on the splat and tablet of the yoke rail'.[15]

Such rhetoric allowed these furniture makers to communicate the tactile qualities of their wares, highlighting the physical connections between material products and the body. Armed with such information, other furniture makers or householders (the two intended audiences of these pattern books) could understand from its form and construction the comfort that a chair might provide to the sitter. They might also be in a position to extrapolate from these explicit comments, inferring physical ease from the material descriptions of sofas – the acme of comfortable seating. Describing a duchesse, Hepplewhite notes that 'the stuffing may be of the round manner as shown in the drawing or low-stuffed, with a loose squab or bordered cushion fitted to each part'.[16] He assumes that his reader (the potential consumer) has some familiarity with the techniques and components described, and might even be able to recognize the ways in which stuffing and cushions would affect not only the appearance but also how it would feel to sit in the sofa. Here, they no doubt brought together their previous corporeal experience of such furniture with the descriptions on the page. In these promotional discourses, didactic and kinetic learning went hand in hand – a point to which we shall return later.

More generally, Hepplewhite and Sheraton showed greater interest in convenience than comfort. They were most effusive when it came to the ways in which design could make familiar pieces of furniture more convenient, often through the ingenuity of the designer or maker. Hepplewhite offered designs for a writing table, desk and bookcase, and dressing table, amongst other things. He describes all of them in terms of their conveniences, which in this case should be understood as their little drawers, shelves and containers; such spaces allowed small items to be stored safely, each thing having its own particular place from which it could easily be retrieved. Sheraton's ladies' writing table, meanwhile, had an ink well and pen drawer that would 'fly out by themselves, by the force of a common spring, when the knob on which the candle-branch is fixed is pressed'.[17] The piece as a whole might be rendered convenient through its fitness for purpose, the scale of construction or the ease with which it might be moved. Thus, Hepplewhite's tambour writing table was a 'very convenient piece of furniture, answering all the uses of a desk with a much lighter appearance', whilst Sheraton's version of the same piece, 'being made for the convenience of moving from one room to another, there is a handle fixed on to the upper shelf'.[18] Being moveable without needing to exert great physical effort made such pieces well suited to the more informal ways of living which, as we shall see, increasingly characterized the English home.

This interest in convenience and comfort was balanced by a persistent concern with fashion and aesthetics. Hepplewhite promised his readers designs that were in the 'newest and most approved taste' and Sheraton described pieces as elegant and handsome as often as he lauded their convenience or ease. If this contradicts Crowley's assertion that comfort emerged forcibly as a mental category and material concern during this period, a subtle change in balance is discernible in the early decades of the nineteenth century. Writing about a dressing table, Brown argued that 'in addition to the *essential* modifications of utility and convenience, the *secondary* objects, elegance and beauty, are indispensably necessary to be studied to render each piece of furniture and graceful and pleasing article'.[19] Consumers are thus invited to consider convenience as their primary concern and, by implication, to bring to the fore all they had read about the ingenuity and practicality of designs.

This rebalancing of taste and comfort is still more evident in the monthly articles published in Ackermann's *Repository* under the heading of 'Fashionable Furniture' – a title which itself highlights the continued importance of fashion and taste to the domestic interior. These articles comprised detailed descriptions of the latest designs and offered readers a clear framework for identifying the materiality as well as the aesthetics of the featured objects. Two examples serve to illustrate the kind of information conveyed. The October 1809 article (Figure 6.1) centres on a Grecian settee (in the form of a window-seat). The materials (mahogany, morocco leather, silk) and construction (lattice-backed, fringed, French-stuffed) are carefully itemized, building a comprehensive material picture. Such catalogues enhanced readers' tactile understanding of the feel of the object, drawing on a shared material literary between maker and consumer. Moreover, the article declares that the settee is tasteful and comfortable, 'not only affording the highest degree of comfort and convenience, but being also an elegant and fashionable ornament'.[20] The next summer, the featured pieces were 'two of the most convenient and comfortable library chairs perhaps ever completed'.[21] Again, the author itemizes the materiality of the chairs before the description turns to the particular features which render them comfortable. The first is fairly conventional, with a moveable reading stand and candle holder, but it promises

A FRENCH WINDOW CURTAIN
& GRECIAN SETTEE.

Figure 6.1 A French Window Curtain & Grecian Settee, *The Repository of Arts*, October 1809, Philadelphia Museum of Art, Library and Archives.

to be 'completely comfortable'. The second offers the gentleman the option of sitting astride the chair, facing a reading desk at the back, or 'when its occupier is tired of the first position' to move the desk around and sit sideways. Importantly, it is the design that assures comfort: 'the circling arms in either way form a pleasant and easy back, and also, in every direction, supports the arms'. Again, the piece draws attention to the materiality and design of the chair, linking this directly to the comfort and convenience of the imagined bodies that might one day occupy it. Another year on, attention focused on a version of the metamorphic library chair which combined an 'elegant and truly comfortable arm-chair, and a set of library steps'.[22] The feature describes the process of metamorphosis, but again identifies specific materials, offering reassurance about the chair's robust character: the steps, we are told, are 'as firm, safe, and solid as a rock'. Both design and materiality work to underscore and reaffirm the chair's comfort and convenience.

All three library chairs – and many other pieces featured as 'Fashionable Furniture' around this period – were available from Morgan and Saunders of Catherine Street, the Strand, not far from Ackermann's own shop. It is possible, therefore, to view these articles as puffs for these furniture makers, but they nonetheless helped to develop the material literacy of potential consumers, allowing them to gain a fuller understanding – or perhaps reinforcing their existing knowledge – of what made furniture comfortable and convenient.[23] For Ackermann and his readers, these qualities were not abstract concepts, but real sensations and experiences. Alongside reassurances of taste and respectability, understanding how materials and design made a chair comfortable and convenient was thus central to choosing the most appropriate pieces of furniture for the home.

Selling comfort? The rhetoric of trade cards and sale catalogues

Pattern books and features in journals helped to establish a crucial framework, for both furniture makers and householders, for communicating and interpreting the language of physical comfort. However, most people bought from more workaday furniture makers and upholsterers, who advertised their skills and products in newspapers and trade cards, or second-hand, from dealers and house sales.[24] These tradesmen and auctioneers were perhaps more directly interested in selling wares rather than promoting designs, which makes their choices in terms of emphasis and language particularly revealing of what made furniture desirable to the majority of consumers. Their descriptions and images articulate how sellers and buyers understood the materiality, comfort and aesthetics of furniture.[25]

Comfort and convenience were not major concerns on trade cards of London tradesmen before the turn of the nineteenth century. Those issued by John Lawrence of Spital Square (1770), Charles Grange & Sons at the Royal Bed, near St Sepulchre's Church (*c.* 1775), and James Brown of St Paul's Churchyard (*c.* 1780) all emphasize fashion, taste and gentility.[26] These claims were often linked with assertions of the quality of the materials and workmanship. As late as *c.* 1810, Abraham Allen announced his new furniture warehouse on Pall Mall with a trade card that promised to supply 'the

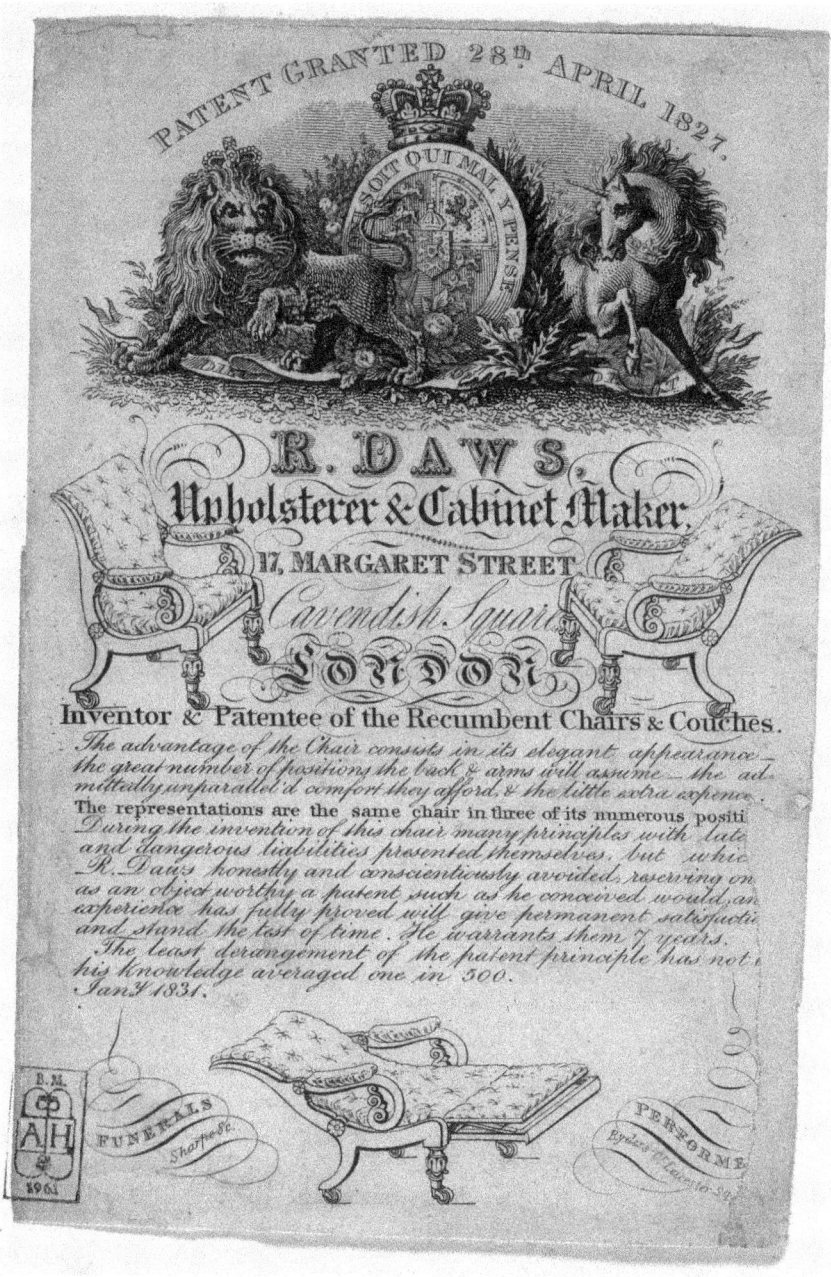

Figure 6.2 Trade card for R. Daws, Upholsterer & Cabinet Maker, 1827, British Museum, 125.25.

most fashionable furniture calicoes' and 'every kind of furniture to which such goods are adapted with an elegance and taste not usually met with'. Around the same time, John McClean & Sons were promoting the 'elegant Parisian furniture' available at their showrooms near Portland Place.[27] This emphasis on taste and fashion matches very closely that seen in contemporary pattern books. However, bodily comfort was implied in both. Pattern books included designs for sofas and arm chairs, even if they were not explicitly presented as affording comfort, and trade cards contained numerous images of such furniture, without it being mentioned in the accompanying text. The sole image on John Read's 1771 bill head, for example, is a low-seated easy chair with a high and angled back, whilst McClean's depicts an idealized interior which includes a similar chair and a chaise longue.[28]

As with the pattern books, convenience and ingenuity of design constitute marketing touchstones on trade cards, demonstrating the material ideals these tradesmen, furniture makers and dealers wished to communicate to their customers. The metamorphizing library chair discussed earlier underlines the key role of innovative construction, but patent furniture offers further evidence of how technical innovation was promoted to consumers.[29] A particularly broad range of patent furniture was featured on a trade card issued in 1806 by T. Butler of Catherine Street, London. This included sofa beds, 'forming an elegant Sofa and may be transformed at great ease into a complete Four Post Bed' and chair beds which similarly converted into tent beds.[30] A similar focus on both the ingenuity of design and convenience of use appears in a series of trade cards and advertisements issued by William Pocock. A double-page advertisement dating from 1814 featured perhaps his most famous piece: the 'Boethema or Rising Mattress', designed to raise invalids to a sitting position in bed and, as shown in Pocock's advertisement, available in the form of a sofa or a chaise longue.[31] Alongside these were his patent sofa beds, which 'make a comfortable and convenient Sofa and Bed' and were aimed less at the infirm and more at those looking for versatility. Whilst convenience came from them being 'very portable by folding into a very small Compass', their comfort was asserted rather than explained. R. Daws was more explicit in expounding the virtues of his Recumbent Chairs and Couches, for which he was granted a patent in 1831. These were essentially a development of earlier reclining chairs, but Daws argued that 'the advantage of the Chair consists in its elegant appearance, the great number of positions that the back and arms will assume [and the] unparalleled comfort they afford'.[32] Daws's customers were thus guided through the features that assured bodily ease; they could then apply these when assessing the comfort and convenience of other (competing) products.

One venue in which customers may well have been looking was the household sale, which remained an important source of second-hand furniture and other household goods well into the nineteenth century. Household sales often offered good-quality pieces at a fraction of the price of new goods, but caution was needed to avoid shoddy replicas.[33] The buyer thus required some skill in assessing furniture and the auctioneer needed to reassure potential customers about the provenance and quality of items for sale. Unsurprisingly, then, the covers of auction catalogues frequently declared the goods as the genuine property of a named individual. Many catalogues contained detailed descriptions of the goods, often focussing on their material and aesthetic qualities. The drawing room sofa available for purchase at Barton Hall in 1784 was 'stuff'd in canvas with the best materials, cover'd with rich crimson silk damask' and came with two bolsters, whilst the 'excellent sofa' in the drawing room of the rectory at Rushton (1809) had a 'mahogany frame, on brass rollers, hair back, mattress, 2 bolsters, covered in canvas, & rich chints cover'.[34] The appearance is emphasized, but so too is the nature of the stuffing and cushions – something which is still more evident in the sale at Crick rectory in 1830 where the drawing room included six lots *en suite*:

A *very elegant 6-ft carved rosewood Grecian chaise longue*, the back, side, and seat cushions stuffed with horse hair, and a reclining pillow, in blue merino damask

A mahogany framed 5-ft 7 sofa, stuffed with horse-hair, thick seat cushion, 3 back cushions, 2 bolsters, and 2 reclining pillows, in blue merino damask

A pair of 4ft 6 ottomans, with thick seat and back cushions, in blue merino

A pair of indulging chairs, stuffed thick with horsehair, in blue merino

A mahogany-framed Woburn chair, stuffed with horsehair, and 2 seat cushions, in blue merino

A *pair of cabriole chairs, a pair of mahogany elbow chairs, and 1 single ditto, in blue merino damask*[35]

The repeated references here to stuffing, bolsters, cushions and reclining pillows highlight the auctioneer's interest in communicating the suite of furniture's comfortable feel alongside its tasteful appearance. Reading the catalogues, potential buyers could both increase their literacy in the terminology used and its material implications and (therefore) assess the desirability of particular pieces.

The filling of beds was also crucial to catalogue descriptions, with distinctions made between feathers, fine feathers, goose feathers and down. The combination determined the price, but also the softness of the bed, allowing potential buyers to judge both the quality and the comfort that different beds might offer. At the sale of Sir Richard Cave's household goods in 1792, for instance, the 'full size goose and down feather bed in a bordered tick, bolster and 2 down pillows' found in the Principal Bed Chamber could clearly be understood as softer and superior in quality to the 'goose feather bed, bolster and 2 pillows' in the Cotton Bed Chamber.[36] Curtains too were described in terms of their materiality and aesthetics. At Crick rectory, there were 'two pair of elegant blue merino damask window curtains, trimmed with silk gymp, deep fringed valens, holders and tassels' to match the drawing room furniture. At Barton Hall, the sale included 'three very rich crimson silk damask festoon window-curtains, lined, with tassels and fringe, 3 1-qr yds. long'.[37] With the latter in particular, the richness of the textile is placed alongside the practicality of their lining and size. Such drapery would communicate the taste of the new owner, but it also afforded them physical comfort by being large and thick enough to cut out drafts as well as light from windows.

In these examples from sales catalogues, the material literacy of the buyer was taken as read: customers were credited with understanding the meaning of textile names and design descriptions, and what this meant for their comfort and convenience, but also for the aesthetics of the pieces. Potential buyers were clearly expected to know the relative softness of regular and goose feathers; the different experience of sitting on French stuffing, as well as its particular aesthetic; how the cushions, bolsters and pillows on a sofa affected its appearance and comfort. They also understood what a rising top might be and could imagine the convenience of a fire screen that converted into a reading table.

Comfortable living: Informality and materiality

What increasingly struck people about homes towards the end of the eighteenth and especially in the early nineteenth century was their informality. Whilst Mark Girouard argues that this first appeared in British country houses from the 1770s, it became a leitmotif around the turn of the century, linked with ideas of relaxation and comfort. This informality was apparent in a growing range of interior illustrations, often produced by amateur artists, but it was also seen in the letters and journals of those who occupied and visited these houses.[38] In 1799, Lady Louisa Stuart wrote to her sister about a house she was visiting in East Lothian. It was, she thought,

> a most excellent one ... all is new and nicely furnished in the most fashionable manner. It wants nothing but more furniture for the middle of rooms. I mean all is set out in order, no comfortable tables to write or read at; it looks like a fine London house prepared for company; quite a contrast to the delightful gallery at Dalkeith, where you can settle yourself in any corner.[39]

We see her understanding of the type of furniture necessary for a comfortable interior, interestingly expressed in terms of what DeJean calls 'convenience furniture' (that is small tables and desks) rather than sofas and easy chairs.[40] But she also makes clear that the arrangement of the room in an informal manner was critical to making it comfortable. Much the same is seen in the account written by Prince Puckler-Muskau when visiting Guy's Cliffe in Warwickshire some twenty-five years later:

> The interior is fitted up with equal attention to taste and comfort ... In the [drawing] room itself sparkled a cheerful fire; choice pictures adorned the walls, and several sofas of various forms, tables covered with curiosities and furniture standing about in agreeable disorder, gave it the most inviting and home-like air.[41]

Here, the sofas combine with convenient tables – arranged in the same informal manner lauded by Lady Stuart and here described as 'agreeable disorder'. The arrangement produced a relaxed and welcoming atmosphere.

Such assemblages of furniture were common in the drawing room furniture advertised in auction catalogues. At Brixworth Hall in Northamptonshire, for example, there were a sofa, four conversation stools and ten upholstered elbow chairs, together with a writing desk, a Pembroke table, a pair of circular card tables and another writing table.[42] The material comfort and convenience of individual pieces are written into the descriptions: the hair squab and two bolsters on the sofa and the 'drawer, slider, and sliding back screen, and a pair of plated patent lamps and shades' of the writing table. The overall comfort of the room comes from the assemblage and its imagined arrangement in the consumer's home.

The same considerations of convenience and comfort, alongside the imperatives of taste, can be seen when home owners bought new goods. After inheriting Stoneleigh Abbey from his uncle in 1813, James Henry Leigh immediately set about refurnishing the principal rooms, which had been largely unchanged since the 1760s. He spent handsomely, laying out £13,862 on furniture, pictures, plate and china over the next ten years.[43] His purchases included five ottoman sofas, two rosewood sofas, a chaise longue, a Grecian couch and an antique sofa.[44] Some were large fixtures around which other pieces would be set. For example, the two ottoman sofas purchased from the London upholsterer David Taylor were 19 feet 3 inches by 14 feet 9 inches, and 17 feet 5 inches by 5 feet 5 inches; both covered with crimson silk and ornamented with gold silk lace. They were clearly opulent items, designed to impress his guests and display his wealth and taste; but they were also built to be comfortable, with 'thick best horsehair quilted squabs & thick back cushions'.[45] This combination was repeated with the antique sofa bought from John Johnstone which had

> mouldings and foliage leaves all out of the solid wood, handsome ornamented legs & on strong castors, the whole richly finished in burnished and mat gold, the ends scrolled to form bolsters double stuffed and bordered in fine brown linen, with thick double stuffed squab seat with feather cushions to the end and back.[46]

What is particularly significant here is the way that this richly ornamented and deeply upholstered sofa was set on castors that made it possible to move even quite a substantial piece around the room with relative ease. Changing the position of sofas as well as chairs, small tables, writing desks and the like further increased the possibility for informal and flexible arrangements, adding to the comfortable feel of the room as a whole.

That James Henry sought to create rooms that were socially comfortable in their arrangement is further underscored by his purchases of convenience furniture. From Chipchase and Proctor he bought two circular loo tables, a backgammon table and two sofa tables, all made from rose wood inlaid with brass; Morel and Hughes supplied a set of rosewood trio tables, a ladies writing desk and a foot hassock (which might raise the feet and further enhance sitting comfort), and John Johnstone provided two more sofa tables and a 'Hope table', alongside the three ottomans noted above.[47] As with the antique sofa, many of these pieces were fitted with castors making them easy to arrange in the kind of informal groupings which Lady Stuart and Prince Puckler-Muskau so much admired.

Other consumers displayed similar close attention to the physical comfort of visiting friends and guests in the less public spaces of their homes. Wilbraham Egerton, for instance, was eager to refit his guest bedrooms and dressing rooms with comfortable and convenient furniture, as shown in the purchases he made from Gillows for Tatton Park in Cheshire.[48] For each, he chose a four-poster bed complete with furniture, hair and flock mattresses, and a feather bed and bolster; a wardrobe, dressing table, wash stand and chest of drawers; a night table, pot cupboard and bed steps, a writing table, usually six chairs, and a Grecian couch. The assemblage comprised individual pieces clearly designed for convenience. The washstand, for example, contained numerous spaces for basins, brush trays, soap dishes, decanters, essence bottles and chamber vases. Having everything in its correct place and to hand might be seen as ergonomically efficient; it also made for a neat and ingenious piece of furniture. It is difficult to know the extent to which Egerton shaped the assemblages, but he appears to have chosen from standard designs illustrated in Gillows's *General Sketch Book*. However, this does not mean that customers lacked knowledge and awareness of the design and material qualities of the furniture that they were commissioning. Egerton chose from a range of possibilities, and many of Gillows's other customers engaged in detailed correspondence about the material and design characteristics of the furniture they required.[49] In the same year that they were fitting out bedrooms at Tatton Park, Gillows wrote to Mrs Winckley of Preston concerning a bureau bedstead that folded into a piece of drawing room furniture when not in use. Of this highly convenient piece, they noted that 'the front may lift up as you describe to form the tester or the doors may be hinged as usual & fall back to cover the 2 wings and have a tester frame to lift up'. The customer has suggested one design and, whilst pointing out a more common alternative, Gillows conclude that they 'shall complete it as you may please to determine'.[50]

Customer correspondence illustrates the material literacy that consumers exercised when ordering furniture to make their homes more comfortable and convenient. Such knowledge was implicit in the instructions given by his housekeeper in 1837 when Matthew Robinson Boulton (son of the famous manufacturer) was searching for a second-hand sofa for a family room at Soho House. His housekeeper forwarded Boulton's instructions to their agent in London, noting that 'it is wanted for an ordinary room and provided it is a good length to lie on say 6ft, no matter for the fashion of it'.[51] The ability to read the material qualities of upholstery comes out more directly in a series of letters written in 1735 by Henry Purefoy of Shalstone in Buckinghamshire to the London draper, Anthony Baxter. His mother, Elizabeth, 'would have one of the new fashioned low beds with 4 posts' and Henry sought advice about the quantity of cloth needed for hangings and also the price of 'such a bedstead that takes all to pieces and goes on 4 swivell wheels to draw about'. Whilst his knowledge here was a little shaky, he was more confident in judging the quality of the quilting needed to finish the bed. Having been sent four different samples, he returned them all, writing 'there are none of the Quiltings will do but that whereon H. Purefoy is wrote on the Edge … If it was finer stitched & as good a cloath I should like it better'.[52] This final remark is perhaps most telling as it demonstrates that he understood the material qualities of good quilting.

Pride was taken in such knowledge and in the ways in which it helped householders to create comfortable rooms. Writing in 1754, Mary Delany reported that:

> I have bespoke four armed-chairs and six other stuffed ... for the drawing-room, and seats low and easy such as we love; but Mr. Dewes [her brother-in-law] shall have a chair of *his own* when he does me the favour to come in *every room*, or at least a cushion to raise him.[53]

Naturally, the chairs are padded, but Delany shows awareness of the additional comfort afforded by low seats and cushions, making them soft and commodious. The attributes of comfort elicited an emotional response from Delany, positioning comfort as something she not only desired, but loved. That she commissioned these things herself suggests an active role in determining the character of the chairs. Much the same can be seen in the detailed knowledge shown by Lady Louisa Stuart when writing to her sister, Lady Caroline Dawson, Countess of Portarlington in July 1797.[54] She first describes in detail the furnishings of her bed, discussing the pleating of the material, the linings required and the way in which they should drop, 'sweeping the ground an inch or two when let to fall straight'. This shows a clear understanding of the materiality and making of bed hangings, but it is her discussion of the bed itself that demonstrates her grasp of comfort: 'my bed is six feet long, and that I find rather too short for comfort; you had better make the couch seven'. Similar to Delany, Stuart is actively involved in shaping her bed, drawing on her experience to determine the nature of the hangings and the dimensions of the bed itself.

Such a grasp of the materiality of comfort was so ingrained that it could be used metaphorically. Complaining about the constant busyness of being in London, Catherine Talbot wrote that ''Tis inconceivable how many vexatious little jobs break in upon one's best hours, and disturb all their serenity'. This was a common enough complaint, even if half in jest. Yet she goes on to wonder 'were but the calm retreat secured in one's own mind, had one a fortress built there with walls of solid philosophy, and a comfortable easy chair, quilted pure and soft with ease of temper, one should enjoy perfect quiet in the midst of a hurricane'.[55] Not only does the comfortable easy chair form an escape from a busy world, it is the material qualities ('quilted pure and soft') that make it such a secure haven.

By the close of the eighteenth century, an understanding of the materiality of comfort was promoted through furniture pattern books, directors and journals. These emphasized the comfort, convenience and ingenuity of individual pieces, but also their alignment with accepted tropes of fashion and taste. They provided readers with a conceptual framework, a material vocabulary and an indication of specific features that enhanced comfort and convenience, such as low seats, reclining backrests, rising table tops and carrying handles. These messages were echoed through the more overtly commercial literature of trade cards and auction catalogues. The latter offered detailed descriptions of the material and aesthetic qualities of objects that allowed potential buyers to construct mental pictures and even imagine the bodily experience of using the furniture described. This was possible because householders could draw on their own experience both to process and refine this information and to formulate a clearer idea of the material qualities of furniture that offered them comfort or convenience – and at times both. The householder brought together diverse information to construct a working knowledge system that enhanced their material literacy. Like Lady Stuart and Prince Puckler-Muskau, they could read furniture and assemblages in rooms to appraise them as comfortable or not, and like James Henry Leigh and Henry Purefoy, they could deploy their knowledge to order furniture that matched their ideals of comfort and taste. There were limits to this knowledge, of course: convenience could be confusing. Having ordered a table desk with a rising top, Purefoy had to admit in a letter written in the summer of 1748 that 'wee can't open the Draw but do suppose it opens in the two Slitts down the Legs. I desire you will let mee have a [letter] next post how to open & manage it'.[56] Nonetheless, an appreciation of the material qualities of comfort was deeply ingrained, even before they were being expressed in those terms, as documented by Purefoy's purchases and Mary Delany's affection for her low easy chairs. In this sense, the ideals promoted by Sheraton, Pocock and Ackermann were merely formalizing a set of value and priorities that were developing much earlier: the language of comfort was perhaps a bit slow to catch up with consumers' growing knowledge about and desire for material comfort.

Notes

1   John Cornforth, *English Interiors, 1799–1848: The Quest for Comfort* (London: Barrie & Jenkins, 1978); John Crowley, *The Invention of Comfort: Sensibilities and Design in Early-Modern Britain and Early America* (Baltimore: Johns Hopkins University Press, 2001); Joan DeJean, *The Age of Comfort. When Paris Discovered Casual and the Modern Home Began* (London: Bloomsbury, 2009). On technologies, see Paul Barnwell and Marilyn Palmer, ed., *Country House Technology* (Oxford: Oxbow Books, 2012); Marilyn Palmer and Ian West, ed., *Technology in the Country House* (Swindon: Historic England, 2016).

2   Crowley, *Invention of Comfort*; DeJean, *Age of Comfort*.

3   William Cowper, *The Task, Book I: The Sofa* (London, 1785), lines 86–8.

4   Marie-Odile Bernez, 'Comfort, the Acceptable Face of Luxury: An Eighteenth-Century Etymology', *The Journal for Early Modern Cultural Studies* 14, no. 2 (2014): 3–21. Some caution is needed here, since DeJean, *Age of Comfort*, argues that comfort was 'discovered' in early eighteenth-century Paris.

5   Jan De Vries, *Industrious Revolution: Consumer Behavior and the Household Economy, 1650 to the Present* (Cambridge: Cambridge University Press, 2009), 44–5. See also Maxine Berg, 'New Commodities, Luxuries and Their Consumers in Eighteenth-Century England', in *Consumers and Luxury: Consumer Culture in Europe, 1650–1850*, ed. Maxine Berg and Helen Clifford (Manchester: Manchester University Press, 2005), 63–85.

6   William Halfpenny, *New and Complete System of Architecture* (London, 1747); DeJean, *Age of Comfort*, 45–66.

7   Shimbo, *Furniture-Makers and Consumers in England, 1754–1851*, 82–93.

8   Thomas Chippendale *Gentleman and Cabinet Makers' Director* (London, 1754).

9   Hepplewhite & Co., *The Cabinet-Maker and Upholsterer's Guide* (London, 1788), 3 and Plate 15.

10 See, for example, Vickery, *Behind Closed Doors*, 217–18; Shimbo, *Furniture Makers*, 88–93.

11 Thomas Sheraton, *Cabinet Dictionary* (London, 1803), 43–4. See also Shimbo, *Furniture-Makers*, 88–9.

12 Ibid., 196.

13 Ibid., 145.

14 Sheraton, *Cabinet Maker*, 388.

15 Richard Brown, *Rudiments of Drawing Cabinet and Upholstery Furniture* (London, 1822), 30.

16 Hepplewhite, *Cabinet-Maker and Upholsterer's Guide*, 6.

17 Thomas Sheraton, *The Cabinet Maker and Upholsterer's Drawing Book* (London, 1794), 396, 389.

18 Hepplewhite, *Cabinet-Maker and Upholsterer's Guide*; Sheraton, *Cabinet Maker and Upholsterer's Drawing Book*, 396.

19 Brown, *Rudiments of Drawing*. Emphasis added.

20 Rudolph Ackermann, *The Repository of Arts*, October 1809, 277.

21 Ibid., August 1810, 182.

22 Ibid., July 1811, 40.

23 Shimbo, *Furniture-Makers*, 88, argues that comfort and convenience together formed a value-set that enhanced the attraction of furniture which combined these qualities.

24 Clive Edwards, *Turning Houses into Homes* (Aldershot: Ashgate, 2005), 55–74.

25 Jon Stobart, 'Luxury and Country House Sales in England, *c.*1760–1830', in *The Afterlife of Used Things: Recycling in the Long Eighteenth Century*, ed. Ariane Fennetaux, Amélie Junqua and Sophie Vasset (London: Routledge, 2015), 25–37.

26 British Museum (BM), Heal Collection, 125.60 (Lawrence), 125.39 (Grange, 125.11 (Brown)

27 BM, Heal Collection, 125.2 (Allen); BM, Banks Collection, D2.669 (McClean).

28 BM, Heal Collection, 28.189 (Read); BM, Banks Collection, D2.669 (McClean).

29 E. T. Joy, 'Georgian patent furniture', *Connoisseur Year Book* (London: The Connoisseur, 1962), 9–11, sees patent furniture as a response to consumer demand for space-saving and portable items that peaked in the first two decades of the nineteenth century. Demand fell after about 1820, a trend that is reflected in the Fashionable Furniture featured in Ackermann's *Repository*, which increasingly focused on taste and aesthetics from the 1820s.

30 BM, Banks Collection, D2.1271 (Butler).

31 Reproduced in E. T. Joy, 'Pocock's – The Ingenious Inventors', *The Connoisseur* (1970), 173, 90–1. Joy's analysis focuses on the ingenuity of design and the manufacturing process.

32 BM, Heal Collection, 125.25 (Daws).

33 Margaret Ponsonby, *Stories from Home: English Domestic Interiors, 1750–1850* (Aldershot: Ashgate, 2007), 86

34 Northamptonshire Central Library (NCL), M0005647NL/9, Rushton, 1809. See also Cynthia Wall, 'The English Auction: Narratives of Dismantlings', *Eighteenth-Century Studies* 31, no. 1 (1997): 1–25.

35 NCL, M0005644NL/19, Crick, 1830.

36 NCL, M0005646NL/11 Stanford Hall, 1792.

37 NCL, M0005646NL/5, Barton Hall, 1784.

38 Mark Girouard, *Life in the English Country House* (London: Yale University Press, 1978), 213–44; Cornforth, *English Interiors*, 11–23.

39   Alice Clark, ed., *Gleanings from an Old Portfolio Containing Some Correspondence between Lady Louisa Stuart and Her Sister, Caroline, Countess of Portarlengton*, 2 vols. (Edinburgh, 1895), II: 281.

40   DeJean, *Age of Comfort*, 131–39.

41   Quoted in Cornforth, *English Interiors*, 65.

42   NCL, M0005646NL/15, Brixworth Hall, 1797, 14.

43   G. Tyack, 'Stoneleigh Abbey in the Nineteenth Century', in *Stoneleigh Abbey: The House, Its Owners, Its Lands*, ed. Robert Bearman (Stratford-upon-Avon: Shakespeare Birthplace Trust, 2004), 120.

44   Shakespeare Central Library and Archive (SCLA), DR18/5/7021, DR18/5/7022, DR18/5/7137, DR18/5/7150, DR18/5/7156, DR18/5/7158.

45   SCLA, DR18/5/7021.

46   SCLA, DR18/5/7137.

47   SCLA, DR18/5/6999, DR18/5/7100, DR18/5/7007, DR18/5/7056. All these tradesmen supplied a great many others pieces as well.

48   Nicholas Goodison and John Hardy, 'Gillows at Tatton Park', *Furniture History* 6 (1970): 1–7.

49   Shimbo, *Furniture Makers*, 101–13; Amanda Girling-Budd, 'Comfort and Gentility: Furnishings by Gillows, Lancaster, 1840–55', in *Interior Design and Identity*, ed. Susie McKellar and Penny Sparke (Manchester: Manchester University Press, 2004), 27–47.

50   Quoted in Shimbo, *Furniture Makers*, 105–6.

51   Quoted in Ponsonby, *Stories from Home*, 89.

52   G. Eland, ed., *Purefoy Letters, 1735–1753* (London: Sidgwick & Jackson, 1931), 100.

53   S. Woolsey, ed., *The Autobiography and Correspondence of Mrs Delany* (Boston, 1879), Vol. 1, 465: letter from Mary Delany to Anne Dewes, 24 November 1754.

54   Clark, *Gleanings from an Old Portfolio*, 296.

55   *A Series of Letters between Mrs Elizabeth Carter and Miss Catherine Talbot, from the Year 1741 to 1770*, 2 vols. (London, 1808), I: 368.

56   Eland, *Purefoy Letters*, 111.

7

Stitching and shopping: The material literacy of the consumer

Serena Dyer

In Frances Burney's 1779 play *The Witlings*, the milliner, Mrs Wheedle, laments that a tippet made by one of her shop girls would 'be fit for nothing but the window, and there the Miss Notables who work for themselves may look at it for a pattern.'[1] Deprived of its inherent commercial value by the poor workmanship of its maker, the tippet's worth becomes bound up in its visible materiality. The object is transformed into a vehicle for material knowledge, as the basic making methods embedded in the design, cutting, stitching and decorating process which brought it into existence supersede its value as a wearable garment. In reframing this tippet as a reference piece, rather than a tradeable commodity, Burney hints at a practice of material exchange which existed alongside traditional commercial transactions. The 'Miss Notables' who seek material education through the tippet take their name from an eighteenth-century understanding of 'notable' to mean industrious domestic activity.[2] The transference of the amateur making skills necessary for 'notable' needlework brought making into the marketplace, and necessitated shopping skills beyond bargain hunting and quality assessment. As Mrs Wheedle infers, such consumer knowledge could be gleaned through the shop's window, before consumers even crossed its threshold (in this case, Mrs Wheedle prefers to provide an inferior model to the women 'who work for themselves', rather than purchase her shop's wares). This chapter proposes a consumption model in which making skills complemented services and material goods as forms of commercial capital within shops. It frames material literacy as a consumer commodity and positions it as a central part of the consumer experience of navigating the eighteenth-century shop.

Historians of retail and consumption are familiar with narratives built around financial transactions for goods or services. A focus on global trade networks and the practices and processes of retail and shopping has led consumption to be positioned as one of the key explanatory frameworks for economic and social change across the eighteenth century. The rich, interdisciplinary literature which has debated the growth of a 'consumer society' has highlighted gender relations, credit structures and consumer behaviours.[3] Genteel consumers, and, especially female consumers, have gained dominance in such histories of consumption, where they have been lauded for their skilled browsing practices. Both Helen Berry's research on the browse-bargain model and Kate Smith's work on the haptic skills of shoppers have celebrated the material skill of browsing, seeing it as a stepping stone towards the financial transaction.[4] The ability to judge quality, suitability and value was required of consumers, but within this model, making a purchase remained the goal. The importance of economic participation in the marketplace dominated satirical denouncements of the disruptive nature of browsing practices, which vilified female shoppers who looked without making a purchase.[5] Yet Burney's brief snapshot of life on the shop floor undermines this traditional model of economic exchange. Mrs Wheedle, her shop girls and the Miss Notables all share and operate within an economy of making knowledge. Scholarship on consumer culture, with its focus on financial transactions of money for goods, has often overlooked the skills on offer within the eighteenth-century shop. Such material skills may not have been tallied in accounts and receipts, yet they remained part of the consumer experience. This chapter turns away from positioning retailers as makers, and consumers only as assessors, purchasers and users, to consider the diverse range of consumer identities that rested on a shared material vocabulary, which was not confined to one side of the shop counter.

Making and acquiring, stitching and shopping: these activities were intertwined for consumers of dress, and especially for women. Stepping away from production processes which took place in factories and workshops, I focus on the role of consumers in the making process, and the means by which they gained knowledge that traversed the producer and consumer binary. Genteel and elite consumers of dress played the role of judge, facilitator, collaborator and maker during the course of garment construction. They engaged in a dynamic process of personal and practical interaction with the goods they consumed, which transformed the material characteristics of the object from fabric to garment. Of course, most garments acquired by genteel and aristocratic women continued to be made bespoke, utilizing the services of a mantua-maker or dressmaker. Such relationships have traditionally been characterized as conforming to familiar divisions between producer and consumers. As I show below, however, the relationships between dressmakers and their clients should instead be read as production partnerships: garment and accessory manufacture constituted a fluid collaboration between professional makers and materially literate consumers. I chart first the acquisition of this material literacy throughout the lifecycle, exploring the links between making and consumption. I then turn to the role of the consumer in the transformative process of turning textiles into garments to reveal the complex array of material literacies possessed by genteel and elite consumers in the eighteenth century. Finally, I draw together this making education and practice to consider the application of material literacy on the shop floor. Together, dressmakers, mantua-makers and consumers were in dialogue over the creation, commissions and making of garments and accessories, bound up in a rich web of material knowledge.

Making consumers

Material literacy was central to the education of girls. As Mary Wollstonecraft notes in *A Vindication of the Rights of Women* (1792), 'females are made women of when they are mere children', replete with a 'contagious fondness for dress'.[6] For genteel girls, their education focussed on the skills they would require as adults to navigate their consumer activities. Within this consumer skillset, material literacy dovetailed with economic literacy. Conduct books, such as the *Polite Academy* (1768), were aimed at mothers overseeing the education of their offspring and contained sample conversations that promoted these combined consumer skills.[7] The apparatus for financial self-regulation – columns for accounting and tables listing values, prices and conversions – was also an integral part of children's pocket books, and has often overshadowed the concurrent training of material literacy in childhood.[8] Yet for the parents of many young girls, economic training was of less concern than their daughters' making skills. Frances and Charlotte Starkie, two daughters of a Cheshire gentleman, attended a school in Twickenham from around 1804. The school was run by Mrs Fletcher and her sisters the Misses Dutton, who were regularly called upon for reports on their pupils' progress. The reports make only brief mention of the girls' grammar and mathematical lessons, and primarily focus on their sewing. Unfortunately, the sisters struggled with their needles. In 1804, the teachers complained of Frances that 'the moment she uses a needle, her hands become so warm and moist that it is with difficulty she can proceed'.[9] She was slow and awkward, and possessed very little skill. Although this lack of talent extended to her other subjects, with the exception of music, Frances's poor sewing is discussed at great length. Her teachers saw her awkward handling of the needle as a barometer for the haptic skill and material literacy Frances would need to navigate consumer culture as an adult.

Although often overshadowed by the needlework sampler, the stitching of dolls' garments was particularly key in developing girls' knowledge of making.[10] Extant examples demonstrate the high level of skill required to make tiny versions of fashionable dress. For instance, around 1805, one girl spent a summer vacation, in which she remained at her school, creating a small, white, cotton dress (Figure 7.1). The girl would not have spent time developing these material skills in preparation for work as a dressmaker in adulthood: as a pupil at a London school for the daughters of the gentry, she would not need a trade. The dress is a perfect miniature copy of a contemporary woman's drop-front gown, and measures approximately thirty-three centimetres in length. It is made from a white cotton, woven with a large check pattern, which out-scales the size of the dress, and the bodice is lined with white linen. A lace trim adorns the neckline and the sleeves and appears again as an insert on the front bib. A closer inspection of the handiwork of this garment shows the young pupil's extremely high level of material knowledge and skill. The lace inserts on the front of the bib are perfectly aligned with the check pattern of the fabric, their edges butting the tiny rolled hems.

The material evidence stitched into this garment testifies that its maker was adept at pattern matching and able to work in tiny, intricate stitches. Knowledge of fabric manipulation and cutting was expertly demonstrated through the sleeves, which were cut on the bias, allowing for greater flexibility of movement – movement beyond that needed by a doll's arms. A back view of the garment demonstrates further skill in the top-stitched construction of the seams, in which the side back is carefully lapped over the back and stitched in place using a backstitch (Figure 7.2). The lining of the garment

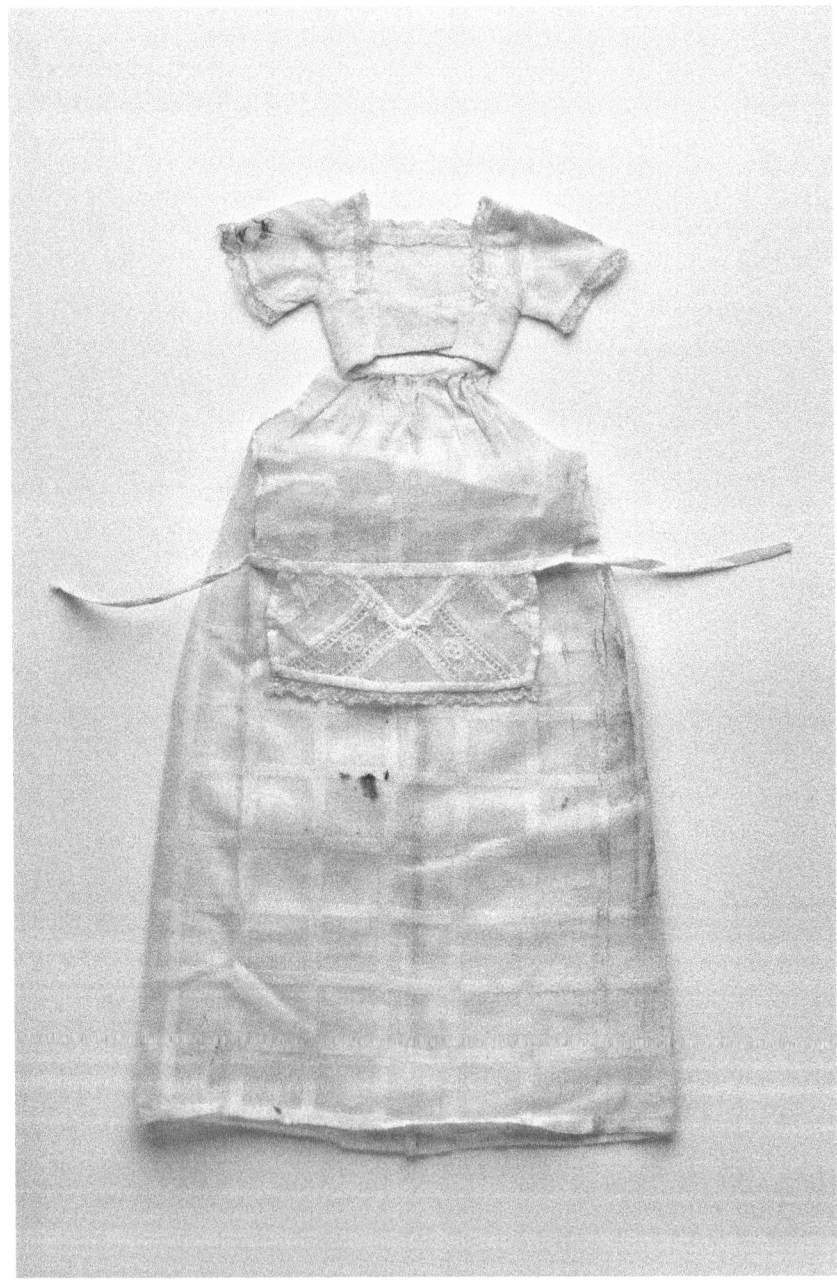

Figure 7.1　Doll's dress, front view, 1805, Museum of London, A21412.

Figure 7.2 Doll's dress, back view, 1805, Museum of London, A21412.

continues to show a careful awareness of the construction methods used in full-sized garments, as all raw edges are either concealed or whip-stitched over to prevent fraying. The child also created the dress with an eye to economy, constructing interior sections from cheaper, plain linen such as the flaps underneath the bib construction. She preserved more costly pieces of textile and trimmings for the outside of the garment, mimicking the economical practices seen in extant garments. The process of making, materialized in each stitch and fold, is identical in its entirety, to the construction practices of full-sized garments.

This small garment invites urgent questions about the development of material literacy, such as why a genteel young lady – the sort of girl who would grow up to become the consumer taking part in Berry's browse-bargain model, and not the dressmaker – would be required to know how to construct a garment in such intricate detail while minimizing costs. Indeed, the skills demonstrated by the maker of this small dress were not unique to her. Emphasis has often been placed upon girls making shirts for their father or brothers – an exercise framed as a comparatively simple task of plain sewing.[11] Yet manuscript sources support the fact that girls were actively involved in the creation of miniature, fashionable garments for their dolls, practices beyond the routine replication of identical men's shirts. Their making was creative, their skills diverse, and their work considered worthy of discussion and epistolary celebration. In the 1770s, Anne King, the daughter of a Gloucestershire baronet, dressed a doll in a full brocade gown, which was praised in letters by her family.[12] Betsy Nutt, of the Northamptonshire gentry, was also commended for her handiwork and the making of dolls garments, after she completed an outfit for her friend's doll; the friend wrote that she had 'made it such a pretty frock and petticoat'.[13] Similarly, in 1830, seven-year-old Caroline Pennant confided to her grandmother that she was pleased with 'the first little pocket handkerchief I have made for my doll'.[14] The work of making dolls' garments was something to be encouraged, admired and celebrated; such practices shored up female friendships and intergenerational family bonds.

At the same time, extant examples of miniature garments confirm that while many young women strived to develop their material literacy, they were not universally successful. Learning to be materially literate was not an inherent feminine quality; it took effort, time and repetition. Some miniature garments show mistakes and faults in the cut and construction of the garments, which indicate the same process of learning by doing discussed by Beth Fowkes Tobin in this volume. Like the Starkie girls, the makers of these garments struggled with their efforts to square their manual skills with their knowledge of garment construction, creating pieces that ultimately showed only a limited literacy in technique. An example of a doll's silk pelisse, which is generally executed to a very high standard, nevertheless contains some inconsistencies (Figure 7.3). The right sleeve is cut correctly along the bias of the fabric, allowing it to cling at the forearm, and puff out at the head. However, the left sleeve follows an imprecise angle – not quite on the straight grain, but also not on the bias – and fails to create the same effect as its twin sleeve. It is possible that this maker was faced with limited fabric pieces for her ensemble and perhaps economy shaped her execution. The sleeve's odd angle would have been evident when placed on a doll and it remains visible in the way in which the garment does not lay correctly when stored. This

Figure 7.3 Doll's pelisse, 1810, Museum of London, A21160 (see Colour Plate 3).

mistake makes it unlikely that this garment was purchased readymade, and highly likely that it was constructed at home. Such misses clearly confirm that girls developed their material literacy through practical trial and error. This approach equipped girls with an intimate knowledge of cut and fit, which they retained and developed through adulthood. Conversant in garment making, they would not only recognize mistakes in the cutting and sewing process of goods, but also would understand why and how those faults had emerged. In turn, childhood practice in clothing dolls enabled women to bring a significant degree of practical material knowledge to their conversations with professional garment makers.

Children's literature, moreover, underlines the central role played by dolls to develop the material literacy of children. In *The Doll's Spelling Book* (1802), Dorothy Kilner encourages young readers to be 'anxious for the welfare of your dear little families, whether they are composed of Wax, Wood, Leather or Rag'.[15] Rather than focusing on the doll's garments and clothing, Kilner centres her didactic storytelling on its material composition. Kilner traces how a girl teaches her doll, using the components of linen, silk, wood and wax as subjects for her lessons. The dialogues mirror the configuration of a mother teaching a child, enacted through play and imaginary conversations:

> MAM-MA: Come, my dear doll, though you can not talk, I will make-be-
>   lieve that you can. So come, my child, sit in my lap, and tell me if
>   you know what you are made of?
>
> DOLL: I am made of paper.
>
> MAM-MA: Oh you sil-ly thing! Do you not know bet-ter than that? You are
>   made of wood. (39–40)

The girl, in the part of 'Mam-ma', trains for her presumed future role through instructions that she is herself in the process of learning. Play as a device was widely used to acquaint children with the world they inhabited, as scholars have noted.[16] The girl links the doll's identity to the material of its form, emphasizing her awareness of the physical world that makes up domestic objects. This extends to the girl's description of wax, which also contributes to the doll's material composition, and an explanation of the processes behind its manufacture. In Kilner's instructive drama, play and the practical enactment of skills call attention to manufacture, materials and making.

Lady Eleanor Fenn's *Rational Sports in Dialogues Passing among the Children of a Family* (1783) offers another example amongst a broad range of texts which bridged education and entertainment. Like Kilner's *The Doll's Spelling Book*, Fenn's work takes the form of a series of dialogues. The first of these is entitled 'Trades', in which the children play out different occupations – primarily retail roles – through conversation. Their dialogue is enacted through a series of questions, and when a child does not know an answer, they have to forfeit. Each vignette opens with a child performing a type of shopkeeper, whose role they begin by describing:

> JANE: I will be a Milliner; and I will sell a thousand things. Jack says, that
>   is the meaning of the name; and I will make caps and ruffles and
>   such things.
>
> GEORGE: And I will be a Haberdasher, and I will sell as many things as you:
>   pins, tape, needles, thread; and I will have a great shop.
>
> WILLIAM: And I will be a Pedlar; and I will buy my goods of George, and
>   carry them a great way about, and call at all the houses; and I will
>   keep a stall at the fair and sell my goods. [...]
>
> JANE: Let Susan be a draper; then what shall she sell?
>
> GEORGE: Cloth to be sure, you know; there are both linen and woollen-
>   drapers.[17]

Facilitated through play, the children learn about different retailers, what they sell, and how they conduct their trade. Not only does the exchange specify the exact goods each retailer sells, such as pins, tape, needles and thread from the haberdasher, it also delineates where the pedlar would acquire their goods, and the location where each retailer can be found, whether a stall at a fair or a shop on the street. Other trades such as a cooper, druggist, stationer and pastry-cook are also included, demonstrating that a wide knowledge of the retail market was deemed desirable. Fenn shows considerable attention to the material properties of each vendor's goods, as in this dialogue about drapers: 'GEORGE: Draper! – When you are asked what your linen is made of, answer hemp or flax. – They are both plants. – You know what the woollen cloths are made of?'[18] Other exchanges divulge similar information about leather, butter, chocolate, cochineal and turpentine, placing the children in the role of instructor. Many of the young characters display a strong grasp of the materials and stock of their shop keeping roles, while others are shown to gather details from their peers. Fenn's subtitle promises that these dialogues were 'designed as a hint to mothers how they may inform the minds of their little people respecting the objects with which they are surrounded'. This explicit focus on material literacy within the context of commerce and retail affirms the significance of the tangible and material world in developing the child as a consumer. Being a capable consumer involved not only adept financial awareness and responsibility, but also a thorough awareness of how goods were made and their material origins.

Making fashion

Training girls in practical making skills and material knowledge prepared them to make informed consumption choices in adulthood, whether in shops or by their own needles. Materially literate women could make fashionable garments for themselves, as well as manage and judge the handiwork of professional makers. As Priscilla Wakefield insisted in *Reflections on the Present Condition of the Female Sex* (1798), 'useful needlework in every branch, with complete skill in cutting out and making every article of female dress … ought to employ a considerable part of the day'.[19] Her advice appears to have been headed by many women, of both the elite and the middling sort. At the end of her life, Catherine Hutton (1756–1846) recalled that she had made 'all sorts of wearing apparel for myself, with the exceptions of shoes, stockings and gloves'.[20] In 1794 Lord Sheffield's daughter was actively engaged in garment making, and boasted of her efforts to relatives.[21] Further down the social scale, Jane Austen recorded in 1798 that her acquaintance, a Miss Debary, was netting herself a gown in worsteds.[22] In references tucked away in letters and diaries, these women record an everyday practice of garment making which is often absent within account books and bills.

Genteel women's engagement with making was intertwined with discussions around women's work. To be industrious was a key attribute of the good and virtuous wife, and of genteel femininity more broadly, as demonstrated in Nicole Pohl's essay in this volume.[23] In praising her daughter Betty Parker in 1780, Elizabeth Shakleton wrote that she was an 'exceedingly Good Wife, she ruffled her Husband a shirt & always is Industrious'.[24] Needlework was regularly framed as industrious labour, and as generative of feminine morality.[25] This industry with the needle involved not only the plain stitching of shifts and shirts (including the whip-gathering of shirt ruffles), but also the cutting out and stitching of full fashionable garments. A watercolour by George Walker, captioned 'Industrious Jenny ever useful miss!! Employs her time in making a pelisse' (Figure 7.4), depicts a young, genteel lady mid-stitch as she creates a fashionable lady's outer-garment. The pelisse is an adult version of the miniature garment discussed above (Figure 7.3), which also dates from the same decade as the watercolour. 'Industrious Jenny' is not a professional maker. Her labour is productive and useful, and her virtue linked to her choice to stitch garments. She is the epitome of the 'notable' domestic maker referenced in Burney's *The Witlings*. Jenny's work reveals that positive productivity, which was often classified for genteel female consumers as participation in the consumer market, could also be transferred to women's engagement with production. While Jan de Vries's 'industrious revolution' model has emphasized that increased productivity was geared towards earning more to consume more, it was also the case that production itself was framed as a form of fashionable consumption.[26]

Letters and diaries record that women used their material literacy to oversee the manufacture of their attire, whether or not they executed the work with their own hands. In 1755 Lady Heathcote made a gown for a masquerade to 'her own design'.[27] This vocabulary of design most likely meant that she actively oversaw the composition and creation of the garment, but that she did not stitch the garment herself. Such composition and production partnerships between consumers and makers were indeed commonplace. In 1767, when organizing the manufacture of her gown for a court birthday, Lady Mary Coke recorded that 'the weather as severe as ever, sent for the lace man & chose some silver lace to trim my gown' and that she also 'sent for the mercer to bring the silk for my Birthday Gown'.[28] Assembling a variety of tradespeople at the same time meant that Coke could compare, contrast and choose the different elements of her final gown. Although these elite consumers did not actively engage in making, they clearly possessed detailed degrees of material literacy and maker's knowledge that were necessary to manage complicated commissions. Hester Thrale's infamous Polynesian court gown of 1781 was constructed through a similar process of design management, yet the creation of this garment has been actively attributed to Thrale as part of her own image-making.[29] In these production partnerships, the materially literate consumer could coordinate the construction of garments, even if professional makers plied their needles to generate the finished piece.

Although many elite consumers relied upon the making skills of professional dressmakers in the construction of garments, their deference was certainly not universal. In fact, some consumers classified their own stitching skills as superior to those of professional makers. In 1781, Lady Carlow wrote to her sister Louisa Stuart about a gown she had made. She wrote, 'My chief amusement since I came from town has been making myself a white polonaise, in which I have succeeded to a miracle, and repent having given one to a famous mantua-maker in Dublin who spoilt it entirely for me.'[30] Carlow's experience of having entrusted the workmanship of a previous polonaise to a mantua-maker was evidently unsatisfactory. Although Carlow perceived the production process partially as an amusement, and her own success and skill

Figure 7.4 George Walker, *Industrious Jenny*, watercolour on paper, 1810s, Yale Center for British Art, B1975.4.974.

in making the gown 'a miracle', she also considered her own remaking to have been more successful than that of a professional dressmaker. Women consumers not only engaged in production, but also felt qualified and able to judge the skill of professional producers in relation to their own. The material literacy required to create fashionable garments was not exclusively the province of mantua-makers and dressmakers but was part of a shared material language and practice: a language in which both consumer and retailer could possess varying degrees of fluency, as also evidenced in Jon Stobart's chapter. In 1774, moralist John Gregory informed his motherless daughters that 'the intention of your being taught needle-work, knitting and such like, is not on account of the intrinsic value of all you can do with your hands, which is trifling, but to enable you to judge more perfectly of that kind of work, and to direct the execution of it in others'.[31] In other words, when women commissioned mantua-makers or milliners to construct a garment to order, they were expected to be able to assess the workmanship of the product being created, and to supervise this process from start to finish.

Whether actively stitching, netting and constructing their own garments, or managing the design and composition of a professionally stitched gown, consumers exercised their material literacy to varying degrees over time and over their lifetimes. As styles changed, construction processes altered too. How such information was conveyed has presented a stumbling block to scholars who have often over-emphasized the role of the fashion plate in the dissemination of fashionable dress.[32] While fashion plates often presented information about changes to style, no practical information about making resided in these images. As sources of visual knowledge, the fashion plate's increasing prevalence from the 1750s onwards in pocket books, and from the 1770s onwards in periodicals, attests to their popularity. Yet their practical use was limited. How to achieve a certain sleeve shape, or to adapt to rising waistlines, could not be communicated in these two-dimensional images. Fashion plates did not in themselves advance making knowledge, yet they managed to promote material literacy in other ways. They offered visual illustrations of the terminology used in fashion journalism, giving shape and style to terms, and thereby developing the material vocabulary of the reader. Fashion plates depicted fashionable garments in action, showing their wearers occupied in genteel activities. They communicated a sociable and polite way of life, which could include taking care of children, playing musical instruments, drawing or engaging in sewing and needlework. While elite periodicals such as the *Gallery of Fashion* offered an exception to this rule by eschewing images of needlework, other popular periodicals favoured by the genteel women consumers of Britain, such as *Journal des Dames et des Modes,* the *Repository of Arts,* and the *Lady's Magazine,* frequently depicted women involved in forms of material activity.[33] As such, fashion plates did not provide practical instruction, but they did promote participation in a culture of material literacy. A fashion plate (Figure 7.5) from the 1802 edition of the *Journal des Dames et des Modes* depicts a fashionably dressed woman at her knitting, her pale hands nimbly working the strikingly long needles.[34] The image advertises not only the fashionable garments she wears, but also her activity as a maker. To be materially literate was to be part of the culture of fashion.

Figure 7.5 *Bonnet garni en Tulle, Fichu sur l'épaule, Journal des Dames et des Modes*, 1802, The Morgan Library and Museum, New York, PML 5687.

Material literacy in the shop

Fashion plates made plain the close ties between the culture of style and women's handiwork, but the eighteenth century also saw many moralists seizing on how shopping for fashionable goods might threaten women's material skills. Amanda Vickery has noted that a decline in homespun goods resulted in a 'moral panic' in the face of 'frenetic shopping'.[35] Widespread evidence of this concern about consumption has led us to overlook how making continued to be maintained, even if women might also be swayed to purchase what they had previously made at home. While preparing for her marriage to a local Yorkshire clergyman, Elizabeth Woodhouse, the daughter of a successful haberdasher, spent a vast amount of money on new clothes and also paid ten pounds to a local milliner, J. Volans, to instruct her in her 'art'.[36] In setting herself up as a genteel clergyman's wife, Woodhouse not only sought to dress herself and equip her house with the necessary tools for married life, but also to arm herself with the knowledge and skills required to make garments for herself, adding skills to her repertoire of material literacies, likely established in childhood. Echoing the practices highlighted in Tobin's chapter in this volume, Woodhouse explicitly framed the acquisition of maker's knowledge as a transaction. Within Volans's millinery shop, making skills were on sale, as well as ribbons, bonnets and lace. Bringing making into the marketplace acknowledged that consumers were also makers and transformed the making practices of customers into an opportunity for commercial gain.

Making activities that took place within the shop offer insight into how such activities were dispersed across the shop interior. Scholars have long identified the shop counter as an important site of exchange for both economic and cultural capital.[37] It was the space upon and over which goods were revealed, browsing occurred and bargaining took place. It was not, however, the only space of exchange within the shop. Our focus on the counter as a place of economic transactions has overshadowed the persistence of making elsewhere on the shop floor. Henry Kingsbury's 1787 satire (Figure 7.6) of an imagined visit by the royal family to a milliner's shop assembles an array of visual tropes of the shop floor and displays the visual tropes used to indicate life upon the shop floor. The engraving's busy composition draws attention to the browsing and bargaining that take place upon the long counter at its centre. The women's vivid, voluminous gowns and hats are echoed by the goods that have been tacked to the top of the wall rising up behind the counter. Yet Kingsbury carefully depicts another kind of activity, located towards the back of the image, within the shop window. Seated at a large round table in the window are two women and a man, in plain view of all who pass by in the street. As a living window display, these three millinery workers present passers-by not with finished goods, but with the processes of making. Their performance of cutting, stitching and making acts as a form of marketing, artfully positioned in the shop window for all to see.[38] The prominence of the makers' position coveys how their manual activities might serve to captivate new customers. Mirroring Burney's depiction of the milliner's shop, where passers-by study construction as much as style, the window space is again the setting for the transmission of making knowledge. This performance of skill both acted as a source of material knowledge for

Figure 7.6 Henry Kingsbury, *A Milliner's Shop*, 1787, Yale Center for British Art, B1978.43.923.

consumers like Burney's Miss Notables and offered up an opportunity for milliners and dressmakers to demonstrate their skill, advertising themselves as makers as well as merchants.

The ability to judge the quality of making was viewed as a key skill to navigate a material world, in which retailers were viewed as untrustworthy. The skill of material judgement was positioned as necessary as part of a discourse which considered retailers as untrustworthy. Some retailers were evidently adept at attempting to dispose of unwanted stock through misleading sales patter and therefore could not be fully trusted as independent arbiters of taste and quality. For instance, Ann Charlton, a London milliner, often attempted to defend her goods to her customer, Lady Sabine Winn of Nostell Priory. Lady Winn, a baronet's wife living in Yorkshire, maintained a correspondence with Charlton, whom she had previously patronized whilst living in the city. This correspondence can be characterized as an epistolary locum shop counter, in which letters, rife with economic and material exchanges, performed as a proxy shop counter.[39] However, this epistolary browsing, initiated by Winn, was frequently interrupted by Charlton's attempt to send Winn goods which would no longer sell in London. Charlton often stated that she had no desire for the return of some of the unwanted goods, and there was certainly no strategy for returns in place. In 1783, Charlton wrote that she would 'take it as a favour if it quite suits your Ladyship if you could dispose of the other' items, which Charlton had sent speculatively.[40]

Charlton was not alone in her attempts to manipulate long-distance, and unknowing, customers who purchased goods through correspondence, by sending unwanted items. Elizabeth Griffith, another London milliner, also sent items which were not requested to Lucy Smythe, along with an existing order, apparently in an attempt to both generate extra sales and dispose of excess and out-of-date stock.[41] These garments may have been of inferior quality or poorly made, out of fashion or simply aesthetically undesirable. The material literacy of the consumer, and the ability to independently assess the fashionability, quality and suitability of goods, was essential if the ploys and schemes of such retailers were to be avoided. Such tools were particularly important to women who lived far from urban centres and their commercial displays of making. The female consumer's material literacy allowed her to judge the workmanship, quality and suitability of the goods she purchased, or was encouraged to purchase. It was to this end that the materially literate consumer developed a careful browsing methodology which, as Kate Smith has demonstrated, was central to shopping practice.[42] Haptic browsing allowed consumers to step back from the rhetoric of retail and physically judge goods, and use their hands to assess and pass judgement on the making process.

Housekeeping and economy skills, and the craft and handiwork ability of genteel and aristocratic women have been dominant in scholarly accounts of consumption, collecting and material culture.[43] However, non-professional women's material knowledge of garment construction, and the role this played in their consumer practices, has often been overlooked in studies of shopping and trade. This chapter has traced how genteel and elite women acquired and deployed making skills throughout the life-cycle, from girlhood to adulthood, and has framed material literacy as a key skill for consumers. The performance of material skills, both at home and on the shop floor, constituted part of the visual rhetoric of the culture of fashion. Similarly, fashion plates and depictions of shops illuminate how the depiction of manual labour buttressed the portrayal of style. Knowledge of production enabled the consumer to effectively collaborate with and judge the craftsmanship of the producer, and to engage with broader discourses on taste, judgement and fashion. It was the shared material literacy between consumers and producers that facilitated these production partnerships and united them in mutual knowledge of the practice of garment construction. Consumers of dress were skilled, knowledgeable and materially literate, and their activities brought making firmly into the marketplace.

Notes

1   Frances Burney, *The Witlings*, in *The Witlings and the Woman-Hater*, ed. Peter Sabor and Geoffrey Sill (Peterborough: Broadview, 2002), 48.

2   For example, the character of Emma Mourtray, in Elizabeth Hervey's novel *The Mourtray Family*, 'had learned to detest all needlework of the notable kind'. See Elizabeth Hervey, *The Mourtray Family*, 2 vols. (London: Faulder, 1800), I: 74.

3   McKendrick, Brewer and Plumb, *The Birth of a Consumer Society*; Brewer and Porter, *Consumption and the World of Goods*; Lorna Weatherill, *Consumer Behaviour and Material Culture in Britain, 1660–1760* (London: Routledge, 1996); Jonathan White,

'A World of Goods? The "Consumption Turn" and Eighteenth-Century British History', *Cultural and Social History* 3, no. 1 (2006): 93–104; Claire Walsh, 'Social Meaning and Social Space in the Shopping Galleries of Early Modern London', in *A Nation of Shopkeepers: Five Centuries of British Retailing*, ed. John Benson and Laura Ugolini (London: I.B. Tauris, 2003), 52–79.

4   Berry, 'Polite Consumption: Shopping in Eighteenth-Century England', 375–94; Smith, 'Sensing Design and Workmanship: The Haptic Skills of Shoppers in Eighteenth-Century London', 1–10.

5   Elizabeth Kowaleski-Wallace, *Consuming Subjects: Women, Shopping, and Business in the Eighteenth Century* (New York: Columbia University Press, 1997), 84.

6   Mary Wollstonecraft, *A Vindication of the Rights of Women*, ed. Janet Todd (Oxford: Oxford University Press, 1999), 192, 275.

7   *The Polite Academy; Or, School of Behaviours for Young Gentlemen and Ladies* (London, 1768), 109.

8   Serena Dyer, 'Training the Child Consumer: Play, Toys and Learning to Shop in Eighteenth-Century Britain', in *Childhood by Design: Toys and the Material Culture of Childhood, 1700–Present*, ed. Megan Brandow-Faller (London: Bloomsbury, 2018), 31–46.

9   Kent Archives, U908/C60/4: 'Letters relating to the education of Frances and Charlotte Starkie, 1804'.

10 For more on eighteenth-century dolls, see Fennetaux, 'Transitional Pandoras: Dolls in the Long Eighteenth Century', 47–66.

11 In practice, constructing a shirt requires vast amounts of material literacy. Economical cutting and a wide variety of stitches are required to complete this task.

12 Gloucestershire Archives, D2455/F1/7/4: 'Letter to Martha Hicks from her daughter Ann, 1770s'.

13 Northamptonshire Record office, B(HH)/148: 'Letter to Mathilda Bosworth'.

14 Warwickshire County Record Office, CR 2017/TP548: 'Letter from Caroline Pennant to Her Grandmother, 1830'.

15 Dorothy Kilner, *The Doll's Spelling Book: Intended as an Assistant to Their Mammas in the Difficult Undertaking of Teaching Dolls to Read* (London: J. Marshall, 1802), v–vi. Hereafter cited parenthetically in the text.

16 Neil McKendrick, John Brewer and John Harold Plumb, 'The New World of Children in Eighteenth Century England', in *The Birth of a Consumer Society: The Commercialization of Eighteenth-Century England* (London: Europa, 1982), 286–315; Andrew O'Malley, *The Making of the Modern Child: Children's Literature and Childhood in the Late Eighteenth Century* (Abingdon: Routledge, 2011).

17 Eleanor Fenn, *Rational Sports in Dialogues Passing Among the Children of a Family* (London, 1783), 18.

18 Ibid., 22.

19 Priscilla Wakefield, *Reflections on the Present Condition of the Female Sex* (London, 1798), 144–5.

20 Hutton, *Reminiscences of a Gentlewoman of the Last Century*.

21 Anne Buck, *Dress in Eighteenth-Century England* (London: Batsford, 1979), 183.

22 Austen, *Jane Austen's Letters*, 22.

23 For extensive discussion of genteel women's industrious activities, see chapter 4 in Vickery, *The Gentleman's Daughter*.

24 Lancashire Record Office, DDB/81/37, 'Elizabeth Shakleton, Pasture House, to J. and R. Parker London, 1780'.

25   Smith, *Women, Work, and Clothes in the Eighteenth-Century Novel*, 13.

26   Vries, *The Industrious Revolution Consumer Behaviour and the Household Economy, 1650–Present*.

27   Bedford Archives, L/30/9a/5, f. 104: 'Wrest Park Papers'.

28   Mary Coke, *The Letters and Journals of Lady Mary Coke* (Edinburgh: David Douglas, 1889), 247.

29   Ruth Scobie, '"Bunny! O! Bunny!": The Burney Family in Oceania', *Eighteenth-Century Life* 42, no. 2 (2018): 56–72; Serena Dyer, '"Magnificent as Well as Singular": Hester Thrale's Polynesian Court Dress of 1781', in *Fashion and Authorship: Literary Production and Cultural Style from the Eighteenth to the Twenty-First Century*, ed. Gerald Egan (London: Palgrave Macmillan, 2020), 43–62.

30   Louisa Stuart, *Gleanings from an Old Portfolio, 1785–1799*, ed. Godfrey Clark, Vol. 2 (Edinburgh: D. Douglas, 1895), 169.

31   John Gregory, *A Father's Legacy to His Daughters* (London, 1774), 22.

32   Aileen Ribeiro, *Dress in Eighteenth-Century Europe, 1715–1789* (London: Yale University Press, 1985), 52.

33   Smith, 'Fast Fashion: Style, Text, and Image in Late Eighteenth-Century Women's Periodicals', 440–57, 448. For more on the diverse forms of women's work, see Maria Ågren, 'Introduction', in *Making a Living, Making a Difference: Gender and Work in Early Modern European Society*, ed. Maria Ågren (Oxford: Oxford University Press, 2017), 1–23.

34   For more on the importance of hands in constructing genteel female identities, see Kate Smith, 'In Her Hands: Materializing Distinction in Georgian Britain', *Cultural and Social History* 11, no. 4 (2014): 489–506. Although *Journal des Dames et des Modes* was a French publication, it had a wide readership within Britain.

35   Vickery, *Behind Closed Doors*, 13.

36   York City Archives, Munby Papers, Acc. 54.19: 'Receipt from J. Volans'.

37   Deidre Shauna Lynch, 'Counter Publics: Shopping and Women's Sociability', in *Romantic Sociability: Social Networks and Literary Culture in Britain, 1770–1840*, ed. Gillian Russell and Clara Tuite (Cambridge: Cambridge University Press, 2006), 211–36.

38   For work on the shop window as a site of display, see Claire Walsh, 'Shop Design and the Display of Goods in Eighteenth Century London', *Journal of Design History* 8 (1995): 157–76.

39   In general, the correspondence refers to finished goods, such as ribbons industrially produced on looms. The correspondence was also accompanied by samples of these ribbons, through which Winn enacted the haptic browsing practice described by Smith. See Smith, 'Sensing Design and Workmanship'.

40   West Yorkshire Archive Service, WYL 1352/C4/9/16: 'Letters from Mrs Ann Charlton to Sabine Winn, 1783–1785'.

41   Surrey Record Office, D641/3/P/21/93: 'Letter from Elizabeth Griffith, milliner, to Lucy Smythe, 1826'.

42   Smith, *Material Goods, Moving Hands*; Smith, 'In Her Hands'.

43   Maureen Daly Goggin and Beth Fowkes Tobin's three edited volumes have been especially influential in this area, see *Material Women, 1750–1950*; *Women and Things, 1750–1950*; *Women and the Material Culture of Needlework and Textiles, 1750–1950*.

8

Stitching the it-narrative in *The History and Adventures of a Lady's Slippers and Shoes*

Alicia Kerfoot

The History and Adventures of a Lady's Slippers and Shoes (1754) is an it-narrative told from the perspective of the slippers and shoes of the title.[1] It-narratives, as Mark Blackwell explains, were also known as '"novels of circulation," "object tales" and "spy novels"' and were 'a subgenre of the novel, a type of prose fiction in which inanimate objects (coins, waistcoats, pin, corkscrews, coaches) or animals (dogs, fleas, cats, ponies) serve as central characters'.[2] The object narrators of these stories offer a window into the circulation of goods in the marketplace, as well as a satirical picture of their various owners. In this vein, the object narrators in *The History and Adventures of a Lady's Slippers and Shoes* emphasize their own circulation, sharing their tales with each other. At the slippers' request, the shoes recount their manufacture, beginning, 'The whole town did not afford a neater work-woman, nor a prettier girl, than she, whose delicate hand, performed the needlework of me' (38). Rather than open their origins tale with the leather and wood that formed the inner structures of eighteenth-century shoes, the shoes start their adventures with the needle and threads that embellished their external appearance. In so doing they align their making with the telling of their tale and their narrative with the embroidery work of women.

Despite their investment in the material states of their object protagonists, many it-narratives of the eighteenth century evade firm details about the very means of production of their titular objects. Instead, they centre on the ways that consumption (including retailing and gift giving) leads to exchange and social mobility; their focus is often on a satire of fashionable society rather than on the objects themselves.[3] For example, both *The Memoirs and Interesting Adventures of an Embroidered Waistcoat* (1751) and *The Adventures of a Black Coat* (1760) gloss over the production of the waistcoat and coat, moving directly to a detailed account of their various owners. The black coat merely refers to 'the æra of my formation' while the waistcoat opens its tale after its production has already occurred, stating only that it 'was originally the Property of a noble Lord'.[4]

In contrast, a number of other it-narratives, including *The History and Adventures of a Lady's Slippers and Shoes*, do reveal a deep knowledge of the production of goods and the relationship between consumption and retailing. These narratives rely on the language of production in their structures, generating a self-reflexive and intertextual treatment of the material realities of their speaking objects. Such it-narratives draw us into the production, consumption, use and disposal of their speaking objects, forming narrative touchstones for material counterparts that survive in collections today. Several of the it-narratives that engage with their speaking objects in this way appear later than *The History and Adventures of a Lady's Slippers and Shoes,* intervening in an already established sub-genre. In 'Adventures of a Quire of Paper' (1779), for example, the quire of paper changes from a thistle to a 'heap of grain' and is converted to a piece of cambric cloth after 'having undergone a variety of new processes, and pains, under the hands of the combers, spinners, skainers, twisters, and weavers'.[5] The cloth is separated into parts almost immediately, but the object narrator is able to follow the several paths of its parts in order to delineate cambric's different uses. Eventually the cloth becomes rags and the narrative moves into a detailed account of paper-making, followed by a parallel examination of the various fates of the quire of paper. These realistic details of production and consumption are placed alongside more fantastical elements that make claims for the paper's ability to reassemble its various parts.[6] Edward Thompson's 'Indusiata: or, The Adventures of a Silk Petticoat' (1773) begins with an origin tale that traffics in myth, as he personifies the silk by giving it the same name as the young woman who produces it and travels with it from Italy to England: Indusiata.[7] He thus offers a more detailed history of production than most it-narratives by acknowledging that the silk must be spun, traded and woven before it becomes a petticoat. However, this depiction is also a symbolic and gendered one because it aligns the bale of silk with the beautiful woman who creates and then sells it to a Spitalfields weaver.[8] Mary Ann Kilner's *The Adventures of a Pincushion* (1784) offers a thorough account of the making of its pincushion by a young girl, but Kilner speaks in the voice of a human narrator before she takes on that of the speaking pincushion: '*Martha* made choice of a square piece of pink sattin, which she neatly sewed and stuffed with bran, and which … was the identical *Pincushion* whose adventures form the subject of this little volume.'[9]

Many of these later it-narratives, however, move more quickly through their production details, than does *The History and Adventures of a Lady's Slippers and Shoes,* in which the object narrators relate their production via their first-person perspectives. This it-narrative is split into two parts, with the slippers the focus of the first part and the shoes the focus of the second (4). The footwear narrators provide more details about their human producers than is typical for the genre. Not only this, but in *The History and Adventures of a Lady's Slippers and Shoes,* references to other literary texts and the material realities of footwear production are intertextual ones that comment on both form and gender identities. In her discussion of 'the bounded text' in *Desire in Language,* Julia Kristeva's definition of intertexuality offers one framework for the intertwining of the spoken, written and stitched in it-narratives. Kristeva's text 'is a permutation of texts, an intertexuality: in the space of a given text, several utterances, taken from other texts, intersect and neutralize one another'.[10] *The History and Adventures of a Lady's Slippers and Shoes* displays an energetic commitment to such varied and intersecting utterances, experimenting with genre and human subjectivity, as it dips in and out of discourses of production, consumption and social satire. Such experimentation renders the fictional slippers and shoes a crucial point of intersection for a number of voices from other texts, including the slipper and shoe objects that exist outside of the narrative in the archive or in the reader's memory.

Narratives of production

To date, several literary critics have mined the tension between the object narrator and the written text that contains its voice. Liz Bellamy, for instance, argues that the sub-genre has a 'subversive potential' because it uses an object rather than a human as its focus and replaces the 'economic for the affective mechanism'.[11] Blackwell similarly notes the tension between written works and commodities by considering the sub-genre's representation of hack writers as 'things, mere instruments of a culture that commodifies texts, thereby "smear[ing] and stain[ing]" writing with the excess of materiality'.[12] For Blackwell and many others, the it-narrative necessarily reflects on the cultural and physical production of the text itself. Christina Lupton has also examined the relationship between the material culture of the book and the power that it-narrators seem to wield, which she argues is 'less through the fictional stance of their knowing objects … than because the fiction of the sentient object gels with the more widely propagated effect of the book as a technology that has the power to trap, enthral, and anticipate its reader'.[13] In *The History and Adventures of a Lady's Slippers and Shoes* these tensions play out through an elaborate correspondence between the production of the slippers and shoes and the creation of a fictional narrative. The story positions the makers of the slippers and shoes as co-authors working alongside the anonymous author of the it-narrative itself, as it delves into detail about the shoemaker, journeymen and pieceworkers who produce the slippers and shoes. The slippers explain that they are 'cut out by the hand of the shoemaker-general to the court-ladies, and delivered up to his neatest workman' who 'having completed me for an honour suitable to the richness of my substance and the exquisiteness of the work, returned me to the master, who after taking a view of me on all sides, clapt me under his cloak, and carried me himself to a lady of quality' (5).

Such stages of production chime with the shift that Giorgio Riello has outlined in his study of eighteenth-century footwear. In the late seventeenth- and early eighteenth-century workshop, the master shoemaker purchased and cut the leather, and finished and heeled the shoes, while the middle stages of production were completed by subcontractors or journeymen who worked under the master shoemaker.[14] D. A. Saguto calls these workers 'all but invisible', noting that they were reimbursed by the piece and laboured either in the workshop or elsewhere.[15] Kimberley S. Alexander emphasizes the difficulty of succeeding at the trade and the long process of apprenticeship; she explains that an apprenticeship lasted seven years 'under the tutelage of a master shoemaker, followed by several years as a journeyman', before the aspiring cordwainer could think about setting up his own workshop.[16] By the 1750s when *The History and Adventures of a Lady's Slippers and Shoes* was published, the master shoemaker still performed the tasks that required greater skill, but 'most of the work was carried on outside the workshop by journeymen who considered themselves very much as independent shoemakers'.[17] This led to a battle between shoemakers and journeymen over access to materials like leather and to the retail market because of the fear that journeymen would become 'independent producers'.[18] The slippers insert themselves into this manufacturing hierarchy when they explain to the shoes why they call the shoemaker, and not the journeyman, their father:

> My saying true father, I perceive, makes you smile; it is certain that 'tis the journeyman often has the greatest hand in things and that the sale of them is all the master minds. How many fathers call those children whose only relation to them is, that they go under their name, as many boast of a piece of workmanship which others have performed for them. But this was not the case with my father; I was formed by his own pincers, from his hand I received my admirable shape. (6)

The slippers articulate the economic dynamics behind the production of footwear, comparing the workshop system with the question of legitimate birth. They acknowledge that networks of production mean that more than one maker can claim authorship of a commodity. Although the shoemaker might attempt to control journeymen by limiting access to materials and the market, the slippers still call the journeyman their 'co-father' and the master shoe-maker their 'chief-father', which attends to conflicts over creative roles in the manufacturing of footwear in the mid-eighteenth century (7).[19]

The gendered position of the fictional slippers and shoes also affirms the complexity of their production and consumption. The shoes and slippers are associated with their female owners, but they also remain the product of both their 'father's' hands and the needlewoman who works their embroidery. The text refers to the slippers as 'it' sometimes, indicating that though there are two slippers they act as one, and sometimes they call themselves 'we' to indicate that there are two: for example, 'we stuck in the mud' (26). The slippers refer to the shoes as 'brother shoe' (4, 38), but the shoes never refer to the slippers in a similar manner. Their indistinct gender and subject positions thus mirror the multiplicity of agents in their production, highlighting the social complexity of such objects.

In the eighteenth century, there were separate shoemakers for women's and men's shoes, and for men's boots. In his *Art of the Shoemaker* (1767), Garsault attributes such divisions to source materials and technique:

> Although the art of the shoemaker generally encompasses every kind of leather footwear, and every master is entitled to fully exercise this, nevertheless he would in practice meet with differences and incompatibilities, finding himself frequently obliged to change his methods, materials, and procedures.[20]

He explains that making men's shoes requires dirty 'resinous materials' while women's footwear requires 'cleanliness, being adorned with silk fabrics' so that to make both could only be perfected 'with great difficulty'.[21] The *History and Adventures of a Lady's Slippers and Shoes* depicts both the gendered division of production (it focuses only on lady's slippers and shoes, just as a shoemaker would specialize in one type of footwear) at the same time that it complicates the idea of an individual subject position with the use of 'we', 'I', 'it' and 'brother shoe' to refer to the slippers and shoes. Although the inclusion of a woman needleworker and the shoes' and slippers' shifting pronouns appear to dismantle the gender binaries of footwear, the it-narrative's details about making practices (in which multiple creators, including women embroiderers and the shoemaker's family, play a role) make a case for the presence, in the production process, of the voices of woman workers and piece-makers, thereby undermining the authority of an individual male maker.

Despite their gender malleability and the way their production contains multiple makers and voices, the slippers' and shoes' portrayal of gender in the tale conforms to stereotypical depictions of women and the shoes they wear as frivolous and sexualized. Bonnie Blackwell has argued that all object narrators should be thought of as feminine, as they 'urge the female reader not to avoid her objectification by men ... but to take care to be possessed by only one male so as to assure her longevity and safekeeping'.[22] In keeping with Blackwell's insight that even gender-neutral objects are feminized by the it-narrative genre, the shoes' and slippers' indistinct gender positions remain temporary and are eventually feminized through use. While the slippers and shoes are manufactured by men and women, their status as women's footwear turns them into commodities for fashionable ladies, connecting them to gendered associations surrounding consumption, display and sexual promiscuity.

The shoes, notwithstanding their title of 'brother', are subject to their owners' whims in a way that mirrors women's subjection to men. In one instance, a housekeeper turns to use the shoes to assault a chamber-maid, when her husband appears and decrees that the shoes be sold. He appropriates the shoes to control his wife, just as she was about to use them to control another woman: 'He indeed at first setting out, knowing her mettle, had disciplined her into docility, so that she was obliged patiently to pocket the misfortune; and ... he took us away to a shoe-maker, in the town, and exchanged us for a plain pair' (50). Although the shoes claim that they do 'not like such implacable severity against a supplicant wife' (50), they possess little authority over the ways that humans use them, apart from being able to voice their opposition to the actions of men.

The conflict that the shoes feel on this and other occasions is visible in the details of their production: made for women by a male shoemaker who folds into his work that of a woman needleworker and subordinate journeyman, the slippers and shoes sometimes further their wearers' agency and sometimes frustrate it. The slippers, for instance, extoll their physical appearance: 'it is certain a neater pair of Slippers never came out of a shop, though I say it myself; for still my remains are respectable; my white bottom, and embroidered flowers are not quite undiscernible' (5). A pair of slippers (Figure 8.1) from about 1760 is constructed from similar materials. Their embroidery includes stem stitch, satin stitch, metallic work, French knots and long-and-short stitch. A closer look reveals a yellow brocaded material where the upper attaches to the sole, demonstrating that these slippers have been revamped (the front part of the upper of the shoe has been replaced with new embroidered material, hiding the original fabric).[23] It was not unusual to transform the external appearance of Georgian footwear. As Ariane Fennetaux has shown, reusing and recycling all sorts

Figure 8.1 French women's slippers (mule) 1760, Bata Shoe Museum, Toronto.

of textiles were common in the eighteenth century, and are central to understanding 'an economic system which rested as much on reuse and resale as it did on novelty and invention'.[24] Kimberly S. Alexander also charts the reuse and repair of shoes in eighteenth-century New England as they appear in cordwainers' daybooks of the period. Such documents show how 'women often brought in textiles or embroidered uppers to be made into shoes', how shoes were remade for other wearers or adapted to changing foot sizes, and how 'older textiles were cut down to use for new shoes'.[25]

The shoes' segment of the it-narrative incorporates such practices into the narrative circulation of the footwear; the shoes are, for instance, sold 'to an old-cloths-woman' who 'soon had us vamped up to qualify us for our original honours' (53). The shoes become subject to each new owner's whims, and we see tensions surrounding their agency at moments in which owners become makers. In this instance, the 'old-cloths-woman' remakes the shoes to support a young man's planned seduction of his married neighbour: 'Here's a pair of Shoes says she to him, buy them, and try what they will do' (54). Such schemes fail, however, when the married neighbour and her husband reverse the intended use of the shoes by playing a trick that teaches the young man to 'leave honest women to their own husbands' (56). The multiple creators and owners of these slippers and shoes make plain the varied uses of footwear by an 'avaricious go-between' and a 'faithful spouse'. Such objects are subject to the hands that make and use them, which complicates the idea of a single gender identity or subject position for them, both in practice and in the it-narrative.[26] The shoe object contains in its form the remnants of workers' stitches and design choices that the wearer in turn interacts with when they shape the object for their own purposes; the multiple uses of the shoes and their variegated fabrics mirror one another in this sense. The question of how to read details of an object's production and consumption, as they are visible in the object itself, is one that historians of dress and material culture have considered. Hannah

Grieg, Jane Hamlett and Leonie Hannan take up the importance of reconstructing such networks of production and consumption: 'by stopping to consider the ways in which objects were made, used, exchanged, lost, adapted and even destroyed, we can reinvigorate our view of historical artefacts otherwise disassociated from human action by the distance of time and place.'[27] *The History and Adventures of a Lady's Slippers and Shoes* shows how a dress object can be just as intertextual as a fictional tale and the two types of material and fictional intertextualities work together in this example of the genre.

Material intertextuality

The shoes' and slippers' competing social functions gain further traction from the tale's changing speakers and narrative modes; both objects and narrative form are shape-shifting things. The tale purports on the title page to be 'written by' the shoes and slippers 'themselves', but the fictional narrator of the piece is, in fact, a traveller who writes to a friend about the conversations he hears between the slippers and shoes. This narrator/editor interjects with footnotes and 'translations' that allude to English authors from Shakespeare to John Dryden, Joseph Addison and Alexander Pope; these appear alongside historical anecdotes about the War of Spanish Succession, common proverbs and notes on fashionable dress. Moreover, in the slippers' and shoes' accounts of their production, the voice of the object narrator is intertwined with the voices of their creators, drawing attention to the relationship between verbal, written and textile creation. The workman who constructs the slippers 'with bended body, lifted knee, and arms alternately closed and extended' chants over them 'many amorous ditties' (5). The needlewoman who works on the shoes' uppers similarly embeds her voice into the construction of the object: 'Whilst she was at work upon us, her tongue moved as nimbly as her fingers, with hymns, and love-songs, stories, jests, and all the effusions of female prattling' (39). The shoemaker and needlewoman both repeat narratives they have heard as they construct the material product of the shoe. This aligns 'amorous ditties' as well as 'love-songs, stories, jests' and 'female prattling' (which are all forms of narrative similar to those that appear in the it-narrative itself) with acts of making and sewing. This it-narrative's material literacy, which is visible in its treatment of textile production, sewing skills and shoemaking, sits comfortably with its awareness of narrative as a fashionable commodity.[28] An intertextuality develops between the shoes and the written or spoken texts, as well as between the multiple authors, makers, consumers and materials that comprise the shoes, slippers and narrative.

In this way, the shoe-object functions as a kind of ideologeme, the meeting point of its literary representation and its material referents, as Kristeva describes: 'The ideologeme is that intertextual function read as "materialized" at the different structural levels of each text, and which stretches along the entire length of its trajectory, giving it its historical and social coordinates.'[29] The fictional representation of objects like the slippers and shoes functions as an intersection that corresponds with the 'historical and social coordinates' that Kristeva identifies. In the case of the slippers and shoes, historical conditions – including that of labour, bodies, voices and social positions of the shoemaker, the journeyman and the needlewoman – meet again with the objects' literary representation. Literary critics have also attended to the way that artefacts such as textiles and dress are part of the larger social and historical narrative in which novels and other forms of literature participate. Elaine Freedgood focuses on 'strong metonymic reading' and acknowledges the importance of the material object when analysing its literary uses.[30] Chloe Wigston Smith makes the point that artefacts are 'tantalizing and often mysterious material objects, whose own stories have been truncated through historical distance', a loss that complicates literary readings of the relationship between texts and textiles and elucidates how 'narrative might diminish and supplant the material realm'.[31]

On the one hand, *The History and Adventures of a Lady's Slippers and Shoes* attempts to 'diminish and supplant the material realm' in the shoes' and slippers' reliance on human bodies in order to realize narrative movement. On the other hand, the detailed account of their production and needlework develops a strong correspondence between textual and material production and consumption, including the shoemaking, stitching, translating, quoting and re-covering that form the narrative's structure. The tale's traveller-narrator displays an anxiety over the relationship between the text and its social and historical milieu when he satirizes the genre of the secret history novel by suggesting that readers will want a key to the shoe's 'place ... of formation' so that they may identify the shoe itself, or that 'some envious incendiaries ... will go about to persuade people that this is all a device of mine, that through a fabulous vehicle I disseminate real scandal' (29–30). He thus refuses to acknowledge that the shoe's description of its own 'formation' is based in fact, even though the details of its production are so specific. Rivka Swenson outlines two kinds of secret-history it-narratives: 'non-it-centred narratives in which the *it* is a titular, otherwise absent, entryway into a mosaic of anecdotes about actual, fictive, and/or allegorical figures witnessed by *it*' and 'the increasingly popular *it-centred secret history*, wherein the secret history of the *it* is a primary focus'.[32] In *The History and Adventures of a Lady's Slippers and Shoes* the emphasis lies somewhere between these two forms. The shoes and slippers introduce anecdotes about fictional (mostly allegorical) figures, but they also explain their own feelings and attitudes about their subject position, give detailed accounts of their appearance and foreground the voices of the shoemaker and needlewoman. While the narrator makes us believe that this will be a 'fleshless' it-narrative, as Swenson calls those narratives that take the *it* as their focus but frustrate the reader's desire for secret details, the shoes and slippers provide enough details about their production and consumption and thoughts and feelings, for us to be able to identify them (or at least comparable slippers or shoes that look like them).[33] It is these very details (especially of the stitching) that draw our attention to the relationship between narrative composition and material composition.

Layers of needlework and text

The haptic, sensory relationship between narrative and material form gains further complexity in the account of the shoe's needlework, where we see the close intimacy between the physical shoes and the voice of the woman who works them. As the shoes recall:

The whole town did not afford a neater work-woman, nor a prettier girl, than she, whose delicate hand, performed the needlework of me, – especially she had not her equal for cross-stitch – and she made her boasts with the lasses of her acquaintance, that she had never done any thing neater ... I am sure, says she, they cost me many a prick'd finger, and broken needle: Ay, said they ... to be sure they must be designed for some bride, but it is well, when the honey-month is over, if mourning-shoes would not become her better; *marry in haste, and repent at leisure*. Whilst she was at work upon us, her tongue moved as nimbly as her fingers, with hymns, and love-songs, stories, jests, and all the effusions of female pratling. The silken flowers, beautifully variegated on an *Isabella* groundlace and the effulgency of spangles, joined to render me worthy to figure in the splendid circle. Not the least bit of canvas, on which I was worked, was to be seen; in all things, a fair outside you know is the main concern. (38–9)

Here the detailed description of production portrays the intertextual, multiple discourses of imaginative creation and how these texts and voices intersect in the production of both the shoes and narrative. This is visible in the types of needlework that the work-woman produces, which are more wide-ranging than that on typical period shoes. The 'silken flowers variegated on an *Isabella* groundlace' might refer to shaded embroidery on ground lace, as seen in an example of embroidered Italian net, which combines colourful variegated flowers in silk with a substrate of net.[34] The embroidery on the slippers in Figure 8.1 also might be called variegated, although they are stitched on silk, not lace. The word 'variegated', however, can mean both 'varied in colour' and 'characterized by variety'.[35] This suggests that variegated flowers on ground lace might also be worked as a needle-lace design, given that 'the tendency was to use the ground rather than the design itself to display the lacemaker's virtuosity', and was associated with Alençon lace.[36] The shoes' surface lace thus stands as a material metaphor for the way that *The History and Adventures of a Lady's Slippers and Shoes* incorporates allusions to many early modern texts. These gestures to texts quite serious in tone often appear in the voice of the narrator in footnotes, or in italics; thus, this textual ground contrasts with the apparently frivolous central motif of speaking shoes.

Moreover, the language of the shoes implies that the needlewoman executes canvas work, one form of which was to fasten canvas onto a background of gold or silk, and then cut away the canvas after stitching to reveal the shimmering fabric underlayer.[37] Such techniques created patterns in which the finished design inverted the original roles of the top and bottom fabrics. This type of design work offers an apt way to think about the genre of the it-narrative itself, in which speaking objects, like the canvas, serve to cut away to the social satire that lurks beneath. The editorial notes of *The History and Adventures of a Lady's Slippers and Shoes* perform a similar kind of cut-away role when they layer the voice of Dryden below that of the slippers and shoes and then replace the traveller-narrator's voice with that of Dryden's. In the first part of the narrative, the slippers begin a critique of Catholic priests with the verse,

Where soldiers do as they think fit,
Where folks t'intriguing priests submit,

Where wives o'er men the breeches wear,
The devil with all his imps is there! (16)

In response, the shoes interject to defend priests, countering,

> I know as much of the world as yourself, and it seems to me that the principles,
> tempers and carriage of mortals, are not a little influenced by their station of
> life, and the company they frequent … Though *one swallow does not make a*
> *summer*, can you, with all your very good authority, produce me a character more
> respectable, more indearing than the following of a priest. (16)

Here the traveller-narrator interrupts the shoe with a note indicating that 'the original
could not be better delivered, than in the following lines from our *Dryden*' (16). He
then quotes a long segment from Dryden's translation of Chaucer's *Canterbury Tales*
that describes the parson. The Dryden passage is included in the body of the text rather
than in the notes, thus displacing the 'following' support of priests that the shoes were
going to cite (indeed this support never appears). The author thus creates an absence
in the text, filling it with a translation, which itself is a translation of an earlier poem.
The intertextual narrative techniques here are similar to those in other material arts,
such as canvas work and needlework because they create texture through layering and
cutting away. The interplay between the voice of the object and that of Dryden, the
poet, also shows how the it-narrative might assemble written and material texts.

The 'effulgency of spangles, joined to render' the shoes 'worthy to figure in the
splendid circle' constitutes both figurative and realistic needlework details that
generate the image of an extremely over-worked shoe. Every rich detail has been added
to this pair, linking it up with excessive court fashions. A pair of mules (Figure 8.2)
from the mid-to-late 1770s exemplifies what an 'effulgency of spangles' might look like.
While the mules were created around two decades after the it-narrative, they are made
from similar materials, including silver thread embroidery and spangles.[38] The word
'effulgency' (suggesting radiance) echoes the 'effusions of female pratling', which pour
forth from the mouth of the work-woman; this aligns the ornamental spangles with the
'hymns, and love-songs, stories, [and] jests' that the work-woman embeds in the shoes
as she embroiders them. To call the woman's narratives 'effusions' suggests that they
are not properly regulated and exceed the space of the text. Hymns, love songs, stories
and jests are also similar to the kinds of narratives the slippers and shoes espouse as
they recount their tales, turning the effusions into a self-reflexive comment on the
composition of both the shoes and the it-narrative itself.

The description of the canvas work below the spangles further positions both
women and shoes as frivolous: 'Not the least bit of canvas, on which I was worked,
was to be seen; in all things, a fair outside you know is the main concern' (39).[39] The
needlewoman proves especially talented at cross-stitch, which is a canvas stitch central
to sampler work used both to create image-based patterns and stitched text. A pair
of shoes (Figure 8.3) from the 1730s–40s is worked with wool in cross stitch and tent
stitch, illustrating a completely covered canvas. This floral pattern was likely designed
specifically for these shoes, as the pattern is not interrupted; instead, it is centred and

Figure 8.2 French women's slippers (mules) 1775–1785 with embroidered pattern on vamp in silver thread and sequins or pailletes, Bata Shoe Museum, Toronto.

Figure 8.3 Pair of women's shoes, linen canvas embroidered with coloured wools in cross stitch and tent stitch, with short heel and latchet fastening, Great Britain, 1730s–1740s © Victoria and Albert Museum (see Colour Plate 4).

finished at the borders of the shoes, rather than being cut off in mid-pattern. The shoes are also initialled on the tongue with what looks like the letters 'A' and 'L'. The needleworker has left a material remnant of individual ownership (and authorship, if the maker and the owner are one and the same) embedded in the fabric of the work. The it-narrative makes a similar corporeal gesture when the shoes note that the work-woman sustained 'many a prick'd finger, and broken needle' as she stitched them. Her identity and experience are part of the shoes' production history (in the form of her pricked finger, broken needle, and songs and stories), just as makers' and owners' identities and experiences are visible in the shoes that remain in the archive, which contain material marks such as initials, choices in stitches and choices in design. A pair of shoes (Figure 8.4), completely worked in queen stitch on a linen substrate, offers another example of canvas work in which the needleworker has inscribed their physical and mental presence in the footwear. Turns in the queen stitch (sometimes called renaissance stitch) mark precise decisions. At points they have turned the orientation of the stitch from horizontal to vertical in order to better accommodate the direction of the design. These shoes also boast a silver lace braid on the vamp, which makes them almost as extravagant as the shoes of the it-narrative.

In the it-narrative, the needlewoman demonstrates her range of skill, evoking the varied techniques of eighteenth-century samplers. The shoes include cross-stitch, lace making (though the text does not explain whether the needlewoman makes or

Figure 8.4 Embroidered green latchet tie shoes with silver lace braid, British, 1710–1740, Bata Shoe Museum, Toronto.

purchases the 'Isabella groundlace' upon which she stitches the 'variegated' flowers), embroidered flowers and spangle work. The 'hymns, and love-songs, stories, jests' recall the verses women included in their samplers, which Crystal B. Lake studies in this volume. The needleworker's varied range of speech and song articulates both the 'inculcation of femininity' and the very disruption of those ideal expectations associated with samplers produced by women, then and now.[40] Just as the it-narrative portrays women both aligned with and able to manipulate the shoes and slippers, the shoes' needlework exists on the border between submissive and unruly femininity.[41] The it-narrative suggests that the songs, jests and stories of the woman needleworker may similarly be in conflict with the hymns she sings. But her industry is also embedded in familial ideals, as she passes the completed uppers onto the shoemaker and his wife:

> This neat girl, delivered us up to a tip-top shoe-maker, who was so pleased with my sightly appearance, that he put no less than a Crown into her hand ... I was finished by his own hand, no journeyman put a stitch in me; his wife ... was called to admire me, and the loving couple turned me about and viewed me with no less joy and pride, than a new author does some fresh product of his brain; reading it over and over, and smiling and laughing at what will cause the purchasing bookseller to sigh and frown. (39)

The shoes proudly explain 'no journeyman put a stitch' in them to emphasize single authorship like the slippers do earlier in the text, but this idealistic image of a single male author is undermined by the needlewoman's role. The shoes' fashionable status requires confirmation – and likely manual input – from both the shoemaker and his wife. Rebecca Shawcross explains of the early shoemaker's workshop: 'wives of shoemakers frequently helped their husbands' "close" the upper. The process involved sewing the upper components of the shoe together and was particularly suited to women at a time when shoe uppers were made mostly from textiles'.[42] Like the production of the slippers, which troubles the idea of a single author, the production of the shoes complicates gendered expectations for women by including the needlewoman in the production of the shoes, even as it later reinforces the idea that women are superficial, false consumers in its account of the shoes' owners. Both the narrative's literary and material intertextuality suggest that communities rather than individuals create commodities such as the book or the shoe. This is true of women's craftwork, which was, according to Amanda Vickery, 'particularly associated with female sociability, female communities and often female retirement', as well as 'productions of supreme individuality'.[43]

The History and Adventures of a Lady's Slippers and Shoes uses the realistic depiction of an object's production and consumption to critique commodity culture and the excesses of women's dress. However, the details of footwear production place pressure on the idea of a single maker as much as they do on the notion that a woman's needlework must indicate submissiveness, in line with similar period tensions surrounding women's samplers. These details also give the shoes a history that includes the bodily experiences and voice of the needlewoman who makes them. Granted, both the needleworker's voice and that of her community offer stereotypical versions of femininity, for example when they suggest that the shoes look like wedding shoes but mourning shoes might work out better: '*marry in haste, and repent at leisure*', the shoes chide (39). But the presence of any maker's voice makes this it-narrative a kind of imaginative work in the recovery of material history. Smith has noted that 'problems of identification are ... vexing to studies of feminine material culture. Hundreds of anonymous artefacts frustrate attempts to match literary usage to extant examples'.[44] The fictional shoes in *The History and Adventures of a Lady's Slippers and Shoes* are an intertext for the stitched shoes that remain in the archives because they animate what has been lost, showing us how intricately worked material objects, rife with hidden seams, stitches and variegated surfaces, might be imaginatively and creatively rendered in prose. As Catherine Richardson argues, 'drawing connections between documentary sources about objects and ... more self-consciously expressive writings can help us to historicize the processes through which objects were cast in words'.[45]

The History and Adventures of a Lady's Slippers and Shoes displays the material literacy of its author. It draws upon the details of the slippers' and shoes' production and consumption to develop an intertextuality that is both textual and material. The author's knowledge of shoemaking and needlework generates a more textured narrative portrayal of the stitched shoe object: one that rejects realism in favour of social satire, but one that also relies on the very realism it rejects in order to structure its tale. Attending to details in footwear production, needlework practices and surviving shoe artefacts alongside narrative devices like footnotes and literary allusion helps to illuminate the literary and material intertexts that the author stitches together in this particular it-narrative. The shoes and slippers talk about their materiality and their producers and consumers while they (as any good pair of shoes should) work together in sympathetic narrative exchange.

Notes

1   *The History and Adventures of a Lady's Slippers and Shoes, Written by Themselves* (London: M. Cooper, 1754). Hereafter cited parenthetically in the text.

2   Mark Blackwell, 'Introduction: The It-Narrative and Eighteenth-Century Thing Theory', in *The Secret Life of Things: Animals, Objects, and It-Narratives in Eighteenth-Century England*, ed. Mark Blackwell (Lewisburg: Bucknell University Press, 2007), 10.

3   As Mark Blackwell explains of it-narrators: 'sometimes these characters enjoy a consciousness – and thus a perspective – of their own; sometimes they are merely narrative hubs around which other people's stories accumulate' (ibid., 10).

4   *The Memoirs and Interesting Adventures of an Embroidered Waistcoat* (London, 1751) in *British It-Narratives, 1750–1830*, Vol. 3, ed. Mark Blackwell and Christina Lupton (London: Pickering and Chatto, 2012), 116. *The Adventures of a Black Coat* (London, 1760), in *British It-Narratives, 1750–1830*, Vol. 3, 113.

5   'Adventures of a Quire of Paper', *London Magazine, or Gentleman's Monthly Intelligencer*, 48 (August 1779), 355–8; (September 1779), 395–8; (October 1779), 448–52, in *British It-Narratives, 1750–1830*, Vol. 4, ed. Mark Blackwell (London: Pickering and Chatto, 2012), 30.

6   'Adventures of a Quire of Paper', 35.
7   Edward Thompson, 'Indusiata: or, The Adventures of a Silk Petticoat', *Westminster Magazine*, 1 (June 1773), 365–8; (July 1773), 432–5; (August 1773), 469–72; (September 1773), 549–51; (October 1773), 598–600; (November 1773), 640–1; (December 1773), 689–91, in *British It-Narratives, 1750–1830*, Vol. 3, 166. See also Chloe Wigston Smith on 'Indusiata' and 'Adventures of a Quire of Paper', in *Women, Work, and Clothes in the Eighteenth-Century Novel*, 73–7, 77–8.
8   Thompson, 156, 166.
9   [Mary Ann Kilner], *The Adventures of a Pincushion, Designed Chiefly for the Use of Young Ladies* (London, c.1784) in *British It-Narratives, 1750–1830*, Vol. 4, 66.
10   Julia Kristeva, *Desire in Language: A Semiotic Approach to Literature and Art by Julia Kristeva*, ed. Leon S. Roudiez, trans. Thomas Gora, Alice Jardine and Leon S. Roudiez (New York: Columbia University Press, 1980), 36.
11   Liz Bellamy, 'It-Narrators and Circulation: Defining a Subgenre', in *The Secret Life of Things: Animals, Objects, and It-Narratives in Eighteenth-Century England*, ed. Mark Blackwell (Lewisburg: Bucknell University Press, 2007), 117–46.
12   Mark Blackwell, 'Hackwork: It-Narratives and Iteration', in *The Secret Life of Things*, 207.
13   Christina Lupton, 'Giving Power to the Medium: Recovering the 1750s', *The Eighteenth Century: Theory and Interpretation* 52, nos. 3–4 (2011): 297.
14   Giorgio Riello, *A Foot in the Past: Consumers, Producers and Footwear in the Long Eighteenth Century* (Oxford: The Pasold Research Fund, Oxford University Press, 2006), 173, 175.
15   D. A. Saguto, *M. de Garsault's 1767 Art of the Shoemaker: An Annotated Translation* (Williamsburg: Colonial Williamsburg Foundation and Texas Tech University Press, 2009), 2.
16   Kimberly S. Alexander, *Treasures Afoot: Shoe Stories from the Georgian Era* (Baltimore: Johns Hopkins University Press, 2018), 19.
17   Riello, *A Foot in the Past*, 175.
18   Ibid., 175–6.
19   Ibid. Alexander also discusses how setting up a shop as a master shoemaker would not have been a simple task and involved collaboration with many others; of the mid-eighteenth-century London shoemaker John Hose she notes: 'Setting up a shop was expensive, however, and required a significant outlay of cash for space, materials, and tools, making it a difficult goal for many to achieve. Other family members no doubt assisted the Hose enterprise in some capacity …' (*Treasures Afoot*, 19–20).
20   Saguto, *M. de Garsault's 1767 Art of the Shoemaker*, 36.
21   Ibid.
22   Bonnie Blackwell, 'Corkscrews and Courtesans: Sex and Death in Circulation Novels', in *The Secret Life of Things: Animals, Objects, and It-Narratives in Eighteenth-Century England*, ed. Mark Blackwell (Lewisburg: Bucknell University Press, 2007), 265–91.
23   The vamp is the front part of the upper, which is the term used to describe 'the entire shoe or boot that covers the top of the foot normally consisting of a vamp, quarters and lining' but not the sole or heel. Jonathan Walford, *The Seductive Shoe: Four Centuries of Fashion Footwear* (New York: Steward, Tabori and Chang, 2007), 281. The vamp is a term usually used when 'two symmetrical side seams exist about midway between toe and heel' (281). Slippers do not often have sides, or quarters, so here the upper (all vamp) has been entirely replaced. The Bata Shoe Museum datasheet indicates that 'the upper is not original as it is slip stitched along the

bottom edge to a yellow brocade substrate' (Ada Hopkins, BSM Conservator) *Bata Shoe Museum, Toronto.* Datasheet for ID No. P85.0181 (Print date: 27 March 2012).

24 Fennetaux, 'Sentimental Economics: Recycling Textiles in Eighteenth-Century Britain', 122.

25 Alexander, *Treasures Afoot*, 11.

26 Chloe Wigston Smith argues objects that seem to gain agency in it-narratives are actually put in their place by the structures of such narratives, which rely on the decay of objects for their plots and thus 'foreclose the subjectivity of sartorial objects, relaying how stories about speaking garments mute the very objects they seek to animate' (*Women, Work, and Clothes in the Eighteenth-Century Novel*, 74).

27 Hannah Greig, Jane Hamlett and Leonie Hannan, 'Introduction: Gender and Material Culture', in *Gender and Material Culture in Britain since 1600*, ed. Hannah Greig, Jane Hamlett and Leonie Hannan (London: Palgrave, 2016), 1–14, 7.

28 Julie Park has noted that 'fashion is inseparable from the project of fiction' because both link to 'the public's need to mediate social relations through objects' (*The Self and It: Novel Objects in Eighteenth-Century England* [Stanford: Stanford University Press, 2010], 27).

29 Kristeva, *Desire in Language*, 36.

30 Elaine Freedgood, *The Ideas in Things: Fugitive Meaning in the Victorian Novel* (Chicago: University of Chicago Press, 2006), 12.

31 Smith, *Women, Work, and Clothes in the Eighteenth-Century Novel*, 79.

32 Rivka Swenson, 'Secret History and It-Narrative', in *The Secret History in Literature, 1660–1820*, ed. Rebecca Bullard and Rachel Carnell (Cambridge: Cambridge University Press, 2017), 117–33, 117.

33 Ibid., 119.

34 Metropolitan Museum of Art, accession number 12.9.6.

35 'variegated, adj', *OED Online*, June 2018, Oxford University Press, http://www.oed. com, accessed 4 July 2018.

36 Edward Maeder, *An Elegant Art: Fashion & Fantasy in the Eighteenth Century* (New York: Los Angeles County Museum of Art Collection of Costumes and Textiles, 1983), 117.

37 According to the *Art of the Embroiderer* (1770): 'one makes each stitch simultaneously go through both the canvas and the gold or silk fabric beneath. One then cuts away the edge of the canvas … The fabric, [which was] originally below the canvas, and which is now totally revealed, becomes the actual background of the needlework. The canvas has only served to organize the stitches' (Charles Germain de Saint-Aubin, *Art of the Embroiderer*, trans. Nikki Scheuer, ed. Edward Maeder [1770; Los Angeles: Los Angeles County Museum of Art, 1983], 53).

38 Spangle was the eighteenth-century word for sequins; large spangles were called paillettes. See Marsh, *18th Century Embroidery Techniques*, 43.

39 Elizabeth Kowaleski-Wallace argues that the characterization of women as particularly frivolous and prone to consumption is part of the way that eighteenth-century writers 'deploy the female body in a debate about the human implications of consumption' (*Consuming Subjects*, 7). For more on fashion and gender, see Jennie Batchelor, *Dress, Distress and Desire: Clothing and the Female Body in Eighteenth-Century Literature* (Houndmills: Palgrave Macmillan, 2005); Erin Mackie, *Market à la Mode: Fashion, Commodity, and Gender in* The Tatler *and* The Spectator (Baltimore: Johns Hopkins University Press, 1997); Smith, *Women, Work, and Clothes in the Eighteenth-Century Novel*; Greig, Hamlett and Hannan, ed. *Gender and*

Material Culture in Britain since 1600; Jennifer M. Jones, *Sexing La Mode: Gender, Fashion and Commercial Culture in Old Regime France* (Oxford: Berg, 2004); and Peter McNeil, *Pretty Gentlemen: Macaroni Men and the Eighteenth-Century Fashion World* (London: Yale University Press, 2018).

40 Parker, *The Subversive Stitch*, 82–109. As Maureen Daly Goggin notes, 'A commonplace exercise was the stitching of hymns, proverbs, psalms, and other sections from the Bible and other moral texts' ('Stitching a Life in "Pen of Steele and Silken Inke"', 33. For a treatment of more recent, global examples, see Brenda Schmahmann, 'Intertextual Textiles: Parodies and Quotations in Cloth', *Textile* 15, no. 4 (2017): 336–43.

41 Amanda Vickery considers women's choice of needlework subjects (such as the natural world, or the mourning widow) as a sign of this negotiation. She concludes that 'women could use embroidery to interrogate and negotiate the constraints of femininity' (*Behind Closed Doors*, 240).

42 Rebecca Shawcross, *Shoes: An Illustrated History* (London: Bloomsbury, 2014), 90.

43 Vickery, *Behind Closed Doors*, 246, 253.

44 Smith, 'Gender and the Material Turn', 154.

45 Catherine Richardson, 'Written Texts and the Performance of Materiality', in *Writing Material Culture History*, ed. Anne Gerritsen and Giorgio Riello (London: Bloomsbury, 2015), 54.

9

Making, measuring and selling in Hampshire: The provincial tailor's accounts of George and Benjamin Ferrey

Sarah Howard

The late eighteenth and early nineteenth centuries saw a shift in men's fashion and tailoring that introduced and consolidated new practices and methods for the ensuing decades.[1] The new styles required tailors to possess flexible and adaptable manual skills, in order to meet the changing tastes, demands and desires of their clients. The story of how tailors adjusted their methods of making at this moment of sartorial change is often lost in archival documents, buried beneath narratives of commercial exchange and economic management. However, the account book of the tailors George and Benjamin Ferrey offers a rare glimpse of the material skills of provincial tailors. These surprisingly thorough, detailed and consistent accounts provide insights into how tailors recorded the material literacies that underpinned successful trade. The Ferreys operated in the rural coastal town of Christchurch, Hampshire, and their surviving accounts date to 1795–1812, a period of significant fashion change in men's dress. Their account book reveals the range of making skills and material knowledge that even provincial tailors required to meet the shifting styles of this era. The Ferrey accounts not only illuminate the economic and business history of the tailoring trade, but also nod to the material competencies and ingenuity required to craft clothing, respond to distinct requirements from clients and also to create non-clothing items for their customers.

The Ferrey account book (Figure 9.1) documents a thriving and successful business that provided an array of clothing to a variety of customers from a wide social stratum, from local tradesmen and professionals of the town to the local gentry and eminent men of the area, their families and servants.[2] Over seventeen years, as their account book affirms, the Ferrey tailors operated as conduits for the circulation of fashionable dress from London and major cities to the provinces. Their accounts show how tailors worked to meet the needs of a rapidly evolving local community and how new opportunities to develop trade were exploited through manual knowledge. The Ferreys included six main areas of work relating to clothing: making, mending, altering, turning, letting out and seating garments. Mending and altering feature extensively across the accounts, especially for certain clients whose professions generated more

Figure 9.1 Title page, Ferrey Account Book, Druitt archive, 10570. Courtesy of Dorset History Centre. The opening page shows the names of the tailors.

intensive forms of wear and tear in their attire.[3] Moreover, the Ferreys cared for clothes across the lifecycle, from childhood to adulthood. Children's clothing is often mended, altered, turned or let out, because of either wear, the passing down of a garment to a younger sibling, alterations for growing bodies or adaptations to changing fashions.[4] In addition to commissions for durable workwear for tradespersons in Christchurch, the Ferreys produced other professional attire, with a particular focus upon military dress. Their accounts reflect the growing military presence in coastal towns in the south during wartime and how tailors could capitalize on political unrest, and gain new custom by applying their maker's knowledge of civilian fashion to the regulations of military attire.

The recent work of Peter McNeil and John Styles has delineated the social and economic impact of men's dress in the eighteenth century, adding to scholarship on the trade and practices of tailors in the period.[5] The Ferrey account book provides further insights into understanding the material as well as social and economic trends of the turn of the eighteenth into the nineteenth century. Although the Ferreys' business functioned in a similar fashion to their peers, this chapter makes a case for the account book as a source of material, as well as economic knowledge, that maps social relations and the demands of style beyond the metropolis.[6] Close scrutiny of the accounts uncovers the surprising variety of activities that the Ferreys undertook to satisfy customer demands and keep a steady flow of income, as they adapted their manual skills to complete commissions for the occasional home decoration project. Across the account, the Ferreys made brief notes about their work and the entries unveil the depth and richness of their material knowledge and skills. Rather than interpret these accounts merely as a list of line items in a ledger, this chapter searches the Ferreys' transactions for evidence of succinct detail that marks the tailor's command of material literacy. The Ferreys' account demonstrates how they applied their material knowledge to a changing landscape of style and how their skills proved flexible enough not only to anticipate how their clothes would fit the body of clients but also to foresee how they would be used, strained and stretched by those bodies. As a result, the Ferrey notations unveil a remarkably rich sense of how tailors' accounts marry information on costs and production with more immaterial forms of manual skill and material literacy.

The Ferrey tailors of Christchurch

The Ferrey account book records, in a neat hand, the transactions for 160 customers, the clothes they were wearing, the fabric, yardage and cost of the garments being made, as well as a range of other activities.[7] For some customers the work undertaken was minimal; for others it was more extensive and spread over several years, indicating regular and loyal custom. Of the known customers in the account book, the majority lived in and around Christchurch. Of the 16 per cent of customers who were visiting the town, many were listed as staying with existing Ferrey clients. The recording of this detail not only affirms customer loyalty, but also indicates the ways in which the Ferreys managed customer credit – the vouching of a visitor to the town by a local member of the community and the provision of an address whereby a new customer could be located and outstanding payment resolved.

George Ferrey established his tailoring trade in Christchurch in 1760 at the age of twenty-three.[8] George's son, Benjamin, entered his father's business, and eventually took over upon George's death in 1804.[9] Travel journals and trade directories from the late eighteenth century document that Christchurch was a small, self-supporting and busy coastal market town where the skills of a tailor would have been in constant demand.[10] Christchurch was known for its production of silk and worsted knit stockings and gloves that contributed to a local culture of sartorial making and manufacture. Similar to the Ferreys, several traders supplied the town's clothing needs, including shoemakers, staymakers, drapers and hosiers. A ready supply of locally purchased fabric existed for Ferrey and his customers to utilize. By the 1790s the town also supported two peruke makers and by 1811 a breeches and glove maker, reflecting the range of its residents' fashionable requirements.

The Ferreys lived in the centre of Christchurch very near to several of the town's drapers and close to the market square.[11] A front window that looked out onto a main street would have made it ideal for passing trade, and location and visibility were both key to any tailor's success in provincial towns.[12] The small size of the house indicates that the Ferreys operated, like many eighteenth-century tailors, in a crowded space where stock was also accommodated, and likely with the assistance of one or two journeymen and an apprentice.[13] In keeping with trade practices of the period, the Ferreys combined the selling of their services with material access to textiles and fabrics on hand; entering the shop would bring consumers into a tactile world of form and function.

The Ferreys' position at the heart of this local community is reflected in the long-term relationships they built with their customers. Certain accounts seem to have taken several months to have been settled, which implies the widespread use of credit, a common practice for the size of business operated by the Ferreys. Credit encouraged trade and ongoing custom, especially in towns where personal networks often supported local business.[14] The Ferreys also used extended billing cycles to accommodate late-paying clients, as shown by their references to accounts being brought over from other account books and the recording of dates when certain amounts were finally paid. This credit system relied upon the business owner's ability to assess a customer's ability to pay for their services and necessitated an ability to read the subtle messages relayed by the quality of clients' clothing. The Ferreys' ability to read visual appearance as an indicator of personal wealth and social status was a form of material literacy in itself. Recognizing each well-placed seam, each perfectly stitched buttonhole offered evidence of previous financial transactions. Operating in a small town, the Ferreys' knowledge of the local community would have supported such visual assessments, although possibly not for first-time customers. Faith in eventual payment wasn't always honoured. For example, in the account of a Mr William Bailey, the comment reads, 'Bill delivered to him at the White Hart. Receipted but not paid owing to his not parting with his change.'[15] The Ferreys' experience as tailors would have allowed them to quickly assess the quality of clothes worn by potential clients, but this did not always translate into a trustworthy future customer.

Apart from the work the Ferreys undertook to make clothing, the account book notes instances where they provided services that extended beyond the production of garments, which made up the repertoire of their main business. The Ferreys itemized sewing work they completed for domestic furnishings at customers' houses and such transactions suggest that visiting customers in their homes was possibly an unusual activity rather than a regular practice.[16] There are twenty-four entries in the account book where the Ferreys mention working in the homes of eleven customers, ranging from a linen and woollen draper, to members of the local gentry, and a politician who was a newcomer to Christchurch. Half of the recorded work was for fitting and mending carpets, with one entry for 'making and putting up three sets of blinds' for a maltster.[17]

These activities demonstrate the Ferreys' ability to apply their sewing abilities and adapt their making knowledge to the specific domestic needs of their customers that reached beyond the usual commission for garments. A relationship between the Ferreys and their customers was robust enough for these customers to trust them to work in their homes, where the Ferreys would have visited their household, met their families and servants, and developed personal relationships with them all to strengthen and sustain ongoing custom.[18]

In addition to when projects were completed outside the walls of their shop, the Ferreys also included some information about who realized the work on certain transactions, signalling their reliance on and employment of additional assistance. The Ferreys organized their employees according to a hierarchy in which seniority was congruent with material skill. The account book notes who completed the work in order to charge for their time, often writing 'me' or 'myself' next to an order or task, 'boy' suggesting an apprentice and 'man' presumably a journeyman. Sometimes the men undertook the work alone, without the Ferreys working alongside them. The amount of time required for the activity is precisely recorded in the account book from three quarters of a day up to five days. The Ferreys charged one shilling for the 'boy' per day, and three shillings per day for adults regardless whether they or one of their men was carrying out the work, marking the hierarchy of haptic skill and expertise within the business. The account book also records the number of men for each job, which ranged from two to four at any one time.[19] The hierarchy of material skill reflected in this price structure intimates the demands placed on tailors by their client base. Higher levels of material literacy and skill equated with a greater financial return.

The Ferrey customers and material demand

The customer base maintained by the Ferreys was diverse, requiring an adaptable array of material techniques to fit a number of professional and social needs. It ranged in gender, age and social status. Of the total number of customers mentioned in the account book 94 per cent were men. Yet this straightforward reading of the account book can be misleading. The Ferreys may have categorized their clients under these male customers' names, but the work they carried out often included their children and where the household was big enough, their male servants. Women represent 5 per cent of the customers listed, although they exclusively purchased clothing for their children or servants. There are seven specific entries for servants, four of whom are named.

Alongside the continuous custom of the local community, the account book also maps the changing social landscape of eighteenth-century Christchurch. The mid-eighteenth century saw the arrival of several eminent men and politicians to the region, who sought the clean air, temperate climate and appealing geography of the south coast.[20] Many of these new arrivals established homes in the Christchurch area as summer and country retreats, and several were customers of the Ferreys who fashioned garments to suit their country, coastal and social pursuits. The presence of an increasingly prosperous and mobile 'middling sort' can be detected in the account book via the orders of clothing for servants, by the men noted as 'esquire', and the making of garments specifically for leisure use, such as shooting jackets.[21] The commissions of these prosperous customers would have brought the Ferreys into direct contact with London and city fashions. The Ferreys not only created new garments for these clients, but copied, mended and altered the garments in their existing wardrobes. This maintenance work provided opportunities for the Ferreys to study London-made garments close up, and to learn new metropolitan making techniques.[22] Their repeat custom suggests the Ferreys fostered good relationships with their customers, showing a level of trust in their skills and standard of workmanship, as evidenced by references to bills being carried from or to other account books.[23]

The majority of the Ferreys' customers were tradespeople from the town (Table 9.1).[24] This dominance of the local trade community in the account book hints at a culture of mutual exchange in provincial communities. These local trading networks were reinforced by the location of the Ferreys' premises in the middle of the town. For instance, the landlord of one of the main hotels in the town was a customer of the Ferreys, as was the landlord of the nearby coaching inn. These contacts would have been of value in gaining custom from local residents, and new and returning visitors to Christchurch. The hotel also housed the South Hampshire militia until a barracks was built in the town in 1794, forming another potential source of trade for the Ferreys.[25] Two other inns were situated nearby, the landlords of whom were Ferrey customers, and all sectors or classes of the community would have variously gathered there to discuss the issues of the day and to conduct trade.[26] Operating a business in such close proximity with other merchants in the town, not least the drapers and hosiers who were also Ferrey customers, provided a source of custom and a reciprocal network of making and trading, promoting the Ferreys' business as well as those who supplied services to them.

Table 9.1 Breakdown of identified Ferrey customers, Ferrey Account Book, Druitt archive, 10570, Dorset History Centre.

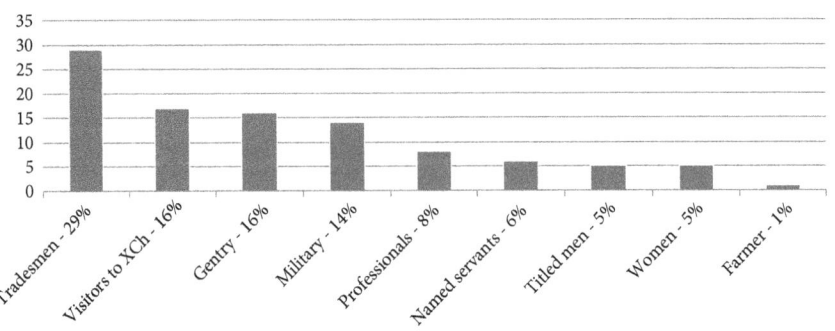

■ Percentage of known Ferrey customers

The account book reveals that the Ferreys explored and utilized opportunities to diversify their business and develop new customers as well as retain existing ones. Military customers, both regular and passing, constitute 14 per cent of the known customers in the account book (Table 9.1). The proximity of the town to the English Channel meant that it was vulnerable to attack from the French during the Revolutionary and Napoleonic wars. The Ferreys sought out this new base of military customers who required the making and mending of Regimental clothing. Although related in cut and fit to civilian dress, working with military clients required that the Ferreys implement a different type of material literacy, one built around regimental identities and military ranks. This new client base, then, would have expanded the material literacy of the tailor beyond the cutting and stitching of cloth, and into a realm of military insignia and sartorial emblem.

The Ferreys' tailoring techniques and commissions

The scope and quantity of garments itemized in the Ferrey account book mark their precise knowledge of materials and range of making competencies. The apprenticeship model was central to the transference of these skills, and it is likely that the Ferreys learnt their own skills through such training. The Ferreys workshops were spaces for the transference of material literacy, knowledge and skill, which occurred alongside and as part of the construction of garments. George Ferrey performed in the role of teacher, training junior tailors and positioning himself as a knowledgeable authority. He took on an apprentice in 1767, only seven years after he started trading, and employed a further two in 1772 and 1790.[27] Benjamin Ferrey, George's youngest and only son to join his trade, continued the business after his father's death in 1804. As a master tailor, Ferrey senior would have passed on his acquired knowledge and sewing skills to his son who would have had the added advantage of being brought up knowing the family trade and observing his father's profession from an early age.

Ferrey father and son constructed a range of garments, from substantial outerwear such as Great Coats to more tailored garments such as pantaloons and frock coats. This garment making sat alongside frequent entries in the account book for altering, turning and seating garments. Although brief in textual detail, the Ferrey transactions reveal the multiplicity of their skills and their rich understanding of cloth and the way in which it could be manipulated and utilized. Their transactions document their aptitude in cutting, shaping and sewing fashionable garments of the day. While the account book cannot reveal how the Ferreys went about this work, what stitches they used or how they transferred the contours of the body onto each length of fabric, it does convey the profusion of skilled knowledge required for each commission. In the language and notations of the account book, we receive hints about the types of technique the Ferreys and their staff employed. References to 'padding' and 'shaping' garments show the tailor's understanding and discernment of how to fit clothing to the shape of the wearer, speaking to the Ferreys' knowledge of the vocabulary of tailoring skill. Such verbal shorthand constitutes another access point for the material knowledge that Emily Taylor identifies, in her chapter, in tailoring manuals and in the stitches, holes and alterations that remain in extant garments from the period. The Ferreys also applied a diverse range of sewing skills to complete numerous tasks beyond garment making, affirming that tailors in the late eighteenth century used their dexterity to undertake a variety of work.[28] The occasional mending of gigs is included under the accounts of two of the Ferrey customers (presumably referring to the repair of carriage upholstery), as is the making of sandbags for a local surgeon, possibly to combat occasional flooding in the town.[29] The Ferreys' notes thus open a window into the range of their material practices and handiwork, even though the garments and objects they made and mended are no longer extant.

The account book reveals that the Ferreys and their staff were able to adapt their material skills to the differing needs of military and civilian clothing. However, such archival sources do have limitations when trying to assess how tailors learnt the skills required to create the latest fashions. This careful recording of transactions does not track the nuances of fashionable change, such as the new shape of a collar or the positioning of buttons on a jacket.[30] However, a significant area where the account book can unlock the materiality of the tailoring trade is through records of fabrics and textiles. In the Ferreys' account book, cheaper and more robust fabrics such as brown Holland, linsey and corduroy are more commonly seen in the garments of tradespeople and those of a lower social standing. Casamere and superfine cloth often falls under transactions for the middling classes and gentry. For instance, the accounts of Ambrose Tucker, a successful grocer in the town and regular Ferrey customer, detail the purchase and mending of a modest variety of fashionable garments.[31] In common with the records of other prosperous Christchurch tradesmen mentioned in the account book, Tucker's accounts hint at his social status and aspirations for self-advancement through the type of cloth used, and the cost and range of the garments ordered and mended. His sense of ambition can be seen further in the commissioning of his portrait (Figure 9.2). Although the portrait is crudely painted, the sartorial skill behind the high-cut collar and well-fitting coat hints at the material skill of the tailor who made them – presumably Ferrey. The range of garments purchased by Tucker (Table 9.2), particularly the pattern and colour of the waistcoats, suggests he was eager to adopt the latest fashions and prepared to pay for them, affirming his fashionable taste by commissioning a portrait that would show off his fine clothes.

The practical requirements of clothing the population of a rural, seaside town can be seen in the diversity of garments and textiles documented in the Ferrey accounts. Nuances in features, colour and cloth, as well as an understanding of the trade, professional and social status of the customers assisted in distinguishing the type of garments created by the Ferreys for their clients, which ranged from the purely practical to the highly fashionable. The Ferreys produced new garments alongside maintaining an active trade in the upkeep of clothes, meeting the material requirements of customers eager to extend the life of their wardrobes. Outer garments appear regularly in the account book and were made throughout the year. In total the Ferreys made, mended, altered and turned (reversing the fabric's surface from inside to outside) ten kinds of outer garments, including spencers and shooting jackets, conveying a familiarity with a wide variety of cuts and styles. Ferrey also made and mended riding habits

Table 9.2 A selection of garments in the accounts of Ambrose Tucker, Ferrey Account Book, Druitt archive, 10570, Dorset History Centre.

Date	Garment type	Fabric	Cost
September 1804	Stock	Velvet	1s. 6d.
December 1804	Pair of pantaloons	Superfine blue cloth	£1 10s. 6d.
October 1806	Pair of pantaloons	Superfine blue cloth	£1 10s. 6d.
January 1808	Waistcoat	Fine Swansdown buff striped waistcoat	15s.
February 1808	Coat	Best superfine blue cloth	£3 4s.
July 1808	Waistcoat	Dark spot Jean	10s.

and pelisses for the daughters and wives of his customers, adapting his knowledge of cloth and cut to the women's dress.[32] The greatcoat was one of the most common coats recorded in the account book, and the outer garment was commissioned by a variety of clients, ranging from the gentry and fashionable elite to their servants. This range suggests the flexibility of its terminology in the accounts; not all greatcoats would have shared the same material qualities, but the Ferreys apply it to describe the cut and general style of the garment. The fabrics mentioned in the account book highlight the multiplicity of 'Great Coats', with some made of moleskin and fine corbeau beaver in olive green, drab and dark blue, and others of cheaper fustian and cloth. The intricacies of fabric and detail also distinguish a box coat mentioned in the account book. In 1807 Richard Norriss, Esquire, ordered a box coat made with '4 whole capes' composed of ten yards of superfine Spanish Impregnable drab cloth and a velvet collar.[33] In this way, the Ferreys' transactions, despite their concision, unveil a material world brimming with coats of varied colours and fabrics in the streets and houses of Christchurch.

The account book's terminology also uncovers that the Ferreys tried their hand at new styles throughout their career, adapting their material literacy to meet fashionable expectations. Pantaloons, which were fashionable from the 1790s, offer a clear example of how tailors continued to expand their repertoire of making.[34] According to their records, the Ferreys made and mended such garments regularly, and used their knowledge of fabrics to adapt to the structural demands of these new styles (echoing Hilary Davidson's findings for dressmakers in the same era). Again, fabric choices affirm, indirectly, this material knowledge. The use of superfine cloth or knitted fabrics shows a recognition of the need for this new garment to mould effectively to the wearer. The production of such items signifies the Ferreys' understanding of the construction of these close-fitting garments and the modification and use of their making skills accordingly. There are many entries in the account book for knitted pantaloons, and the supply of these garments at the height of their popularity offers evidence of the Ferreys' ability to adapt their material skills quickly to changing styles.

The Ferreys' response to new styles of the nineteenth century is made clear by their production of trousers.[35] This garment originated as part of the garb of sailors,

Figure 9.2 Portrait of Ambrose Tucker, *c.* 1820, Hampshire Cultural Trust, FA2006.491. Tucker was a local grocer and Ferrey customer.

and the Ferreys' coastal position no doubt fed their knowledge of how to construct trousers, long before they were adopted by the fashionable elite. Of the total number of trousers which appear in the account book, 54 per cent were made for the sons of Ferrey customers, reflecting the fashion for them, from the late eighteenth century, for boys wear.[36] However, in spite of their coastal location, 80 per cent of trousers in the account book were ordered and mended between 1807 and 1808. This correlation with the fashionable wearing of trousers from 1807, a trend first noticed in Brighton that year, suggests a speedy adoption and adaptation of this garment in response to the needs of fashionable clientele.[37]

The account book clearly itemizes the garments the Ferreys sold in three different ways, with each method offering distinct degrees of the tailor's material agency over the finished product. Where the cost of the making of a garment is given in isolation, it can be assumed that the customers supplied their own fabric and that the Ferreys were charging for their time and skills alone. As Table 9.3 shows, the Ferreys standardized the cost of making different garment types. Customers would have had opportunities to purchase fabric from several Christchurch drapers or from visiting journeymen; in such cases, the Ferreys would have had to adapt their techniques to the fabrics supplied by clients.[38] The Ferreys' prices for the making up of garments match those of other tailors in the area. They correlate, for instance, with those in the account book of Robert Mansbridge, a fellow Hampshire tailor working in Basingstoke around the same time.[39] The Ferreys charged 7s. 6d. to make a Great coat, as did Mansbridge to make a coat, presumably the same type of garment suggesting a regional match in the pricing of work.[40] For other garments, the Ferreys wielded complete authority over the garment from start to finish, supplying the fabric and trimmings themselves. In these cases, the Ferreys indicate such oversight in textual ways through meticulous notes on the fabric type, costs and quantities used alongside the cost to make up the garment. Such cases presumably would have been the result of consultations between tailor and customer, in accordance with the making dialogues documented by Serena Dyer, Emily Taylor and Hilary Davidson.

A final method of selling a garment detectable in the account book is where neither the cost of making nor the amount of fabric is given and only the garment description provided, such as 'Pair of blue pantaloons'. Such notes confirm that the Ferreys were selling ready-made garments (Table 9.3 shows the average costs for garments made by the Ferreys and those they sold ready-made).[41] Sometimes the word 'complete' is added at the end of the garment's description suggesting the clothes were fully made up and did not require additional work. This third method of making granted the Ferreys complete autonomy, without any input from their customers, except at the point of purchase. The sale of ready-made garments alongside bespoke tailoring services substantiates Miles Lambert's assertion that eighteenth-century Britons were familiar with purchasing ready-made clothing.[42] Almost half of the Ferreys' customers were purchasing ready-made goods. However, the purchase of ready-made clothing did not come at the expense of bespoke wear. Ferreys customers paid for ready-made and other tailoring services at the same time. Even eminent local men purchased ready-made wear, although these garments were likely for their servants and children rather than for themselves.

The Ferreys produced ready-made garments across all the clothing types they supplied, from gaiters and breeches to coats and pantaloons. Gaiters, drawers and flannel waistcoats were popular ready-made garments – small, easy to store flat and to keep in stock. The percentage of ready-made garments to bespoke clothing includes 40 per cent of all breeches, 37 per cent of all pantaloons and 28 per cent of all trousers. This indicates a slightly lower overall percentage of ready-made garments than those

Table 9.3 Details from the Ferrey account book of the prices of bespoke and ready-made garments, Ferrey Account Book, Druitt archive, 10570, Dorset History Centre.

Type of garment	Standardized cost of making a garment	Average cost of a made-up garment including fabric	Average cost of a ready-made garment
Waistcoat	3s. 6d.	Flannel waistcoats, 8s. 4d. Other fabric waistcoats, 11s. (two @ £1 10s.)	Flannel waistcoats, 8s. 2d.
Breeches, pantaloons and trousers	3s.	Breeches, £1 6s. Pantaloons, £1 10s.	Breeches, £1 6s. Pantaloons, £1 10s. Trousers, £1 10s.
Spencer	5s.	£3 9s.	None mentioned in the account book
Shooting jacket	7s.	Shooting coat and long gaiters, £2 4s.	None mentioned in the account book
Coat	7s	Between £1 and £4	£2
Great coat	7s. 6d.	Between £2 and £7s6d	None mentioned in the account book

sold by other tailors such as Jeremiah Murray and Robert Mansbridge.[43] Of the total number of ready-made garments purchased, 10 per cent were bought by named visitors to the town. One such visitor, Mr Bosanquet, Esquire, purchased almost exclusively ready-made goods for himself and his son. These included a pair of light drab corded Casamere pantaloons, a pair of best India Nankeen trousers and several pale coloured, lightweight waistcoats which would have been appropriate in terms of fabric type, colour and fashion for a stay in the area.[44] The Ferreys thus recognized the importance of supplying the instantaneous demand for appropriate clothing for visits to Christchurch, especially from new clients staying with existing customers, and they saw the financial benefit of having a small stock of ready-made garments on hand. Either they made up certain garments in advance during quieter periods in the business as other tailors of the time did, or they outsourced their production to journeymen at busy times, as intimated by the quantity of ready-made and bespoke garments sold during 1807–8, a particularly bustling time in the accounts.[45] A comparison of prices of made-up and ready-made clothing in Table 9.3 shows that the Ferreys offered both for approximately the same amount, variations of which depended on the type and quantity of fabric used. This appears to be in contrast to the pricing practices of Murray and Mansfield, who differentiated between such garments, possibly indicating that the Ferreys made these garments themselves rather than outsourcing their production.[46]

Provincial tailors were required to maintain a flexible array of haptic and fashion knowledge. Deploying a range of textual details, the Ferrey account demonstrates the ability of eighteenth-century tailors to develop, adapt and alter their making skills; to use their knowledge and experiences to cultivate loyal custom; and to satisfy the fashionable and social aspirations of their customers as well as their own commercial ambitions. Relying on their material literacy, the Ferreys were able to embrace and promote new fashions, using their material knowledge and specialist discernment to ensure that the figure-hugging garments of the time, such as knitted pantaloons, would fit the shape of the wearer. The Ferreys broadened their client base and developed their business by taking advantage of a changing and burgeoning local economic and social scene, as evidenced by the number of visitors to the town who relied on their services. Similar to other tailors in the eighteenth century, the Ferreys possessed a range of sewing skills and techniques to satisfy diverse demands for the provision of practical and fashionable clothing for men of the town as well as visitors to the area, and for their children and servants. Their ability to apply this material knowledge to additional sewing activities such as the fitting of carpets in the houses of customers shows how they expanded their business outside of the walls of their shop.

The Ferreys also capitalized on clothing the military located in the area during the Napoleonic wars, and catered for the influx of eminent men to the region. They adapted their making knowledge and experiences from civilian clothing to military conventions. Finally, the account book provides a rich resource that details the construction and dissemination of a range of late Georgian garments at a pivotal time in men's fashion. It reflects the tastes and aspirations of an English provincial coastal town, whose residents participated avidly in fashion trends, from cut to colour and fabric. The supply of trousers to adults starting in 1807 marks the desire for this garment at the earliest stage in its fashionable status in Britain. In addition to its documentation of new styles and colours, the accounts also record everyday garments and those worn for work by traders in the town, including the use of overalls and stout fabrics for hard wearing clothing. The presence of ready-made clothing alongside bespoke garments shows how tailors attempted to balance commercial demand with customer requirements during periods of busy trading. The Ferreys utilized a range of material techniques and knowledge to dress the men and boys of Christchurch, whether residents or visitors, demonstrating how they applied their hands to meet and even anticipate the tastes, preferences and choices of their customers and community. Their handwritten transactions may remain concise and succinct to our eyes, but they nevertheless record the richness and variety of the material literacies tailors relied on to create new garments, mend old ones and extend the lives of clothes.

Notes

1   Madeleine Ginsburg, 'The Tailoring and Dressmaking Trades, 1700–1850', *Costume* 6 (1972): 64–71.

2   Trade directories of the time indicate the existence of other tailors working in Christchurch, although these businesses do not appear to be as long-standing as that of the Ferreys. In 1784 two other tailors were operating in the town but ten years later neither of them was listed and there was only one other tailor in business apart from Ferrey. Baileys British Directory, 1784, 56–7; The Universal British Directory 1792–8, Hampshire extracts.

3　Of note is a local maltster and regular Ferrey customer whose accounts are predominantly for mending clothing, including overalls; see Ferrey Account Book, Druitt archive, 10570, Dorset History Centre, 58.

4　Buck, *Dress in Eighteenth-Century England*, 205.

5　McNeil, *Pretty Gentlemen*; Styles, *The Dress of the People: Everyday Fashion in Eighteenth-Century England*; Miles Lambert, 'Bespoke versus Ready-Made: The Work of the Tailor in Eighteenth-Century Britain', *Costume* 44 (2010): 56–65; Ginsburg, 'The Tailoring and Dressmaking Trades'.

6　Christine Fowler, 'Robert Mansbridge, a Rural Tailor and His Customers 1811–1815', *Textile History* 28, no. 1 (1997): 29–38. The Ferrey account book extends some of the research into London and urban-based tailors; see Lambert, 'Bespoke versus Ready-Made'; Beverley Lemire, 'Developing Consumerism and the Ready-Made Clothing Trade in Britain, 1750–1800', *Textile History* 15, no. 1 (1984): 21–44; Ginsburg, 'The Tailoring and Dressmaking Trades'.

7　The account book is probably in Benjamin Ferrey's own hand, as indicated by the entries that mention work undertaken by 'self'.

8　Date taken from a Ferrey business letterhead of 1911, from a private collection.

9　The account book covers a period which spans both George and Benjamin Ferrey's time managing the business. Throughout the chapter, they are referred to as the Ferreys.

10　Richard Pococke and James Joel Cartwright, *The Travels through England of Dr Richard Pococke, during 1750, 1751, and Later Years* (Burlington: Tanner Ritchie Publishing, 2015); T. A. Baker, *Companion in a Tour Round Southampton*, 1794, cited in Sue Newman, *The Christchurch Fusee Chain Gang* (Gloucester: Amberley Publishing, 2010), 186.

11　A blue plaque produced by the Christchurch Historical Society identifies the Ferrey house in the town today which remains little unchanged since the eighteenth century.

12　Nancy Cox, *The Complete Tradesman: A Study of Retailing, 1550–1820* (London: Routledge, 2016), 125.

13　Ginsburg, 'The Tailoring and Dressmaking Trades', 65. By 1816 Ferrey's business was successful enough to warrant a move to larger, more prominent premises in the nearby High Street closer to the market square.

14　Cox, *The Complete Tradesman*, 146; Margot C. Finn, *The Character of Credit: Personal Debt in English Culture, 1740–1914* (Cambridge: Cambridge University Press, 2003), 6.

15　Ferrey Account Book, 53.

16　Buck, *Dress in Eighteenth-Century England*, 171.

17　Ferrey Account Book, 63.

18　Cox, *The Complete Tradesman*, 127.

19　This indicates the number of staff working for the Ferreys, or possibly journeymen they would hire in to assist. See Ginsburg, 'The Tailoring and Dressmaking Trades', 66.

20　*Holden's Annual London and Country Directory* (London, 1811) lists, for example, a bathing house at Christchurch.

21　Finn, *The Character of Credit*, 4.

22　Ginsburg, 'The Tailoring and Dressmaking Trades', 65.

23　The means of ensuring repeat custom is further discussed in Ginsburg, 'The Tailoring and Dressmaking Trades', 65; McNeil, *Pretty Gentlemen*, 41.

24　To date 60 per cent of the total number of Ferrey customers has been identified by trade, profession or title. The number of tradesmen reflects the ease in which it is possible to identify Ferrey customers via trade directories. Further research will alter the statistics as additional customers and their professions are identified.

25 Newman, *The Christchurch Fusee Chain Gang*, 33.

26 Ibid.

27 Entries in the Register of Duties Paid for Apprentices' Indentures 1710–1811, accessed via www.ancestry.com. A Christchurch charity founded in 1734 after the death of a wealthy local mercer John Clingan supported several apprentices across the town in the eighteenth century including those assigned to tailors. It is possible that the Ferreys benefitted from this assistance either directly or through those they employed. The property rented by the Ferrey family from 1819 to 1936 was owned by the Clingan charity indicating a close link with this charity. See Newman, *The Christchurch Fusee Chain Gang*, 28.

28 Lambert, 'Bespoke versus Ready-Made', 56; Styles, *The Dress of the People*, 154.

29 The account book also records that the Ferreys were making black cloth covers for the coffins of named individuals in the accounts of a local carpenter and builder.

30 McNeil, *Pretty Gentlemen*, 78; Styles, *The Dress of the People*, 192.

31 Tucker's grocery business was successfully operated by successive generations of the family until its closure in the town in 1958. Hodges, *Christchurch: A Short History* (Christchurch: Natula Publications, 2003), 46.

32 Lambert, 'Bespoke versus Ready-Made', 60; Styles, *The Dress of the People*, 155.

33 The Ferrey Account Book, 34.

34 Ribeiro, *Dress in Eighteenth-Century Europe*, 214.

35 Buck, *Dress in Eighteenth-Century England*, 205.

36 Ibid., 93.

37 Jeremy Farrell, *Costume Accessories: Socks and Stockings* (London: Batsford, 1992), 48.

38 Jane Tozer and Sarah Levitt, *Fabric of Society: A Century of People and Their Clothes 1770–1870* (Wales: Laura Ashley, 1983), 107.

39 Fowler, 'Robert Mansbridge', 29–38.

40 Ibid., 33.

41 Similar observations regarding the supply of ready-made goods can be found in the accounts of a London tailor Jeremiah Murray, and Robert Mansbridge. See Fowler, 'Robert Mansbridge'; Lambert, 'Bespoke versus Ready-Made'.

42 Lambert, 'Bespoke versus Ready-Made', 56; Lemire, 'Developing Consumerism', 30.

43 Fowler, 'Robert Mansbridge', 33; Lambert, 'Bespoke versus Ready-Made', 57.

44 The Ferrey Account Book, 73, 76.

45 Lemire, *Developing Consumerism and the Ready-Made Clothing Trade in Britain, 1750–1800*, 39; Lambert, 'Bespoke versus ready-made', 58.

46 Lambert, 'Bespoke versus Ready-Made', 57; Lemire, 'Developing Consumerism', 34.

10

Gendered making and material knowledge: Tailors and mantua-makers, *c.* 1760–1820

Emily Taylor

The closing decades of the eighteenth century saw significant social, political and industrial changes, which were reflected in the *métiers* of fashion. Cotton fabrics came to dominate women's dress, with the waistline moving to directly underneath the bust; men's dress increasingly incorporated cotton fabrics, in heavier weights and monochrome weaves.[1] Early dress historians cited mechanized textile production, the pursuit of global markets and the extremities of the 1789 French Revolution in explaining these changes, but some have recently uncovered a more nuanced view of long-term causes and regional variations.[2] Kimberly Chrisman-Campbell, for instance, has discussed the influence of place and social group in the dress of the French royal court and post-revolution Paris.[3] Similarly, in her study of late eighteenth-century Swedish tailoring, Pernilla Rasmussen concludes that Swedish making shared principles with Germanic practices.[4] Following Rasmussen's example, this chapter reconnects material evidence with makers, investigating how the knowledge of tailors and mantua-makers remains embedded in the garments they created. Tailoring and mantua-making were the core trades behind main garment production in this era.[5] In Britain, each held legally independent positions (e.g. in Scotland, mantua-making was recognized as a distinct trade from 1763).[6] These positions were, in some cases, hard won as craft and trade protection generated conflicts around skills, as well as around economic rights.

In the late eighteenth and early nineteenth centuries, gender ideologies shaped not only garment styles, but also the choice of materials and makers' working practices. In elite and intellectual discussions of taste, women were regarded as arbiters of domesticity through natural gentility, whereas female makers were popularly associated with 'pride, social ambition and a lack of moral refinement'.[7] Such perceptions were reinforced by period guides to trades, with *The London Tradesman* (1747), for instance, minimizing the competencies of mantua-makers by insisting on the male stay-maker's role in moulding shape. By contrast, mantua-makers were credited for possessing the patience, discretion and dissimulation to carry out the 'innumerable Whims' of clients, with no mention of their material skills.[8] The mantua-maker's primary skill was seen as coping with 'perpetual Vicissitudes' in fashions, implying a greater need for knowledge of trends than for creating structure and fit.[9] Despite this early categorization as a

'Servant of the Ladies', the mantua-maker expressed extensive skill in working with stays to prevent strain on materials and using methods that allowed for reconfiguring garments as fashions changed.[10] Many of the debates about the establishment of mantua-making as a trade in its own right underscored how female makers' business practices might stand outside of patriarchal workshop structures.[11] Female makers turned gender distinctions to their benefit: using ideas around modesty and propriety to argue for women's right to the option of a same gender maker. Despite the efforts of female makers to establish a clear professional identity – in their view, 'Tailors are not *Mantuamakers*; that the two Professions are altogether distinct' – mantua-makers often experienced an uncertain career trajectory.[12]

Women's dresses, with their re-worked pleat creases, pin marks and stitch-holes, reveal how demands for alteration helped to configure mantua-makers' skills differently from those of tailors. Much menswear survives largely unaltered. In contrast to trade literature on mantua-makers, tailoring theories emphasized form, proportion and disproportion. Tailors sought to accurately fit garments with an understanding of the elasticity and fragility of the materials they worked. They made more moderate allowance for physical change, building in generous seam allowances, but otherwise fully finishing hems and seams with strong stitches. Contemporaneous tailoring publications describe a preoccupation with accommodating proportions, not just of different men but of women in opposition to men. Their resistance to the complete reconfiguring of garments emerges as a cultural gender distinction based on notions of stability associated with masculine fashion.[13] This chapter traces the making practices of mantua-makers and tailors by placing objects and archives located in Scotland into conversation with contemporaneous publications designed for a national readership, to show how makers implemented and understood changes in fashion. In so doing, it addresses three main themes of making practices: how materials affected making and influenced gendered practices; how makers adapted to the technical demands of new styles at the turn of the nineteenth century; and finally, how client-maker communication influenced construction practices.

Materials

The practices of tailors and mantua-makers were affected by the fabrics they worked. Different material qualities demanded distinct making skills, and men and women favoured particular fabrics that were viewed as suitable to their separate social roles, positions and occasions. David Kutcha has argued that in the wake of the Glorious Revolution elite men adopted plain, high-quality cloth in their dress: 'English aristocratic masculinity was newly defined in opposition to the luxury fabrics worn by women, the French, and social upstarts.'[14] Kutcha suggests this sobriety of fabric continued to define masculinity in different manifestations into the mid-nineteenth century.[15]

Collections of eighteenth-century menswear show a broad adoption of plain cloth fabrics across the spectrum of social classes, alongside continued use of patterned luxury materials by elite men for specific social circumstances, such as attendance at the Royal Court. For example, wool cloth samples affixed in a notebook started in 1736 by William Young, a merchant tailor from Aberdeen, consist of shades of grey, dark blue, brown and red, with one purple sample at a higher cost.[16] A man's double fronted waistcoat in a pale brown silk belonging to the Earls of Haddington in the 1760s combines muted colour, practical function and luxury fibre.[17] The Marquis of Tweedale's bills, from around 1800, prominently feature fustian, cambric, linen, shag and cloth, but also finer fabrics such as 'a Superfine Olive Clothd Coat with velvet neck, & platted buttons' made for £3 7s. 3d., and 'hair shag breeches for 3s. 0d., and 'Gray Cassmire breetches' for 2s. 7d.[18] Sobriety of colour and simplicity of texture are mixed with extremely high-quality fibres. Deceptively unfussy, the Marquis's velvet collars, decorative buttons and cashmere constituted an overt demonstration of luxury and social status that would have been quickly recognized by his contemporaries. He is neither giving up ornamentation nor seeking functionality alone. That any of these fabrics were in opposition to feminine luxury is less clear. In a century transfixed by ideas of polite (mixed gender) sociability, public and private display, gender opposition in dress materials was tempered by a complex matrix.[19]

In 1796 *The Taylor's Complete Guide* published details about the typical fabrics used for bespoke menswear. The text offers a wealth of information about the behaviour of materials during the making process, alongside advice on how to treat them. It also includes comments on suitability: velveteen has 'strength, beauty, and convenience'; Weymouth satin should 'never be used but upon the most captivating figures'; the 'barbarous' use of shag has caused it 'to have entirely bid adieu to the lively imagination and reason of the Trade'.[20] The tailor's comments show that gentlemen sought to distinguish themselves from the labouring classes through qualities of fabrics, ceasing to use anything that was popularized for the broader public and retaining luxury finishes such as satin and velvet.

The nexus between material, maker and gender that played out in late eighteenth- and early nineteenth-century garments is usefully illustrated by women's riding habit jackets. Choosing fabric for a habit was more complex than an expression of gender and class; it was equally affected by the properties of the material, knowledge of which relied heavily on the maker. These garments were often commissioned from tailors, were usually made with plain wool fabrics and required a specific cut to accommodate their active use. In 1776 Louisa, Lady Stormont, wrote at length to her sister Lady Mary Graham from London about commissioning a riding habit:

> I sent to desire [Katy] to make you a stone colour riding habit lined with green & a green waistcoat, as it is quite the fashion now to have them lined with different colours I thought, I should not keep an apointment with him if I desire him to come here for me to chuse the colour, so I sent him word to make it of a good colour that would last well. Excuse my having executed your commission in so slovenly a manner.[21]

The tailor must already have had the required measurements and Louisa clearly trusted him to select a fabric colour. Both sisters were likely to have had a standing relationship with their tailor; Louisa's concern that the colour of the garment will 'last well' indicates that she anticipated fabric colour to be affected by fashionable change ahead of shape. She leaves further decisions about the fabric's suitability entirely with the tailor. Louisa's instructions for a two-tone colour scheme match three similar habit jackets in different collections (Figure 10.1).[22] They are all a shade of pale brown, faced and lined with a pale blue-green silk, two of which differ only in the placement of their pocket flaps. This raises questions around whether they were bespoke, ready-to-wear or made by the same tailor.[23] Each one uses the same buttons as the double-fronted waistcoat, mentioned above, and another habit jacket with a yellow waistcoat, whose related colour tone is suggestive of the same era of fashion, if not the same tailor.[24]

The fine, hard wool of the habit jackets could be described as masculine 'plain cloth'. It would have been light-weight to wear, with some natural water repellence suitable for riding. The construction stitching of the two-tone jacket (Figure 10.1) is small and even throughout, using elongated buttonholes raised over a gimp thread for decoration.[25] Threads complement both the blue lining and pale brown cloth. The lining is fully fitted and hides all raw edges with stab and whip stitching; the jacket possesses the high-level finishing of a skilled maker, which, again, closely resembles the double-fronted waistcoat. As the habit jackets and waistcoat are all dated to at least a decade before Louisa's letter, 'stone' colour was evidently a long-standing choice for active-wear, despite Louisa's concerns to the contrary. The practical function of the jackets made tight construction necessary, which corresponded to material longevity in fashion and tailors' unhesitant finishing.

Although the establishment of mantua-making had involved some contestation of makers' rights, publications at the end of the eighteenth century suggest that many tailors were happy to leave women's dress out of their business. *The Taylor's Complete Guide* asserted: 'Habit-making is a neat and delicate piece of practice, and understood but by few; seldom practiced, and unknown to thousands of Taylors; ... quite dissimilar and as different as joinery and cabinet-making.'[26] J. Golding seconded this: 'There are many in our trade whose practice is very extensive, and whose abilities in cutting men's clothes is indisputable; who nevertheless consider the Riding Habit an object entirely above their comprehension, and have refused to take an order.'[27] In refusing to make riding habits, tailors were exhibiting nervousness about their skills in relation to female proportions. The 'rising prominency of the breast ... [and] artificial, though very useful appendage, viz. the stays ... ' introduced different technical requirements.[28] The *Complete Guide* noted that the 'protuberance in front ... will baffle anything but great experience' by requiring a long forepart, and that 'Ladies through custom have a manner of holding their arms more upon the bend than men', so that 'the friction of the arm against the stays very soon wears them out'.[29] It is likely that stays were in part responsible for the position of the arms, and possibly for a posture the author believed made women proportionally wider across the shoulders and 'up to the back of the neck' than men.[30]

In 1815 Benjamin Read promised that his proportionate table 'will answer correctly for the difference or void space in the upright position of the back, for habits, pelisses, dresses and spencers'.[31] However, Golding claimed that,

Figure 10.1 Woman's riding habit, *c.* 1760s, National Museums Scotland, A.1978.419. Image © National Museums Scotland.

[A]ll the plaits and foldings, all the curves and fullness, puffings and drawings, must ultimately be brought down to the standard of the human shape … to a proficient in the art of Cutting, it is of little consequence what the prevailing taste may be, his business is to fit the body, in the symmetry of proportion, to which may be added, the taste, grace, and finish of the reigning mode.[32]

The confusion some tailors felt around accommodating stays may have been caused by the transfer of what they perceived as a male garment, like the riding habit, onto a female body. Their familiarity with men's coats of all descriptions potentially exaggerated a sense of difference when they had to change their pattern for a woman. Making riding habits was perceived as a specialist skill, which belonged to the tailor's remit. The problems it presented were not created by unfamiliar fabrics or garment function, but by unfamiliar proportions that confused the tailor's known world of masculine posture and movement.

The repetition of the materials and features seen in the women's riding habit jackets are not isolated examples. When set against two matching *caraco* jackets, the details of their construction offer information about makers, duplication and gendered materials (Figures 10.2 and 10.3).[33] The *caracos* are outwardly made to the same pattern in the same material, but are lined in different wool fabrics.[34] Some of their construction details almost match, to within millimetres. For example, the length of the front openings differs by only four millimetres, and the distances between their hook and eye fastenings are ten millimetres (Figure 10.3) versus thirteen millimetres (Figure 10.2), roughly the width of a small finger. Sometimes the *caracos* offer slight variations, such as the forty millimetres difference in the waistline position that causes one to have ten fastenings at the front (Figure 10.3) and the other eight (Figure 10.2). Other small differences include Figure 10.2's lining, which is sewn with all raw edges turned in with the outer material stitched over the lining, similar to the tailored items above. Its counterpart, Figure 10.3, has the armhole seam allowance facing out into the garment, a whip stitch securing its raw edge, and the lining is over the outer material on the front neckline. The armhole and neckline differences might be explained by later alterations, as they use a different thread. Such alterations would have potentially been made by a female maker, as out-turned armhole seam allowances are more common in women's dresses than in tailored garments. These changes, however, explain neither the different linings nor the shared inaccuracy between armhole circumferences: both garments have an approximately 160-millimetre difference between the left and right, but one with a larger right armhole, the other a larger left armhole.

Such construction details generate questions around the making of the garments.[35] The *caracos* possess fine stitching, with a degree of inaccuracy, although not as fine as the habit jackets above (Figure 10.1). Their wool linings are unusual in extant British womenswear (except for ready-to-wear petticoats), although common in Scandinavian dress.[36] Similar *caracos* were worn in the Netherlands and France, and the silk design of these examples was more often found in British menswear than womenswear. It was, however, a familiar pattern in continental European womenswear.[37] The application of hooks and eyes over the lining, with some of the similarities and inaccuracies in cut, suggests speed in construction and standardized production. If not quite in the category of ready-to-wear, it is probable the *caracos* were jointly produced.

Juxtaposed against one another, the habit jackets and *caracos* tell of different material challenges. The neatness of the habit jackets illustrates the concerns exhibited by tailoring publications to transplant their exacting skills onto a female form. The *caracos* instead present an easier fit, with small tucks and changes. They are constructed from thick silk lined in more hardwearing wool, as opposed to fine wool lined with

Figure 10.2 Woman's caraco, *c.* 1775–1780, Manchester City Galleries, MCAG 2003.171. Image courtesy of Manchester Art Gallery.

Figure 10.3 Woman's caraco, *c.* 1775–1780, Dalgarven Mill Museum, R+MF 86.22. Image courtesy of Dalgarven Mill Museum.

twill silk. Their elbow-shaped sleeves, low necklines and short fastenings indicate they were not expected to endure the same physical rigours of a riding habit nor fit so precisely. While they use fabrics that were associated with masculine dress, they present feminine shapes and making techniques, but lack the 'foldings', 'puffings and drawings' more often found in garments by mantua-makers. The *caracos* are, instead, situated in a middle ground of production, mixing techniques used by both tailors and mantua-makers.

The 'puffings and drawings' dismissed by Golding were a staple part of the mantua-maker's skill and allowed for variety and individuality in garments that followed the same fashion trend. A cotton dress from 1797 provides an example of how puffing and

Figure 10.4 Woman's dress, 1797, National Museums Scotland, A.1977.820. Image © National Museums Scotland (see Colour Plate 5).

drawing exploited the properties of muslin (Figure 10.4).[38] Gossamer thin, the dress uses a jamdani fabric, which allows extremely tight gathers into the rear waistline, which then billow out into a trained skirt.[39] Minute pleats in the sleeve heads provide shape to the sleeves and bodice back, which have been sewn onto a linen lining for strength and structure. The lining of the bodice front has loose panels which overlap and would have been pinned to the wearer's stays.[40] Over these are panels of the outer fabric, cut square they gather loosely around the neckline under loops and buttons.

Muslin demanded a very different attention from makers to wool and silk. The wrong tension of stitching could easily tear the light fabric; seam allowances and hems were visible through the outer fabric; and unlike denser weaves, muslin could create cloud-like puffs and drifting movements when worn.[41] The bodice of the dress (Figure 10.4) has been sewn with four threads per stitch, using a thread-counting technique that was practised in shirt-making and linen work for strong, discreet construction.[42] However, the regularity of stitching on the rest of this dress is generally not counted, instead varying according to the function and visibility of the seam. Internally, the dress's arm-hole seam allowances have a turned-out raw edge, similar to the *caraco* above, with all other seams being neatly pressed and felled. The mix of precise and fluid techniques visible in the dress was determined by the different features of the garment.

Professional and amateur women makers, in Britain, had long held a traditional association of working with linen and cotton, which may have enhanced the quick adoption of these fabrics in women's fashion.[43] Married and genteel women undertook the making of shirts, in some instances as outsourced labour, the materials for which changed from fine linen to cotton fabrics in the nineteenth century.[44] In eighteenth-century Edinburgh millinery work was deemed a suitable employment for gentlewomen: milliners dealt with gauzes, muslins and lace, and in the latter decades they often combined business practice with mantua-making, but employed trade-specific artisans.[45] Although some of the materials milliners and mantua-makers used around this time were the same, the making skills were different.[46] As the dress in Figure 10.4 shows, the mantua-maker worked with the wearer's form and proportions. The use of the stays as a support required skill in fitting the lining anatomically, to enable play with the outer fabric. It demonstrates that for a mantua-maker stays were less a problem to overcome and more a welcome asset on which to anchor external fabric manipulations.

The ability of mantua-makers and tailors to combine the properties of different materials in one garment was clearly central to their practices. The construction of the riding habits and dress were prompted by the materials and function of the garments: their construction shows how tension, length and type of stitch changed between seams, material combinations and different activities. The differences between mantua-makers' and tailors' skills, as expressed in the garments, lie in the gendered associations around the fabrics and vocational trends.

Adaptation

The core skills highlighted by Golding's *Tailor's Assistant* were the abilities to mould a garment to different proportions and to meet the fashion. Such material literacy had been radically tested in the decades preceding his text, which found tailors and mantua-makers adapting to style changes and standardizing methods to enable future modifications. As early as 1756 the mantua-makers of Perth had argued that readiness for change was central to their profession:

The Question then is not, Whether the Fashion of female Dress ought to be changed? but, Whether its perpetual Vicissitudes do not require the peculiar Attendance of Mantuamakers? The various Refinements in Dress may be termed Extravagance; but it is an Extravagance which Tailors and Mantuamakers will not probably reform.[47]

By the end of the century their 'various Refinements' had been upgraded to the complete restructuring of the body of women's garments, as Hilary Davidson discusses in the following chapter. The *Book of Trades* (1804) suggests, 'The mantua-maker must be an expert anatomist … she must know how to hide all defects in the proportions of the body and must be able to mould the shape by the stays.'[48] This descriptive stress on a mantua-maker's anatomical skill mirrors the technical emphasis in trade manuals for tailors, which positioned them as a skilled elite that could best fit to physical form. Golding, for instance, argued that the tailor's business was primarily 'to fit the body', fashion remaining a secondary concern.[49] Golding implies that, unlike mantua-making, tailoring was rational and scientific, engaged with fashion but resistant to its dictation.

Mantua-makers do not seem to have expressed the same resistance to change. In part, this was no doubt due to the long-standing tradition of accommodating short-term fashion needs in women's dress. Meeting these needs was a skill in itself. Running stitches and out-turned seams enabled quick alterations to garments, suggesting the intentional impermanence to the construction of gowns. As women's waistlines moved from a natural position to under-bust by the late 1790s, old dresses were not immediately dispensed with: surviving examples testify to the trials, successes and errors of refreshing old forms to new. A pink taffeta silk dress of *c.* 1798 has been skilfully made from a much earlier item (Figure 10.5).[50] The original bodice has been shortened and extended across the front with piecing that closes over the bust. The skirt, which starts behind the hips like a train, has been attached higher than the front waistline to exaggerate the bodice shortening from behind. Inside, the original low centre back has been unpicked up to the skirt seam, and tucks in the sleeve linings indicate they have been tightened. The skirt uses wide box pleats, with a sharp double fold at each side – a feature associated with *sacque* dresses, which fell out of fashion in the early 1780s. These pleats indicate the dress was reconfigured from a much earlier incarnation.[51] The newly modelled dress has the hexagonal back panel that became a regular feature in the following decade.[52] However, the quality of the silk is much stiffer than the twill fabrics featured in pelisse dresses of the 1800s; it creates a slightly billowing over-dress that would not have suited the newly figure-hugging fashions.[53]

Another silk dress that struggled to adjust into the new waistline is made from a green shot taffeta (Figure 10.6).[54] This dress features a collar, bodice back and sleeves that are unaltered from the 1780s.[55] The skirt, however, has been re-pleated to a higher waistline and two additional ruche panels have been added to the bodice front. These panels end at an under-bust waistline, but beneath them the original bodice fronts and lining have been stitched together to create a second, longer layer. This layer tucks into the new waistband and would have been affixed to unaltered stays.[56] Overall the dress's silhouette resembles illustrations in the *Gallery of Fashion* from 1795 (Figure 10.7).

Figure 10.5 Woman's dress, *c.* 1798, National Museums Scotland, A.1977.821. Image © National Museums Scotland (see Colour Plate 6).

Waistlines rose quickly and at first superficially: mantua-makers altered the exterior of dresses over long-waist stays, with stay-makers remaining slower to meet fashionable demand.[57]

The neat stitching and ingenious adjustments to the green dress (Figure 10.6) show a high level of skill, suggesting they were undertaken by a professional maker. However,

Figure 10.6 Woman's dress, *c.* 1795, National Museums Scotland, A.1975.6. Image © National Museums Scotland.

the way the ample skirt billows from under the bust would have been unflattering on the wearer, unless she was pregnant. The maker obviously attempted to retain as much fabric in the skirt as possible and avoided cutting or unpicking long seams, but the quantity and quality of the fabric resist a straight drape.[58] Experimentation in forming a flattering under-bust line went hand-in-hand with the development of new

Figure 10.7 Fashion print 'Fig. 51, Fig. 52.' In *The Gallery Of Fashion*, published by Nicolaus Heideloff, London, 1 May 1795, National Museums Scotland, H.Rhi 15.1. Image © National Museums Scotland.

underpinnings, and by the early 1800s the two layer bodice system was superseded by a closed front with a centre back fastening.

Tailors generally worked with tighter margins and finishing than mantua-makers: their garments were sewn to withstand greater stress against the body and, in some instances, a greater range of movement. At the same time, the *Taylor's Complete Guide* counselled tailors to leave a margin for error: 'The reason of our hint for laying in the side seam is to provide you with the means on any occasion of making the Coat

Figure 10.8 Man's riding coat, *c.* 1805–1815, National Museums Scotland, K.2018.15. Image © National Museums Scotland.

wider upon the breast, either from the fault of your own practice, or by rectifying the errors of others.'[59] Such allowances in side seams remain visible in surviving coats. An unlined example of a hunting coat, made for Captain Robert Cathcart *c.* 1805–1815, has minimal seam allowances, except for the body side seams, which are around two-thirds of an inch, or 150 millimetres (Figure 10.8).[60] More allowance is also given on the inside of the collar and the head of the coat tails, where extra fabric is used to smooth over interlining and complex joins. To avoid bulk where the excess seam allowance meets the sleeve, the tailor has cleverly made a small cut to enable the sleeve seam to poke through the allowance. The extra length is then folded back from under the sleeve lining, improving the smooth line and fit of the coat.

Communication

The relationship between professional makers and their customers fostered a two-way exchange of material knowledge. This reciprocal relationship was a core part of garment and accessory making practices, as Hilary Davidson, Serena Dyer and Elisabeth Gernerd also note in this collection. Correct measurements were central to the success of a commission and relied on both parties, as emphasized by Golding:

> Be very careful in taking your measure, for upon that will depend in a great degree, your success in fitting, observe the natural gait and position of your figure ... together with every particular relative to the article in question, which you receive from your customer, and enter them carefully into your order book ... careful attention ... will not only assist your memory, but will also enable you to make the necessary allowance for all disproportion.[61]

This description stresses the importance of observation, which could only be made in person. The importance of accurate measurements, and the place of anatomical and biomechanical knowledge in securing a successful fit, would have long-term impacts as tailors often retained measurements for their clients. For instance, in 1829, Elizabeth Sackville West, Countess De La Warr, wrote from Buckhurst Park, Sussex, to a Monsieur Reynault in Paris. She requested a black velvet decorated dress, hoping that a dress she had commissioned from him four years previously would suffice for a pattern.[62]

Conversely, without accurate observations and reliable measurements, makers could be at a loss. In 1772, J. Peter Schein wrote a frustrated letter from London regarding an order for several suits on behalf of the Duke of Gordon, complaining that 'no man can Gues the mining of that taylor which took the musers', and requesting new measurements to his specifications. Schein states that he has already 'Bespock the Silver Lace to be made with all expedition for all the 13 suts so I promes you that nixt weak the silver suts which I have the musur of shall be finished'.[63] He finishes his letter by asking whether the Duchess of Gordon is satisfied with a silk gown. Schein's confidence of having thirteen suits made within a fortnight, his commissioning of trimmings and his production of men's and women's wear, suggest he had a sizeable business. However, his frustration with poor measurements was not unique: surviving trade cards from the late eighteenth-century evidence experiments in assisting remote customers in taking comprehensible and accurate measurements.[64]

Knowledge and communication were also vital to the maker's understanding of a client's needs, as Sarah Howard discusses in the case of provincial tailors. A maker might serve a broad populace, all influenced by different factors, not all demanding the latest fashion.[65] Makers and their suppliers travelled widely to gain understanding of materials and supplies, with women 'actively operating in the same world as their male counterparts'.[66] For instance, the merchant tailor William Young travelled to Manchester, Halifax and Leeds for thread and cloth. He notes memorandums from conversations with other tradespeople, including lists of recommended button makers in Stockport, ribbon weavers in Coventry, water-borne transport costs, material qualities and a supply of trimmings at different rates.[67]

Consumers equally pursued information on materials and fashionable styles, which enabled their reciprocal interaction with makers. For those outside urban areas knowledge was gained by proxy, while for aristocrats socializing in the *beau monde* direct contact with fashionable centres could be crucial.[68] Louisa, Lady Stormont's letter about her riding habit also mentions: 'a little cap, which is made after one by Barrison, Straight from Paris' and 'a pattern of an Italien night gown'. She continued, 'I fancy you would have time to have one made & if they should remonstrate at Edr its being not enough dressed tell them there is nothing but Polonaises worn here at Balls & even at Assemblies'.[69] Among other details, Louisa informs her sister about the sources of trends to enable her to persuade Scottish society and makers of their stylishness and suitability.

Numerous distinctions existed between the techniques used by tailors and mantua-makers in the later eighteenth and early nineteenth centuries. The origins of these distinctions do not lie in relative makers' abilities, however, but reflect society's conceptions of masculinity and femininity, which divided garment trades by gender and consequently the methods used for adapting garments to different physical proportions. Women's dress was made in expectation of changefulness, whereas men's dress was expected to follow ideals of restraint, stability and scientific fit. While dresses might be significantly remodelled, the tailor only allowed for refining to perfection. However, both tailors and mantua-makers adapted their techniques to suit the properties of their materials. They similarly combined fabrics and proactively sought out knowledge about materials and styles. The high-end riding habits and muslin dress clearly demonstrate different expectations of finishes and treatments; the combination of these finishes in the *caracos* raises questions about speed of manufacture and hand-making duplication.

At the turn of the eighteenth into the nineteenth century, skilled mantua-makers and tailors assisted and met the demands of changes in fashion through multiple means. Publishing theories on proportion and cutting practices to hone their craft, tailors became openly conversant with one another as well as potential clients. Surviving garments document that mantua-makers were similarly engaged in creating new proportions. To achieve change, partnership with clients and supporting trades was crucial: shared knowledge created a foundation of understanding around materials, value and suitability. The variables in hand-making reveal how the materials, seams, pinholes and pleats of garments articulate their construction histories, and how mantua-makers in particular reworked clothes over time. These internal workings speak of a rich world of material engagement forged both through gendered practices and expertise gained by trial, error and experience. By balancing quality of craft, fashion and the demands of clients, mantua-makers and tailors occupied an unusual position to both catalyse and contain trends in dress. Their community was in constant pursuit of material knowledge and new applications for their material literacy.

Notes

1   Anne Hollander, *Sex and Suits* (Brinkworth: Claridge Press, 1994), 7; Christopher Breward, *The Culture of Fashion: A New History of Fashionable Dress* (Manchester: Manchester University Press, 1995), 122–3.

2   James Laver, *A Concise History of Costume* (London: Thames and Hudson, 1969), 148–53; Doreen Yarwood, *English Costume from the Second Century, B.C. to 1967* (London: B. T. Batsford, 1952), 196–8.

3   Kimberly Chrisman-Campbell, *Fashion Victims: Dress at the Court of Louis XVI and Marie-Antoinette* (London: Yale University Press, 2015), 286–303.

4   Pernilla Rasmussen, *Skräddaren, sömmerskan och modet: Arbetsmoder och arbetsdelning I tillverkningen av kvinnlig dräkt 1770–1830* (Stockholm: Nordiska Museets Handlingar, 2010), 136.

5   'Main garment' refers to the larger and visible layers of an outfit, excluding shifts, shirts, accessories and unstructured outer-layers.

6   Elizabeth Sanderson, 'The New Dresses: A look at How Mantuamaking Became Established in Scotland', *Costume* 35 (2001): 14–23; Mantua-making and tailoring are also listed as distinct trades in R. Campbell, *The London Tradesman: Being a Compendious View of All Trades, Professions, Arts, Both Liberal and Mechanic* (London, 1747).

7   Robert W. Jones, *Gender and the Formation of Taste in Eighteenth-Century Britain* (Cambridge: Cambridge University Press, 1998), 106; see also Vickery, *The Gentleman's Daughter*, 1–12; Batchelor, *Dress, Distress and Desire*, 52.

8   Campbell, *London Tradesman*, 224, 227.

9   *Answers for Mary Lyon, Rachel Currie, and Agnes Ramsay, Mantuamakers in Perth; to the Petition of George Robertson, and Others, Tailors in Perth* (Edinburgh, 1756), 10.

10 Campbell, *The London Tradesman*, 227.

11 For discussion of women as makers in tailoring businesses, see Clare Haru Crowston, *Fabricating Women The Seamstresses of Old Regime France 1675–1791* (Durham: Duke University Press, 2001), 96–8; Beverly Lemire, 'Redressing the History of the Clothing Trade in England: Ready-made Clothing, Guilds, and Women Workers, 1650–1800', *Dress* 21 (1994): 61–74; Sanderson, 'The New Dresses', 14–23.

12 *Answers for Mary Lyon, Rachel Currie, and Agnes Ramsay, Mantuamakers in Perth*, 9.

13 Masculine stability gained consensus at the end of the eighteenth century; in 1757 a tailor was said to 'change shapes as often as the moon', ibid., 192.

14 David Kutcha, *The Three-Piece Suit and Modern Masculinity, England 1550–1850* (Oakland: University of California Press, 2002), 122–24.

15 Ibid., 133–72.

16 National Museums Scotland, SAS MS.620

17 High fastening with a double layer, these waistcoats are associated with outdoor pursuits such as riding and shooting; National Museums Scotland's collection, K.2002.495.

18 National Library of Scotland, MS. 14691/173.

19   Jon Mee, *Conversable Worlds Literature, Contention, & Community 1762–1830* (Oxford: Oxford University Press, 2011); see 7–12 for a summary of polite conversation and domesticity; see also Annie Richardson, 'From the Moral Mound to the Material Maze: Hogarth's *Analysis of Beauty*' and Elizabeth Eger, 'Luxury, Industry and Charity: Bluestocking Culture Displayed', in *Luxury in the Eighteenth Century Debates, Desires and Delectable Goods*, ed. Maxine Berg and Elizabeth Eger (Basingstoke and New York: Palgrave Macmillan, 2003), 119–34 and 190–206.

20   *The Taylor's Complete Guide* (London, 1796), 32–3, 37, 34–5.

21   National Archives of Scotland, GD155/857/15.

22   National Museums Scotland A.1978.419; Victoria and Albert Museum, T.198–1984; Metropolitan Museum of Art, 2011.72.

23   The origin of A.1978.419 is unknown, purchased at Christies, South Kensington, 'Important costume, embroidery, textiles and fans', 25.05.1978, lot 118; it is also worth considering the possibility that these items might relate to the blue and buff colours adopted by the Whig party, see Hannah Greig, *The Beau Monde, Fashionable Society in Georgian London* (Oxford: Oxford University Press, 2013), 126.

24   Manchester City Art Galleries collection, 1947.2802.

25   Gimp thread is thicker and stronger than standard sewing thread; for illustration of use in buttonholes, see Claire B. Schaeffer, *Couture Sewing Techniques* (Newtown: The Taunton Press, 2011), 95.

26   *The Taylor's Complete Guide*, 110.

27   J. Golding, *Golding's New Edition of the Tailor's Assistant, or, Improved Instructor, Containing a Synthesis of the Art of Cutting to Fit the Human Form* (London, 1817), 72.

28   Ibid., 73.

29   *The Taylor's Complete Guide*, 119–20.

30   The bent elbow is particularly visible in the cut of women's sleeves from the later 1770s, when a longer and tighter fit became fashionable, see Figures 12.2 and 12.3; *The Taylor's Complete Guide*, 116.

31   Benjamin Read, *The Proportionate and Universal Table* (London, 1815), 12.

32   Golding, *Tailor's Assistant*, 89.

33   Manchester City Galleries, MCAG 2003.171 (Purchased by Manchester City Galleries from Sotheby's Castle Howard auction, 07.10.2003 lot 232; purchased by Castle Howard from Christie's South Kensington, Costume, embroideries and textiles auction, reference TOSO910, 09.10.1979, lot 38); Dalgarven Mill Trust, R+MF 86.22. The term *caraco* is French and there is ambiguity around what was defined as a *caraco* in mid-eighteenth-century Britain. Here the term refers to a fitted jacket, cut with a skirted section extending to the upper thigh and with a low neckline similar to contemporaneous women's dresses.

34   For comparative pattern analysis, see Emily Taylor, 'Women's Dresses from Eighteenth-Century Scotland: Fashion Objects and Identities' (PhD Thesis, University of Glasgow, 2013), 127–36 and appendix 3.1.

35   The provenance here is opaque: R+MF 86.22 belonged to the Blair family, of Blair Hall, Ayrshire. It is possible that both garments belonged to Agatha and Jane, sisters to the 24th Laird, William Blair of Blair, and daughters to Major Hamilton Blair, whose military service took him to the European continent (during the War of the Austrian Succession, the Royal Scots Dragoons were sent to Low Countries in 1742 and deployed to the Battle of Wilhelmstahl in North West Germany, 24 June 1762; Battle of Minden, 1 August 1759; Warburg 31 July 1760; and Vellinghausen 16 July 1761). The sisters left Ayrshire for Dorset with their mother Jane, daughter of

Sydenham Williams of Herringston, Dorset, in 1782 after the death of their father. See *Burke's Peerage*, Vol. 3 (London: R. Bentley, 1836), 171–2. If the *caracos* belonged to the sisters, the sudden move might explain their separation. The Blair *caraco* is accompanied in the Dalgarven Mill Trust collection by a striped wool and silk jacket, probably belonging to William Blair. See David Wilcox, 'Cut and Construction of a Late Eighteenth-Century Coat', *Costume* 33 (1999), 95–7. Family history notes are from a transcript of 'Sayings of Aunt Mary dedicated to her niece Cecily Blair', private collection, and with grateful thanks to Caroline Borwick for her generosity in time and detailed knowledge of the Blair family history.

36   Lemire, 'Redressing the History of the Clothing Trade in England', 61–74; a red coat item in Kulturhistorisk Museum Randers uses a wool said to be from East Anglia, England. Information via email correspondence with Kirsten Toftegaard, Design Museum, Denmark, 2 November 2016: [www.tojpaakroppen.dk/], accessed 14 June 2018.

37   Pascale Gorguet Ballesteros, *Modes en Miroir, La France et la Hollande au Temps des Lumières* (Paris: Paris Musées, 2005); Taylor, 'Women's Dresses from Eighteenth-Century Scotland', 129–30.

38   A.1977.820, National Museums Scotland.

39   Jamdani fabrics are very fine cotton muslin fabrics, traditionally spun and woven by hand in the Bengal area of India, with supplementary weft patterns passed into the warp by hand on the loom; see Sonia Ashmore, *Muslin* (London: V&A Publishing, 2012), 20.

40   This is indicated by pinholes in the material.

41   Traditionally, the finest muslins were Indian imports and this dress captures their production just as muslin weaving was beginning to be observed as in decline at the export epicentre of Dhaka, Bangladesh; see Ashmore, *Muslin*, 11–23, 29.

42   With gratitude to Susan North for first drawing my attention to thread counting; Luca Costigliolo and Jenny Tiramani, *Seventeenth-Century Women's Dress Patterns Book One*, ed. Jenny Tiramani and Susan North (London: V&A Publishing, 2011), 9; examples include National Museums Scotland, A.1966.314 and A.1979.406.

43   Lemire, 'Redressing the History of the Clothing Trade in England', 61–74.

44   For an example of a soldier's wife undertaking outsourced work, see Elizabeth C. Sanderson, *Women and Work in Eighteenth-Century Edinburgh* (Hampshire: Macmillan Press Limited, 1996), 32. For example, Janet Richardson of Pitour supplies her Brother Robert in London, 25 March 1782, Perth and Kinross Archives, MS101/Bundle 29; see the mixture of cotton and linen in Thomas Coutts wardrobe in David Wilcox, 'The Clothing of a Georgian Banker, Thomas Coutts: A Story of Museum Dispersal', *Costume* 46, no. 1 (2012): 17–54.

45   For example, Miss Montford and Calder entries in Peter Williamson, *Williamson's Directory, for the City of Edinburgh, Canongate, Leith, and Suburbs; from June 1778, to June 1779* (Edinburgh, 1778); Elizabeth Sanderson, 'The Edinburgh Milliners 1720–1820', *Costume* 20 (1986), 18–28, 23–4.

46   Milliners were generally understood to provide linen garments, headdresses and small accessories. Although they might provide outer-wear, these were usually unstructured garments, with fitted womenswear being provided by a mantua-maker, see entries in A. Bell and C. Macfarquhar, *Encyclopaedia Britannica* (Edinburgh, 1797) and Campbell, *The London Tradesman*.

47   *Answers for Mary Lyon* (Edinburgh, 1756), 10.

48 Janet Arnold, *Patterns of Fashion Englishwomen's Dresses and Their Construction 1660–1860* (London: Janet Arnold, Wace and Company, 1964), 9.

49 Golding, *Tailor's Assistant*, 89.

50 National Museums Scotland, A.1977.821; a similar pink can be seen in John Wilkes and his daughter Mary Wilkes Johann Zoffany, 1782, National Portrait Gallery NPG 6133.

51 *Sacque* may also be spelt 'sack'. *Sacque* is preferred as the style of dress originated in France and it is the contemporaneous French term used by Louisa, Lady Stormont and her counterparts.

52 *Fashions of London and Paris*, 1803, plate 62.3, 'Paris Dresses' and plate 66, 'Paris June 1803', Museum of London Library MOL 62.153.4.

53 For example, Catalani silk pelisse in plate 113.2 'London Walking Dresses', *Fashions of London and Paris*, 1807, Museum of London Library 62.153.8; Pelisse 1805–9, Glasgow Museums E.1977.1

54 National Museums Scotland, A.1975.6.

55 Taylor, 'Women's Dresses from Eighteenth-Century Scotland', 123–6.

56 A similar bodice disjuncture between a narrow over-bust ruche and a deeper, earlier lining can be found in a muslin dress of *c.* 1795–1800, National Museums Scotland, H.TM 8.

57 Anne Bissonnette, '"Dessiné d'après nature": Renditions from Life in the *Journal des Dames et des Modes*, 1798–9', *Journal for Eighteenth-Century Studies* 38, no. 2 (2015): 213–37.

58 Achieving a clean drape was equally challenging to tailors working with heavier fabrics and ample riding skirts: see *The Taylor's Complete Guide*, 124.

59 Ibid., 103.

60 National Museums Scotland, K.2018.15; the same allowance and seam technique are found on a coat that belonged to Thomas Coutts, *c.* 1810, National Museums Scotland, A.1915.225 A; see Wilcox, 'The Clothing of a Georgian Banker', 17–54.

61 Golding, *Tailor's Assistant*, 16; Golding acknowledges M. Cook on page 6, suggesting the fundamentals of tailoring were little changed between 1787 and 1817. See M. Cook, *A Sure Guide against Waste in Dress* (London, 1787).

62 Translated from French with assistance of Julien Willems, Institut Français Écosse, Edinburgh; National Museums Scotland, H.RHI 142.21.

63 National Archives of Scotland, GD44/43/72/25.

64 Miles Lambert, '"Sent from Town": Commissioning Clothing in Britain during the Long Eighteenth Century', *Costume* 43 (2009): 66–84.

65 Vickery, *The Gentleman's Daughter*, 172–7.

66 Sanderson, *Women and Work*, 2 and 34; Sanderson, 'The New Dresses', 17.

67 National Museums Scotland, SAS MS.620.

68 For reference to the importance of high fashion client knowledge, see Greig, *The Beau Monde*, 118–19; and Clare Haru Crowston, *Credit, Fashion, Sex: Economies of Regard in Old Regime France* (Durham: Duke University Press, 2013), 162.

69 National Archives of Scotland, GD155/857/15.

11

Dress and dressmaking: Material evolution in Regency dress construction

Hilary Davidson

'I have got over the dreadful epoch of Mantua-making much better than I expected,' wrote Jane Austen to her sister Cassandra in 1799.[1] Penned at the start of the last year of the eighteenth century, Austen's letter marks a moment of transition for the profession of mantua-making: over the next two decades, it would undergo dramatic changes, even losing its name to become 'dressmaking'. This new nomenclature coincided with radical changes in women's fashion at the end of the eighteenth century. From the mid-1790s, waists rose from their natural position to settle underneath the bust, and gown construction techniques altered with them. During the first two decades of the nineteenth century, a period that can be characterized as the 'long' Regency in Britain (c. 1795 to the early 1820s), gown construction incorporated new front opening methods, back openings, separate bodices and skirts, and shaped skirts, amongst other innovations such as net fabrics and the dominance of cotton fabrics in textiles for clothing.[2] These developments were not an abstract process of fashionable change, but constituted responses to the physical aspects of new styles. Makers, professional and amateur, established new methods to construct fashionable dress. From the professional mantua-makers who morphed into dressmakers to the home sewer, makers of Regency clothing had to address frequent variations, but they also helped to create them by adopting, inventing and exploring new material literacies through their garments. Sewing guides, surviving dresses in museum collections, archives and visual culture together illuminate the processes of transforming a gown, from length of fabric to final garment. By examining a collection of understudied sources, this chapter shows how style was grounded in the physical by examining the manual techniques that accommodated the new fashions.

I look at some previously undiscussed aspects of the making of women's gowns in Britain during the long Regency, adding to scholarship that has sought to detail some specific aspects of dressmaking in the period; these studies often focus on extant garments, as well as the practice of taking patterns.[3] This research has helped to establish general construction changes, such as Winifred Aldrich's survey of tailored women's jackets, from the 1800s onwards, and Madeleine Ginsburg's coverage of Regency shifts in the tailoring and dressmaking trades.[4] In addition, David E. Lazaro and Patricia Campbell Warner have examined techniques for pleating a bodice onto a lining, which demonstrate the rapid transitions in dressmaking between 1780 and 1805.[5] All of these approaches, however, focus on a particular aspect of the history of dressmaking, rather than the striking dependency between style and technique of the Regency period.[6] Cassidy Percoco's book of patterns comes closest to linking style to structure, illustrating how the construction processes for gowns changed during this era.[7] In this chapter, I further examine how the Regency gown was fitted to the body, how the labour of making was broken down and the effect of this on price and style, and how trimmings increased in importance to the creation of fashions. My examination of the material literacy of Regency dressmakers and their fashionable clients is shaped by my own experience of reconstructing a pelisse-coat once belonging to Jane Austen.[8] This project and others have helped to underpin my examination of archives, visual sources and surviving gowns with an embodied practice of experiential research that illuminates the key place of remaking to understanding from within the technical challenges that faced Regency makers. My re-creations of Regency gowns and garments have helped me access the unwritten body knowledge and material literacy implicit in Regency gown construction and the sewing skills that period makers required to shape and style finished gowns and garments.

Regency dressmaking was finely tuned to the connection between external surfaces and the internal structures of gowns, particularly in the crucial area of the bust and upper torso. Professional and amateur makers adapted their manual techniques to respond to developments in fashionable silhouettes, especially in handiwork that would support light and diaphanous fabrics, as well as intricate trimmings. Alterations to individual garments and the creation of trimmings occurred in both workshops and homes: a large proportion of the dressmaker's work of trimming and remaking paralleled domestic work. While professional makers remained largely responsible for the under-structures of gowns, surface embellishments could be designed and produced by women with the requisite skills. Similar to the support of stays and linings to the gown, new methods and materials for making trimmings encouraged innovations in specific technical skills to fit fashion to form. The proliferation of new construction methods, fabrics and trimmings, imported and domestic, encouraged consumers and makers to apply their material literacy to the relationship between structure and fashion in novel ways.

Construction and fit

In 1795, a mantua-maker was a woman who made women's outer clothes.[9] These professionals ranged from those with formal apprenticeship training to experienced makers to women setting themselves up to earn from their sewing abilities in a reasonably respectable trade.[10] They occupied the middle ground between seamstress and tailor, and could be found on almost every street in London. The word 'dressmaker' came in with the nineteenth century, was in use by 1799 and established by the early 1820s, as the Sun Fire Office insurance records show.[11] The term 'mantua-maker' continued in common usage to about 1811, with the term 'dressmaker' increasing after

Plate 1 Mary Emilia Cecil, Marchioness of Salisbury, *The Misses Van*, 1791, Yale Center for British Art, Paul Mellon Collection.

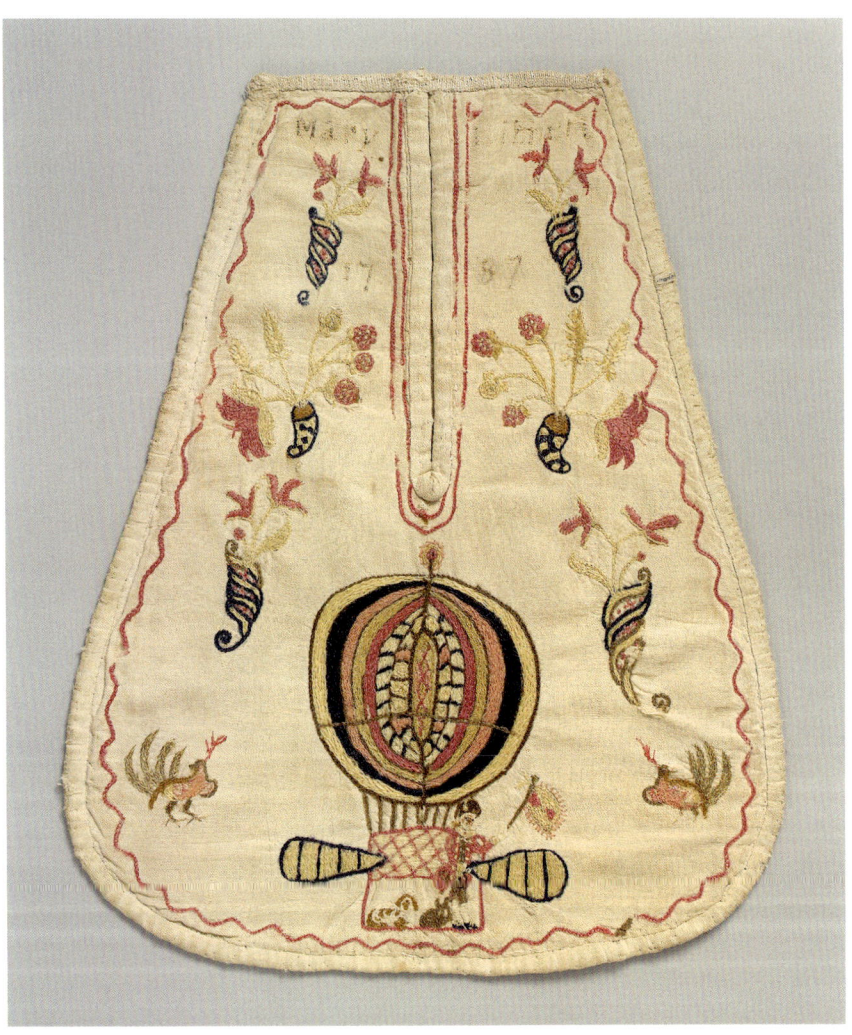

Plate 2 Embroidered pocket made by Mary Hibberd (or Hebbert), 1787, Museum of Fine Arts, Boston, 40.80. Gift of Mrs Samuel Cabot.

Plate 3 Doll's pelisse, 1810, Museum of London, A21160.

Plate 4 Pair of women's shoes, linen canvas embroidered with coloured wools in cross stitch and tent stitch, with short heel and latchet fastening, Great Britain, 1730s–1740s © Victoria and Albert Museum.

Plate 5 Woman's dress, 1797, National Museums Scotland, A.1977.820. Image © National Museums Scotland.

Plate 6 Woman's dress, *c.* 1798, National Museums Scotland, A.1977.821. Image © National Museums Scotland.

Plate 7 Dress probably worn at her wedding in 1807 by Ann Deane, Upton Pyne, Devonshire, The Springfield Collection, National Museum of Australia, 2005.0005.0141.

Plate 8 Box of shells, *c.* 1800, given in memory of Mrs Joan Griffith, W5·1 to 4–2010 ©
Victoria and Albert Museum, London.

1800; until around 1806 it outnumbered mantua-maker by two to six times. Maria Edgeworth employed the new phrase with a certain asperity in 1813: 'I have half an hour before the cursed mantuamaker, I beg her pardon dressmakers appointment.'[12] A 'gown' starts transforming into a 'dress' during this same period. As 'dress' is a noun and a verb, and both singular and plural, it can be hard to pinpoint the earliest instances of 'dress' meaning the women's main body garment, instead of the whole ensemble. The first unambiguous usage I have found dates to 1802, in the caption to a fashion plate (Figure 11.1). By the 1820s, 'gown' is more usually 'dress' at about the same rate as mantua-maker to dressmaker usage. Both senses are in liberal use throughout this liminal period of dress history.

Dressmakers rapidly developed new techniques as their craft and fashion mutually influenced the structure and appearance of clothing.[13] Changes to fit and construction in Regency gowns – the term 'gown' was understood to include the bodice, skirt and sleeves – reveal a period of diverse experimentation in design that played out in manual techniques. During the Regency period, the female waistline went up to under the bust around 1794–6 to reach its peak height *c.* 1800, lowered a bit by 1810, rose again to create bodice widths (the measurement from the waist seam to the neckline) as narrow as 2.5 inches in 1817 for evening wear, while at the same time dropping in daywear to return to the natural waistline by the early 1820s. When fashionable styles were at their most uncertain and mutable since the end of the fifteenth century, gown makers pushed their existing material literacy into new forms by finding solutions to construct and support those styles. The eighteenth-century straight panelled skirt acquired gores and more triangles to create bell-shaped skirts. Dress openings gradually moved from the front placement of eighteenth-century gowns to the back, through a variety of experimental cuts including the fall- or apron-front of the 1800s. After 1800 skirts were generally cut separately to the bodice, instead of being cut in one long piece through the back. The ensuing back bodice panel opened up new options for gathering and shaping skirts. From the 1810s, bust darts (short shaping seams) and bias-cut bodices increased in popularity, enabling a smoothly fitting bodice front, without the use of gathering. This was made possible by the gown's new back opening as the material tension created by fastening no longer dictated construction methods over the bust. The subtle bodice shaping methods emerged from refining earlier experimentation with fabric manipulation in the form of pin tucks, gathers, ruching and the piecing together of trims and fabrics.

The material record shows uncertainty about how a gown ought to be constructed and what it should look like when finished. From the late 1790s, makers appear to explore techniques and possibilities to achieve the latest style, inventing it as they go. This sense of experimentation about how a front closes, if a back can close, how deep an armhole is cut, whether a seam is straight or curved, if a bodice is lined, if a skirt is pleated or gathered, how it attaches to the bodice, how gathering and tucking create the shape of a bodice, continues through the 1800s and early 1810s in extant garments. By around 1814–15, makers seem to have achieved some consensus about basic construction and techniques became more uniform. This haptic confidence emerged with the establishment of the new styles and the decreasing costs of fabric. The consensus introduced structural changes: earlier adjustable gathers on skirts and

Figure 11.1 *Morning Dresses*, 1802, Original publication unknown, Private Collection. Photo © Hilary Davidson.

bodices gave way to form-fitting cuts with greater fabric wastage and makers exploited, with greater frequency, the possibilities of bias, the elastic quality of fabric cut at 45 degrees to the straight edge.

Most dramatic perhaps were the changes that occurred underneath the gowns themselves, as fashion's foundations, the stays, transformed fundamentally during this time. Heavily stiffened eighteenth-century stays had flattened the breasts into a round monobosom, an easily covered, neat and geometric shape. Softer, less boned supporting garments called 'corsets' – French for 'little bodies' – appeared in the late 1790s alongside stays, helping to introduce into fashion a bosom with two separate breasts. The cut of gowns now had to account for this new sculptural roundness, as well as the space between the breasts that was created by the busk, a straight piece of wood held by a pocket running vertically down the front of the corset or stays. Percoco connects the smoother busts of 1810s dresses with the settling of stays into this new line.[14] As Rasmussen observes, 'The achievement of the three-dimensional form over the bust remained the most demanding aspect of constructing women's clothes during the nineteenth century.'[15]

Some women could and did go without separate bust support, using the internal overlapping fronts of the lining to hold up the breasts before about 1810 (Figure 11.2), or relying on the petticoat bodice underneath.[16] Portraits with nipples faintly visible through diaphanous muslin reinforce this practice, bemoaned but exaggerated by contemporary commentators on women's dress, and therefore in subsequent histories of Regency dress. Other portraits that depict an apparently natural bosom demonstrate the centrality of stays in achieving this look. In order to make the gown contour around the bosom, the fabric of the bodice was pinned between the breasts (Figure 11.3) to the firm foundations of the stays underneath.

The oft-quoted description of a dressmaker from Souter's 1818 *Book of English Trades* emphasizes the maker's intimate knowledge of the body: 'The Dress-Maker must be an expert anatomist … she must know how to hide all defects in the proportions of the body, and must be able to mould the shape by the stays, that, while she corrects the body, she may not interfere with the pleasures of the palate.'[17] Some studies of Regency dress have identified direct and indirect methods that dressmakers used to cut, drape and fit gowns.[18] In a direct method, the maker created a pattern by fitting fabric or paper to the figure, shown in the illustration to Souter's text (Figure 11.4). This method required previous knowledge of making to do successfully, skills that were passed on through apprenticeship or regular domestic practice. The indirect method used an existing dress to make a template or copy, by replicating its shapes or unpicking it to take a pattern.

Such 'pattern gowns' – either one's own that fit, or a friend's for style – appear throughout Regency texts as a way of communicating material literacy, the knowledge of an effective fit embodied in its shapes. Before commercial paper patterns, such techniques ensured proper fit and consistency in patterning, and bypassed the need for drafting skills. It was also much easier to make a gown that one had made before, in contrast to working through a gown's inner construction for the first time. The earlier gown could be used as a template and exemplar of the stitches and stages of construction, and depth of turnings, as affirmed by my remaking of the Austen

Figure 11.2 Dress probably worn at her wedding in 1807 by Ann Deane, Upton Pyne, Devonshire, The Springfield Collection, National Museum of Australia, 2005.0005.0141 (see Colour Plate 7).

Figure 11.3 Sir Thomas Lawrence, unknown sitter (formerly called Miss Lamb), oil on canvas, 1815–1820, Birmingham Museum of Art, Alabama, 1966.143.

pelisse twice.[19] Previous mistakes can be corrected for or avoided. Both now and in the past, the pattern gown acted as a reference text 'explaining' how to make itself for a materially literate maker, usually an experienced needlewoman.

To start a Regency gown, the needlewoman would first make the linen body lining as the fitting piece (continuing eighteenth-century practices) and then finish the main fabric over it, as Rasmussen outlines. Rasmussen's summary of the construction of British dresses 1800–30 only holds for the period up to *c.* 1810: 'If the dresses were lined … [t]he lining kept the bust in firm shape and the fit was regulated by a flexible

Figure 11.4 'The plate is a representation of a mantua-maker taking the pattern off from a lady by means of a piece of paper, or of cloth. The pattern, if taken in cloth, becomes afterwards the lining of the dress,' *The Book of Trades*, 23 Oct. 1804, *Ladies Dress Maker*, etching © The Rijksmuseum.

closing of two trapezoid pieces of linen or cotton pinned over the bust', as seen in Figure 11.2.[20] Rasmussen asserts that Regency making practices 'are a typical result of shaping directly on the body'.[21] Lazaro and Warner also explore folding and shaping cloth over this foundation, explaining that 'the mantua-maker had to drape the dress fabric on the client to achieve the proper cut, the proper alignment of the pleats, and the overall fit'.[22] These assessments do not account, however, for some physical aspects of Regency dress. Indeed, the various constructions of surviving gowns reveal that for every generalization there is an exception until making practices started to coalesce around 1814–15. Fitting and sewing simultaneously, as Rasmussen suggests happened, is time-consuming and difficult to achieve within the limited frame of a dressmaker's appointment, where fit would have been the primary concern. Arranging cloth around a body shape was only possible with a person present, as the body substitute of dressmakers' dummies, which modern makers rely on, did not appear until the 1830s.[23] Arranging pleats can also be done off the body once the lining bodice is correctly fitted, as the lining becomes a substitute body on top of which the dress is built.[24] In addition, trying to sew straight lines on a three-dimensional moving form like the human body is much harder than construction on a flat plane. These material challenges, coupled with the constraints of time, suggest that Regency makers likely spent a good part of the production process stitching garments off the body of their clients.

In the Regency gown, the bust area including breast, back and armholes required the most accurate fit, as the sleeves and skirt took their proportions from, and attached onto, this foundation.[25] The bodice's importance to a gown's success is consistently reiterated. Virtually all visual depictions of women's dress from the period show a snug, firm bosom, no matter how capacious the actual flesh underneath. In the front-opening bodice styles, this effect was created during dressing with the help of undergarments. With back-opening gowns, achieving a snug fit at the bosom depended on the cut of the bodice lining or petticoat bodice. References to 'body linings' and 'bodies' – meaning the plain linen or cotton fabric made into a lining for the outer gown – as separate gown pieces in documents are plentiful. However, one of the period's technical evolutions was the disappearance of sewn-in body and sleeve linings on transparent muslin, net or gauze gowns – a change distinctive to the Regency period. Innovations like this exemplify the more experimental approach and fluid form of Regency garment construction as the constraints posed by procedural eighteenth-century dressmaking were questioned materially. The mutable shapes of stays and corsets – and in some cases, nothing under the gown – removed the certainty of the stable, solid geometric shape created by earlier stays, and which eighteenth-century mantua-makers had relied on as a structural foundation. This rapid evolution of long-established practices to incorporate new methodologies characterizes Regency dress and the responsive material literacy required in changing dressmaking practices.

Fashioning fashion

We currently lack a definitive understanding of how clothing clients and makers decided on a gown's final style: the 'fashioning' of the garment. The final article seems to have been a balance between being 'entirely my own thought', as Isabella Thorpe in *Northanger Abbey* (1817) proclaims of her new sleeves, and the professional expertise in new styles and current taste, filtered through tact and flattery regarding what would suit the customer's figure and budget.[26] Judging from contemporary tailors' complaints about their customers – and vice versa – and the number of returns of ill-fitting garments or those which the customer decided against, dressmakers would have spent considerable time balancing their clients' whims, ideas and demands with their own production capabilities.

Appointments would have been a negotiation of competing material literacies about what was possible in a gown. An unexpected, rare source on the process of dressmaking and fitting comes from French-English language lesson books. Changes to *Conversations of a Mother with Her Daughter* between the 1803 and 1823 editions document a simplified version of client-maker negotiations, as well as changes in dress construction over a twenty-year period.[27] The texts present a dialogue about the fitting process between the mother and a clothing provider who visits her at home. Recalling the Souter image, the 1803 edition finds Mrs Melville being fitted in her stays and petticoat by a 'mantuamaker' (Figure 11.4); her fitting attire for the appointment with her 'dressmaker' is excised by 1823. The dressmaker uses her time to fit three dresses of different shapes, from which she can extrapolate three other dresses required by her client, including a 'high-necked' dimity gown, for morning wear with fashionable tassels (in both years) and two 'frocks' of lighter muslin and similar fabrics. In 1803, trains are the fashion, which Mrs. Melville decides she will only wear in town; by 1823, they are long out of fashion and the lady is glad of their inconvenience being gone. Other styles such as tunics and shorter over gowns remained in style for twenty years, as fashion plates also attest.

In both editions of *Conversations of a Mother with Her Daughter*, the armholes receive the most attention at the fitting. This fitting was enacted through cutting, and presumably pinning, as no sewing is mentioned. The sleeves (possibly the lining sleeves) are 'too strait' in 1803, seemingly a result of their tightness. By 1823, when fuller puffed sleeves had become fashionable, it is the constriction of the armhole that causes discomfort. While Mrs Melville queries a similar narrowness at the back, she is assured, and accepts, that this cut will not 'hurt' or 'confine' her. Both these points reinforce the importance of the bust region as the key place to achieve fit. One dress fits perfectly, and from this success the dressmaker can 'take it for a pattern and the others will do well'. It was thought reasonable in both editions to fit three dresses on a Monday and have six made from that fitting delivered the next Saturday. The vignette stresses the conversation and mutual influence between both parties for the final garments: the dressmaker advising and agreeing, the client expressing her preferences and bodily experience.

The subtle differences between the white cotton fabrics such as muslin, cambric muslin, percale and dimity listed in the *Conversation* were of particular importance to the Regency consumer. The client – and sometimes the dressmaker – needed to select the fabric first, using an experienced haptic knowledge and material literacy to inform her choice. The length of uncut cloth required for the purpose was also called a gown, as Jane Austen, for instance, illustrates in an 1813 letter: 'I shall take the opportunity of getting my Mother's gown–; so, by 3 o'clock in the afternoon she may consider herself the owner of 7 yds of B[lac]k Sarsenet.'[28] The garment and its material were synonymous for Regency shoppers, who selected carefully the fabric for their chosen style. Textiles were handled and tested by shoppers, their qualities imparted through the sensory experience of materially educated buyers, as Serena Dyer has shown.[29] As I have argued elsewhere, Regency shoppers, faced with a proliferation of retailers and fabrics, required a near-intuitive haptic sense, which deployed sight and touch to differentiate which fabrics would wash and wear well, and keep their colour from those that would fray and fade.[30] Sight and touch were supplemented by smell, especially for imported goods: some manufacturers even added the scent of spices to their imitation Indian textiles.[31]

The proliferation of new fabrics, and the increased range of their qualities and varieties during the Regency, demanded an advanced material literacy from consumers. Fine muslins, for instance, were increasingly distinguished between handwoven Indian and machine-woven British products – differences that Henry Tilney is famously able to identify in *Northanger Abbey*.[32] Improvements to the quality of British muslin allowed them to compete with imported textiles by the early nineteenth century.[33] Cotton fabrics in general increased in quality and quantity, requiring consumers to understand not only the quality of the fibres and weave, but also the effectiveness and fastness of their printed patterns. John Heathcoat's 1808 invention of a net-making machine that reproduced the twist-net (diamond-shaped) ground of bobbin lace, in cotton or silk, was perfected in 1809.[34] Other advances followed, bringing further technological innovations to the marketplace. Airy silk gauzes, tulles, bobbinet(te)s and cheaper patent nets appear frequently in extant garments in the 1810s.

All of these novel and more cheaply available fabrics required improved material literacy to understand their structural, constructional and wearing qualities on the body. Diaphanous net textiles in particular needed petticoats – such as full underdresses called 'slips' – to wear under them, which were made either of white or coloured silk. These slips could affect the appearance of the gown's outer colour. A net dress's fragile tensile strength also called for construction techniques that provided structural integrity along seams and stress points, to avoid any ripping or stress during wear. Gauzy fabrics increased in gowns from about 1814, at the same time as longer stays that would afford a firmer under-shape to reduce the strain of physical movement on the external fabric followed suit. Print publications such as Rudolph Ackermann's *Repository of Arts* also called attention to the central importance of the consumer's material literacy, by including fabric samples in its pages.[35] The quality of the fabric needed to be felt to be valued. Descriptions accompanying engraved fashion plates detailed the images' specific weaves, fibres, colours and trimmings, tapping into the reader's existing material literacy as an essential part of bringing the image to life. To understand the qualities of a dress in a fashion plate, a Regency woman would need to know its component fabrics. The white morning gown in Figure 11.5, for example, is described in the magazine as

> a petticoat and bodice of fine jaconet muslin, finished round the bottom in vandykes and small buttons. The Rochelle spencer composed of the same material, appliquéd with footing lace down the sleeve, and trimmed at each edge with a narrow, but full border of muslin. Double fan frill of muslin round the neck, very full, continuing round the bottom of the waist where it is gathered on a beading of needlework.[36]

The text is directed to those who not only possess the literacy to read and interpret the description, but who also know how clothing is made. A vocabulary of specific materials and technical sewing skill is deployed, demonstrating the style and manual knowledge that readers brought to publications to understand text and image.

MORNING DRESS.

April 1812

Figure 11.5 Fashion plate, 'Morning Dress' in *The Repository of Arts*, 1 April 1812, hand-coloured engraving on paper, Los Angeles County Museum of Art.

Making trimmings

The accounts of Regency women highlight the importance of haberdashery and trimmings as an essential part of making fashion. The cost of all the components of an ensemble was decipherable to a precise, informed eye, combining material and economic literacy.[37] Contemporary sources repeatedly distinguish between gown and trimmings as separate aspects of fashionable dress, and the rapid increase in trimmings and handwork applied to the base gown remains an under-appreciated element in Regency dressmaking. This was a transitional period, during which the production of textile and haberdashery items was becoming increasingly mechanized; technological innovation both increased the variety of goods and made them more readily and cheaply available. 'Trimming' as a verb or noun is as ambiguous as 'dress', so items in maker's bills and accounts could either be the task of applying the decoration or what is being applied. The accounts of Eliza Jervoise (b. 1770) of Herriard Park in Hampshire, between 1798 and her death in 1821, document one gentry woman's ongoing relationship with her dressmakers and the costs involved.[38] She first employed Esther Crowther of Marylebone, before changing by 1806, to Mrs Killick (either Crowther's mother or daughter), also in London. One 1808 example of Mrs Killick's bills shows how most costs come from the material components, including the gown fabric which she supplies, and a handkerchief worth nearly twice the cost of making a gown. The account itemizes the structures and linings underneath the gown, and its decorative exterior: the sleeve lining is a separate cost. Killick also itemizes ribbon and tape separately: these 'trimming' costs could cover the labour of trimming the gown she has made, or the cost of the materials she used. In another bill, Mrs Killick itemizes '2 bodies', again emphasizing the separate nature of this part of a gown's structure.

There is a distinct increase in haberdashery and trimmings on gowns after 1815.[39] In line with the techniques that Emily Taylor discusses in the previous chapter, Figures 11.6 and 11.7 show gowns using the thrifty practice of remaking older silk gowns into more up-to-date styles. Comparing construction details between the two gowns shows that alterations to the 1820s dress required more labour and components than those to the earlier garment. In the later dress (Figure 11.7), the seams are trimmed with piping made of cotton cord covered in the gown fabric, where the earlier gown, made from a silk patterned with green and pink sprigs on a pale green ground (Figure 11.6), has cream silk braid simply tacked on top along the bustline and sleeve. The piping on the 1820s gown (Figure 11.7), which refashioned a blue and cream figured silk, required at least two more actions of cutting strips of fabric and wrapping it around the cord, plus the careful, slower stitching of the seam to ensure the inserted trimming was straight and even. Tacking braid is quick work done in larger stitches. Likewise, where the sleeves of the 1800 gown are merely turned and hemmed, the 1820s gown shows bias-cut satin trimming forming cuffs and decorative bands on the sleeves, and the neck is bound separately in the same fabric. Creating these decorations required additional fabric, accurate bias-cutting (which is more difficult than straight), manipulation and folding of the bias strips, and extra stitching to apply them over the top of the already constructed sleeve and neckline. The alterations and trimmings on the 1820s gown

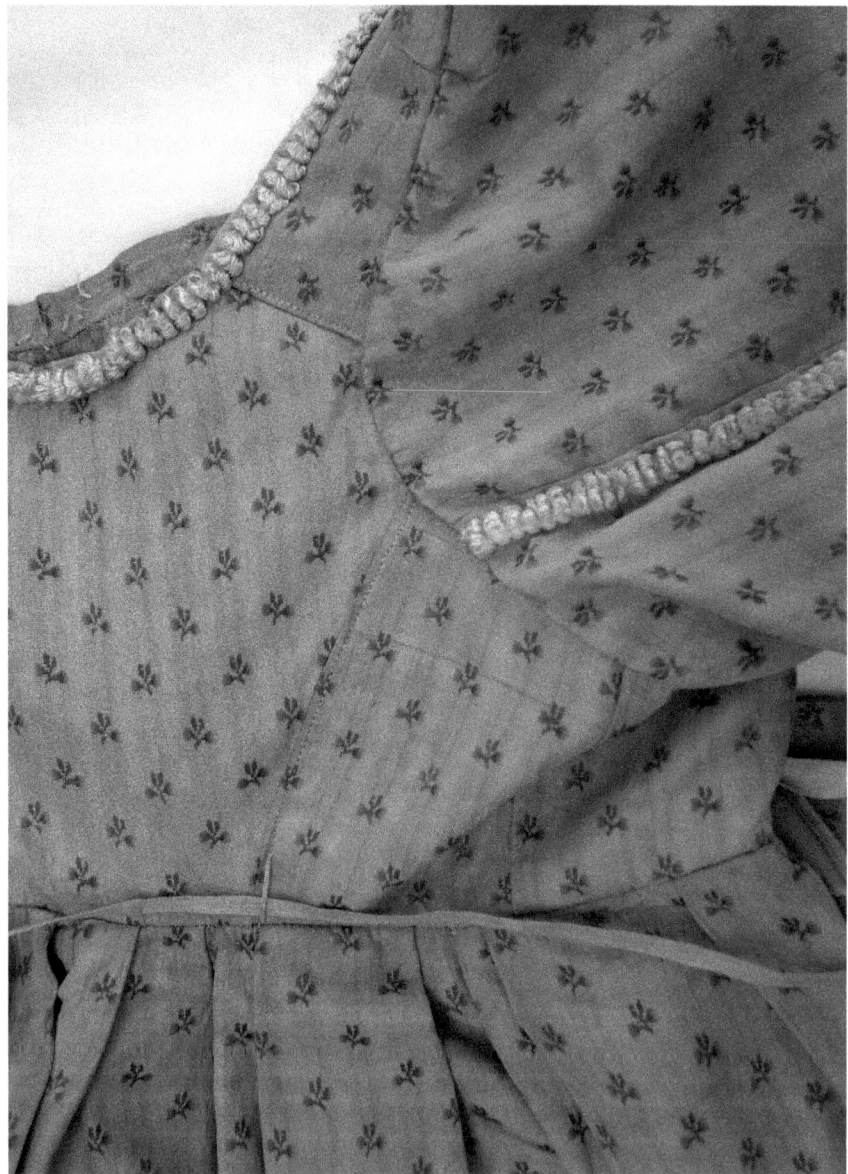

Figure 11.6 Gown, 1800–1805, remade from late eighteenth-century silk, T.761–1913
©Victoria & Albert Museum, London.

required similar basic needlework knowledge. However, these skills needed to be
wielded more carefully, and took a longer time to complete.

The fashion from the late 1810s for rouleaux decoration – narrow bias-cut tubes
made of the main gown fabric and applied in decorative patterns – required even

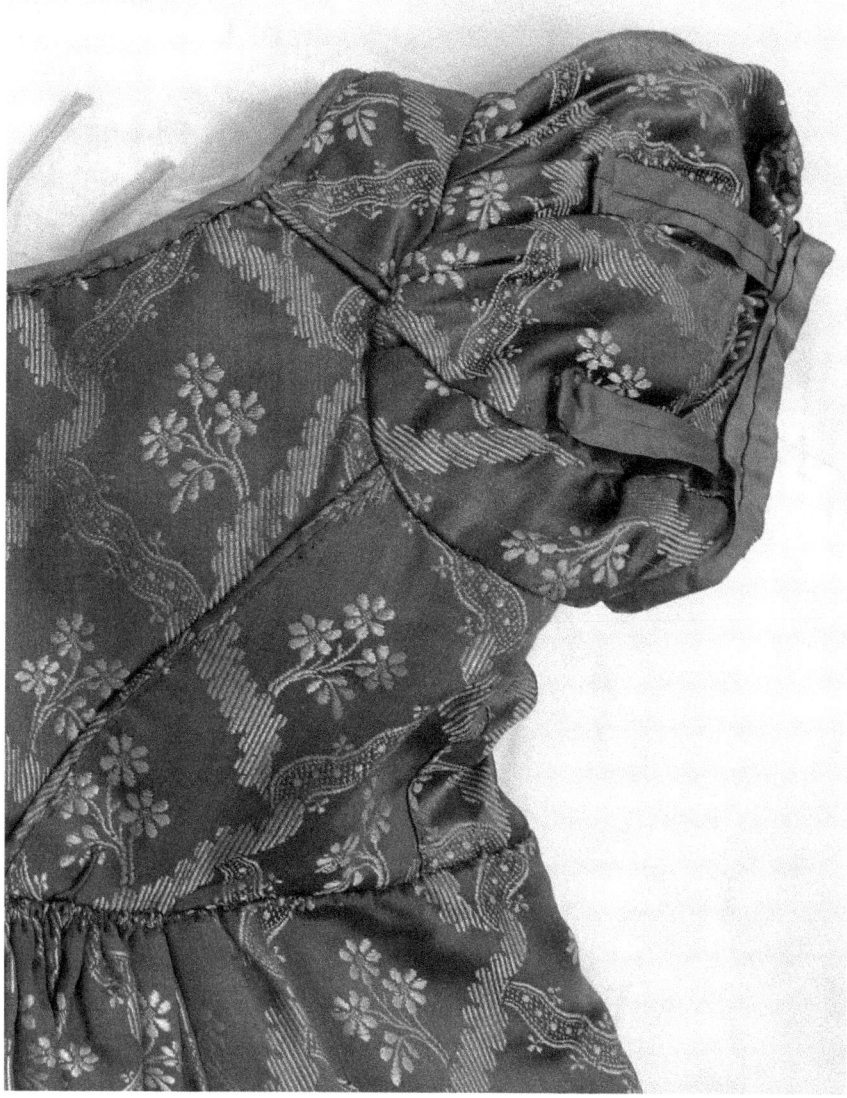

Figure 11.7 Evening dress, 1820, remade from late eighteenth-century silk, T.220–1957
©Victoria & Albert Museum, London.

more labour and fabric investment than most Regency trimmings. The anonymous
authoress on dressmaking in the *Guide to Trade* (published in 1840 but based on the
dressmaking practices of the late 1820s and 1830s) comments that dressmakers 'are
edging with cording hundreds upon hundreds of little vine leaves ... while they are
spending whole days in embroidering with piping of their own making ... They must
eventually sigh for a change of fashion ... so that they may again be able to make a

dress in a day'.[40] By about 1818, it was possible to confect a gown almost entirely of trimmings, using gauze, net, blonde lace, bias strips, ribbon leaves and flowers, and cartisane (decorative shaped cards wrapped in silk floss).[41] The ability to construct a gown mostly of trimmings constituted a significant change from even the lightest muslin gowns of 1800. Then, and through the early teens, many fashionable muslins came already sprigged, spotted, tamboured or embroidered.[42] The decorative labour was incorporated into the cost of the uncut gown length before purchase, and had often been worked in a different country. By the 1820s, the effort of decorative labour had largely transferred to the fashioning of the gown after it was made, a more mobile response site for fashion styling than relying on a textile that was purchased with its decoration already sewn or woven into the fabric. Even where trimmings might not increase sewing time overall, they were always an extra expense, as dressmakers' accounts demonstrate. The material and economic value of trimmings, although obvious to a contemporary observer, has become almost invisible now as secondary to their decorative value.

The tradition of eighteenth-century milliners' fashion styling continued in the Regency period but styling became more crucial as trimming formed the vehicle for much of the style work in later Regency dresses.[43] These trimmings put 'fashion' onto the foundation garment, while the foundation itself became redefined and multi-layered. The 1810–25 account book of London widow and heiress Mrs Mary Topham (*c*.1755–1825) illuminates a fuller picture of the costs involved in making fashion.[44] Her accounts show that extras such as lace, flowers, ribbons, pins, fringing, shoe binding and roses, gimp and galloon (types of braid), cord, tape, wadding and buttons formed 47 per cent of the quantity of textiles and clothing-related items bought by her between 1811 and 1825, although fabrics constitute the greater cost. Ribbons, in particular, form the overwhelming majority of purchases at around 42 per cent of the haberdashery, or around 20 per cent of the total number of textile items, bought in every colour, width and quality on a weekly basis throughout the accounting period.

In her letters, Austen often documented the end uses of trimmings such as her purchases of ribbons: 'I have determined to trim my lilac sarsenet with black sattin ribbon just as my China Crape is, 6d. width at the bottom, 3d. or 4d., at top'.[45] Here she classifies ribbons by price, rather than width, and knows that her sister Cassandra (her correspondent) shares this knowledge and understanding. Mary Topham also documented minute expenditures on ribbons: she buys ribbons according to price, based on width and quality, and weaves (taffeta, satin, embossed), in small repetitive lengths reflecting the immediacy of requirements in everyday life and the frequency of ribbon purchases. This quantification again reflects a nuanced consumer literacy of specific textile and trimming qualities, which is hard to reconstruct from surviving gowns. We see ribbons; they saw sixpenny ribbons. The ability to read both material and economic information in objects was inherently interlinked, as shown in Elisabeth Gernerd's and Jon Stobart's contributions to this volume. Austen's letters also illustrate the ways in which small details created style; in the same 1814 letter, she records 'ruining myself in black sattin ribbon with a proper perl edge; & now I am trying to draw it up into kind of Roses instead of putting it in plain double plaits [pleats]'.[46] After buying the ribbons, Austen used her manual skills to manipulate them into a more fanciful shape. However, applying trimmings could take as long or longer than sewing a gown, with needlework ranging from quick stitches to complicated precise or repetitive sewing to make a decorative surface with applied trimmings. Trimming labour added extra value to the economic worth of clothing's material components.

Home dressmaking

Once the gown structure existed it might be refashioned for as long as the fabric endured. Many surviving garments in museum collections show evidence of careful alterations to improve or update an earlier style. The point of changing surface trimmings and altering the gown structure was to extend the life of a garment and presumably prevent the wearer from becoming bored with her limited wardrobe. Professional bills frequently include this work of refreshing and remaking to renew existing gowns, and such work extended to the wardrobes of men, as Sarah Howard discusses in her chapter. At the same time, such labour could be accomplished at home, as remaking and trimming sat comfortably within the realm of domestic work. Surviving garments and evidence of making practices convey that dressmaking was a task of two halves: foundation structures and surface embellishment. Austen and her contemporaries, who changed trimmings themselves, were doing the work of dressmaking even if they did not construct the entire garment.

Print sources suggest that most amateurs relied on the skills of professional needlewomen, even as they supplemented their own embellishments and sometimes produced gowns at home. *The Female Instructor* (1811) observed that 'Many young ladies make almost every thing they wear; by which they can make a genteel appearance at a small expense'.[47] Accessing the experience of genteel domestic making is challenging. While Janet Arnold notes that the simplified Regency silhouette invited attempts by amateurs, formal occasions would have required professional assistance: 'probably a lot of muslin dresses made at home were worn for mornings, while the dressmaker, or mantua-maker ... would have been called in to tackle the ones worn on more public occasions'.[48] The loaning of completed garments to use as patterns suggests amateur needlewomen were capable of making cloth templates, but that they also gave the pattern gowns to dressmakers or maids to copy. Harriet Smith in Austen's *Emma* (1815) is a young woman with needlework skills, and time on her hands, yet she still outsources her gown construction to another young woman in the neighbourhood.[49]

The regular domestic application and amendment of trimmings further blurred the lines between professional and domestic dressmaking. *The Lady's Economical Assistant* (1808) provides examples of what was regularly produced by genteel and middling women. The manual emphasizes linen ancillary garments for adults and children such as stays, drawers, petticoats, caps, shirts and shifts; outer clothing for children such as coats and spencers; and clothing for the poor. These articles appear constantly as items women made, embellished or mended. Caroline Fry's *The Assistant of Education* (1825) includes a detailed and arresting vignette of home dressmaking:

[T]he mercer came, and his bale of goods came, and the yardwand came – and there was measuring of breadths and measuring of lengths, and many very intricate calculations beside, to make the least possible quantity do the greatest possible service. In the issue, it appeared to me, that the materials selected were simple, tasteful, and very little expensive ... It would be quite superfluous to describe the whole process of dress-making – every lady who has made her *entré* into the gay world, without a long purse at her command, knows what ensues upon wanting a ball-dress in a hurry, and can picture to herself the state of the apartment, during the first stage of the proceeding – the various articles of apparel consigned to the backs of the chairs – the piano converted into a measuring board – the attendance of all the females in the house, except the cook, with thimbles on their middle finger – the trying on, and cutting out, and fitting in. It was impossible not to admire the skill and ingenuity of the young ladies ... And much I heard of the comparative merits of full fronts, and plain fronts, and high backs, and lowbacks, and circles, and squares, and Vandykes, and scollops, and straight-ways, and cross-ways, and long-ways ... Time, with its usual malevolence, sped the quicker for the need there was of it – night came, and the ladies stole some hour or two upon its wintry length, and rose but the earlier to renew their labours: and like to the first day was the second.[50]

The author's assumption of her readers' familiarity with the material literacy and technical knowledge of home dressmaking suggests it was quite common. Her dismissal of a fuller description as 'quite superfluous' perhaps explains why it is hard to find evidence for it: women were too busy sewing to write down what they did. The scene conveys a multiplicity of knowledge and skills. The 'various pieces of apparel' suggest gowns to copy for fit or fashion. The 'comparative merits' of different structural and decorative elements are heard, rather than read about. Fashion information and material literacy were being transmitted directly, from personal experience, to arrive at a collective decision.

A combination of ability, willingness, finances, time and necessity probably determined whether the amateur needlewoman made gowns for herself or others. Arnold highlights, for instance, how *The Economical Assistant* shows that 'cutting out presented the biggest problem to the dressmaker', due to the hazard of wasting fabric.[51] In Fry's vignette, home dressmaking requires certain material conditions: a clean, flat space to lay out fabric and to work, and the appropriate tools. One feminine accomplishment, the piano, is literally covered by another, the skill of dressmaking. Fry's gentry dressmaking commandeers a whole room and nearly all the female staff, prioritizing making as a different form of sociability than the polite kinds usually enacted in drawing rooms. These resource requirements would have hampered opportunities for poorer women to make their own clothes even if they knew how. Further issues of mobility affected sewing projects held in work-bags, such as what could fit easily on a lap or small table to be taken on the move, or to social events. A gown skirt was too large to move about easily, but a bodice or sleeve was of a more mobile size.

The Regency period saw the introduction of a fit that focused on the bust instead of the whole upper torso; new fabrics and greater, or cheaper quantities of established textiles; rapidly changing styles; different methods of putting a gown together; and experiments with new trimmings. By the 1820s, these changes and evolutions meant that women's fashions had a continuing but fundamentally transformative relationship to the fashions of the previous century. Professional and domestic makers alike confronted these transformations, adapting and altering their material literacy to the way they made and trimmed clothes. Shifting our perspective from Regency fashion to its makers and materials answers different questions about how this period of great stylistic change was embodied.[52] Broader technological innovations affected the materials of clothing and how makers manipulated them, shifts which can be observed in the fabric of objects. Exploring modes and processes of making emphasizes the ways makers' and wearers' bodies interreacted with materials, and their bodies within clothing. The material details of fashions and fashionability had equal weight to their visual qualities. The dramatic transition in style between mid-eighteenth century and Regency fashion was indivisible from the changing manual dressmaking and trimming habits of both professional and amateur makers, as together they found new ways to create and respond to technical and structural change through their material literacy.

Notes

1   8–9 January 1799, Letter 17 in *Jane Austen's Letters*, 34.

2   Hilary Davidson, *Dress in the Age of Jane Austen* (London: Yale University Press, 2019).

3   Janet Arnold, *Patterns of Fashion 1: Englishwomen's Dresses and Their Construction. c. 1660–1860* (London: Macmillan, 1972). Arnold also examined the dressmaker's craft over a longer timescale, as well as the classical influence on long Regency fashion; see 'The Classical Influence on the Cut, Construction and Decoration of Women's Dress *c.*1785–1820', *Costume* 4 (1970): 17–23; 'The Dressmaker's Craft', 29–40. For patterns from museum objects, see Norah Waugh, *The Cut of Women's Clothes, 1600–1930* (New York: Theatre Arts Books, 1968); Cassidy Percoco, *Regency Women's Dress: Techniques and Patterns 1800–1830* (London: Batsford, 2015).

4   Winifred Aldrich, 'The Impact of Fashion on the Cutting Practices for the Woman's Tailored Jacket 1800–1927', *Textile History* 34 (2003): 134–70; Ginsburg, 'The Tailoring and Dressmaking Trades, 1700–1850', 64–71.

5   David E. Lazaro and Patricia Campbell Warner, 'All-Over Pleated Bodice: Dressmaking in Transition, 1780–1805', *Dress* 31 (2004): 15–24.

6   See also Natalie Garbett's history of the Regency's high-waisted gown, which draws on her background as a re-enactor to bring an understanding of the material qualities of these styles in *A History of Women's Fashion from 1790 to 1820* (Havertown: Pen and Sword, 2020).

7   Cassidy Percoco, *Regency Women's Dress*.

8   Davidson, 'Reconstructing Jane Austen's Silk Pelisse, 1812–14', 198–223; 'Jane Austen's Pelisse Coat', in *Jane Austen: Writer in the World*, ed. Kathryn Sutherland (Oxford: The Bodleian Library, 2017), 56–75.

9	The name derived from the 'mantua', formal French courtwear introduced in the 1690s, which rapidly became the template for a gown throughout the eighteenth century.
10	See Jessica Collins, 'Jane Holt, Milliner, and Other Women in Business: Apprentices, Freewomen and Mistresses in The Clothworkers' Company, 1606–1800', *Textile History* 44, no. 1 (2013): 72–94.
11	Records of Sun Fire Office, London Metropolitan Archives, MS11936/350-492.
12	16 May 1813, Maria Edgeworth, *Maria Edgeworth: Letters from England 1813–1844*, ed. Christina Colvin (Oxford: Clarendon Press, 1971), 59.
13	This information is summarized from research across eight museum collections in Australia and Britain, and extensive research amongst digitized collections online, looking at hundreds of surviving Regency gowns.
14	Percoco, *Regency Women's Dress*, 9.
15	Pernilla Rasmussen, 'Creating Fashion: Tailors' and Seamstresses' Work with Cutting and Construction Techniques in Women's Dress, *c.* 1750–1830', in *Fashionable Encounters: Perspectives and Trends in Textile and Dress in the Early Modern Nordic World*, ed. Tove Engelhardt Mathiassen, Marie-Louise Nosch et al. (Oxford and Philadelphia: Oxbow Books, 2014), 50–71.
16	I have successfully tested this theory of using tightly pinned bodice linings to hold up my bosom in a re-created 1798 gown throughout a night of period dancing.
17	John Souter, *The Book of English Trades: And Library of the Useful Arts: With Seventy Engravings* (London, 1818), 224–45.
18	Aldrich, 'The Impact of Fashion'.
19	This insight was gained by making a second copy of the Austen pelisse in 2018 for the Jane Austen House Museum. I referred constantly to the first replica in the way a Regency maker would have used a pattern gown, and it was easier than using my photographs.
20	Rasmussen, 'Creating Fashion', 68.
21	Ibid.
22	Lazaro and Warner, 'Dressmaking in Transition, 1780–1805', 20.
23	Alison Matthews David, 'Body Doubles: The Origins of the Fashion Mannequin', *Fashion Studies* 1 (2018): Article 7, https://www.fashionstudies.ca/body-doubles, accessed 22 May 2020.
24	I have confirmed this on my recreated 1798 gown.
25	I tested this approach successfully with a group of costume making students in 2016.
26	Jane Austen, *Northanger Abbey*, ed. Barbara M. Benedict and Deirdre Le Faye, *The Cambridge Edition of the Works of Jane Austen* (Cambridge: Cambridge University Press, 2006), 68.
27	The conversation with her milliner is similarly enlightening. J. Franceschini, *Conversations d'une mère avec sa fille en français et en anglais* (Paris, 1803; *Conversations of a Mother with Her Daughter, and Some Other Persons*, 4th ed. (London, 1823).
28	Austen, 20 May 1813, Letter 84, *Jane Austen's Letters*, 219.
29	See Dyer, 'Shopping and the Senses: Retail, Browsing and Consumption in 18th-Century England', 694–703.
30	Davidson, 'Jane Austen's Pelisse Coat', 64.
31	See, for example, the advice proffered to early American merchants in the anonymous *Valuable Secrets Concerning Arts and Trades: Or, Approved Directions, from the Best Artists, for the Various Methods* (Norwich: Thomas Hubbard, 1795), 50.

32   Austen, *Northanger Abbey*, 20.

33   Davidson, *Dress in the Age of Jane Austen*.

34   See Santina M. Levey, 'Machine-Made Lace: The Industrial Revolution and After', in *The Cambridge History of Western Textiles*, ed. D. T. Jenkins, II vols. (Cambridge: Cambridge University Press, 2003), II: 846–50.

35   See Serena Dyer, 'Fashioning Consumers: Ackermann's *Repository of Arts* and the Cultivation of The Female Consumer', in *Women's Periodicals and Print Culture in Britain, 1690–1820s*, ed. Jennie Batchelor and Manushag N. Powell (Edinburgh: Edinburgh University Press, 2018), 474–87.

36   *Repository of Arts, Literature, Commerce, Manufactures, Fashions and Politics*, Vol. XI, April 1814 (London: Rudolph Ackermann), 241.

37   For discussion of the relationship between material and economic literacy, see Dyer 'Training the Child Consumer: Play, Toys, and Learning to Shop in Eighteenth-Century Britain', 31–45.

38   Jervoise of Herriard Collection: Family and Estate Papers, 44M69/E13/13, Hampshire County Record Office.

39   This is probably connected with the new free trade of textile items around Europe after Napoleon's defeat and the end of his economically unsuccessful Continental System.

40   *The Guide to Trade: The Dress-Maker, and the Milliner* (London, 1843), 68.

41   My thanks to Gina Barrett for identifying cartisane.

42   See Ashmore, *Muslin*.

43   For a discussion of trimmings in relation to eighteenth-century fashions, see Kimberly Chrisman-Campbell, 'The Face of Fashion: Milliners in Eighteenth-Century Visual Culture', *Journal for Eighteenth Century Studies* 25, no. 2 (2008): 157–71.

44   Lady's Account Book, 6641, Chawton House Library.

45   Austen, 5–8 March 1814, Letter 98, in *Jane Austen's Letters*, 269.

46   Ibid., 260.

47   *The Female Instructor; or, Young Woman's Companion* (Liverpool, 1811), v–vi.

48   Arnold, 'The Dressmaker's Craft', 31; Dyer also addresses the degree to which genteel women engaged in garment construction in this volume.

49   Jane Austen, *Emma*, ed. Richard Cronin and Dorothy McMillan, *The Cambridge Edition of the Works of Jane Austen* (Cambridge: Cambridge University Press, 2005), 190.

50   Caroline Fry, *The Assistant of Education: Religious and Literary* (London, 1825), IV: 98.

51   Janet Arnold, '"The Lady's Economical Assistant" of 1808', in *The Culture of Sewing: Gender, Consumption, and Home Dressmaking*, ed. Barbara Burman (Oxford: Berg, 1999), 229.

52   My recreation of Jane Austen's pelisse transformed how I understood the inner structures of Regency dress. Fry's dressmaking scenario notably omits 'the whole process of dress-making' because it was so fundamental: her readers know what a dress comprises, and the sewing skills necessary to realize it. Recreating Regency dress through the practice of experiential research has increased my access to what Regency makers knew about making clothes from the inside out. For me, the process of reconstruction returns an embodied experience to the understanding of history, allowing for a more comprehensive picture to emerge of the intimate relations among form, function and fashion in the Regency era.

12

Fancy feathers: The feather trade in Britain and the Atlantic world

Elisabeth Gernerd

On 25 April 1775, Matthew and Mary Darly published a pair of satirical prints entitled *The Breeches in the Fiera Maschereta* and *The Petticoat at the Fieri Maschareta* (Figures 12.1 and 12.2). As their titles suggest, one portrays a figure in a giant pair of breeches and the other in a giant petticoat. In *The Breeches*, a face pokes out in profile from the unbuttoned front flap and tiny heels protrude from the legs. Nine ostrich feathers project upwards from the waistband, encircled by a crown atop the figure's head. *The Petticoat* mirrors this composition with a face visible through an opening in the fabric of the head-to-foot petticoat, tied in place under the chin. A smaller ducal coronet tops the figure's head while larger shoes stand beneath the skirt and a gloved hand protrudes from the pocket slit at the side. The shoes of each figure, as well as their facial features, convey that a woman is wearing the breeches and a man is wearing the petticoat.

The date of the prints, 25 April 1775, places their publication at the beginning of the mid-1770s feather craze. Early April saw a flood of newspaper reports and satirical prints heralding the ostrich feather's prominent appearance.[1] Towering plumes crested the high coiffeurs of fashionable *têtes*, saturating print culture with feathers. Hairdressers, or *frizeurs*, constructed these hairstyles with cushions of horsehair, natural and false hair, and wire, which were held in place with pomatum (a scented greasy paste made of animal fats and flour), adorned with feathers and other addendums, and dusted with white, grey or coloured hair powder.[2] The 1770s coiffeur was predominantly tall and narrow in shape, some reaching as high as thirty inches.[3] Ostrich feathers added extra height, further elevating the coiffeurial silhouette. As Caitlin Blackwell's work on the appearance of ostrich feathers in satirical prints in the mid-1770s has demonstrated, on the printed page, high heads and ostrich feathers were a tandem trend that dominated women's visual representations in portraits and satirical prints.[4]

Like tiny-heeled feet, ostrich feathers atop a high coiffeur became a visual signature of the Darlys' female figures in the mid- to late 1770s.[5] In contrast, the figure in *The Breeches* diverges from their typical feathered women. In addition to her unusual attire, she lacks the archetypal high coiffeur, which characteristically featured three to five plumes. These differences can be attributed to the satire's source. While the specific inspiration of many graphic satires remains lost to the modern print scholar, the

Figure 12.1 *The Breeches in the Fiera Maschereta*, 25 April 1775, published by M. Darly, etching, 17.4 × 12.6 cm. Courtesy of The Lewis Walpole Library, Yale University.

THE· PETTICOAT·
at the Fieri maschareta

5315

Figure 12.2 *The Petticoat at the Fieri Maschareta*, 25 April 1775, published by M. Darly, etching, 17.4 × 12.6 cm. Courtesy of The Lewis Walpole Library, Yale University.

source material for this pair was published. The 'Fiera Maschereta' was a masquerade ball held at the Opera House in London on 24 April 1775. The following day, an article detailing the event was reported in four London newspapers: *The Morning Post and Daily Advertiser*, *The London Chronicle or Universal Evening Post*, *Middlesex Journal* and *Lloyd's Evening Post*.[6] In addition to detailing the decoration of the theatre and the inclusion of 'elegant little shops, in which gloves, ribbons, *feathers*, jewels and toys of all sorts were to be sold', the columnist observed 'Mat. Darly selling off his old stock, and taking Caricatures to lay in a new one for the spring trade.'[7] This observation that the satirist, Matthew Darly, not only attended the ball, but also sold prints and 'took Caricatures' suggests that this pair was drawn from life.[8] It also speaks to the speed that a print could be created – from design to publication overnight – and to the intimate involvement of London satirists with those they satirized.

The attendance of this couple dressed in these masquerade costumes is confirmed further down the page. The columnist observes:

> We cannot close this account, without mentioning two very singular masques, viz. a Lady in a very large pair of breeches that reached from her feet to the top of her head, where the waistband was fastened, and crowned with a prodigious bunch of Ostrich feathers, and a Gentleman in a petticoat, that covered his whole figure with a ducal coronet, ornamented with jewels on his head.[9]

The newspaper's detailed description of these masquerade costumes could nearly be a description of the Darlys' prints, a rare match which confirms that the satirical image was not a construct of imagination or exaggeration, but of direct observation. The report concludes, 'This petticoat and breeches, which we hear was intended as a satire on the Duke and Duchess of ——, afforded much diversion to the company throughout the whole of the evening's entertainment.'[10] While the newspaper chooses discretion here, the couple satirized was the Duke and Duchess of Gordon, whose inverted relationship of marital power was well-circulated society gossip.[11] Although the trope of a woman wearing the breeches was a recurrent visual device of an overbearing wife and a cuckolded husband, this satire of the Gordons' marriage may be one of the sharpest.

The satirical masquerade costumes, including the 'prodigious bunch of Ostrich feathers', worn on 24 April 1775 and, in turn, the public perceptions of the Duke and Duchess's marital relationship, were printed in newspapers and graphic satires in the capital and further afield. Reaching the readers of the just-then rebelling American colonies, the 24 June 1776 edition of *The Virginia Gazette* republished the masquerade costumes' description, along with other items of social news from London.[12] The geographically extensive circulation of these costumes' visual and discursive representations demonstrates the widespread public consumption that print culture facilitated, highlighting how far impressions of fashion could travel.[13] The visual and written reception in newsprint and graphic satires, as well as paintings, represents the final stage of the material article's fashionable lifecycle and the culmination of a series of tradesmen, makers, consumers and artist's material knowledge and skill.[14] Their portrayal in print culture expresses their ultimate incarnation – available for public consumption long after the original material artefact has been lost.[15]

Widely visible to eighteenth-century and twenty-first-century viewers alike, the Darly prints depict the end of the feather's fashionable lifecycle, as the satirical accessory of print culture that documents natural materials no longer extant. However, the thread I wish to follow in the feather's story unearths the stages that preceded its portrayal in print: its sourcing, making, selling and wearing. Feathers, particularly goose, were quotidian materials of eighteenth-century life, used as quills and as the stuffing of feather beds.[16] As accents of interior decoration and dress, both fashionable and military, a wide range of plumes were used, including heron, egret, peacock and swansdown, with perhaps the most prominent being ostrich. Previous scholarship on ostrich feathers has primarily addressed the industry in the nineteenth and twentieth centuries or, in the eighteenth century, has focused on the feather's social and culture significance, particularly in relation to perceptions of Indigenous feathers.[17] Similar to the transcultural tomahawks and knives discussed by Robbie Richardson in this collection, feathers were freighted with competing cultural connotations. This chapter will address the feather's material trajectory in the late eighteenth century, highlighting its journey, as well as the makers behind its production and sale. The British feather trade has remained overshadowed by the more prominent clothing trades.[18] To fill in the contours and context of feathers to British consumers, this chapter first establishes the source of ostrich feathers and the means by which they were imported to Britain. It then turns to the process, business and consumption of the feather trade, emphasizing the material practices and skills that underscored the feather's production. Though subtle in its treatment, material literacy permeated each stage of the feather's material lifecycle – from an understanding of the feather's acquisition, transportation and handling to the dexterous skill of the maker to knowledge and expertise of the buyer and wearer. Together the market for and fabrication of feather accessories demonstrate not only the plume's broad geographic footprint, but also the material processes and understanding required to transform the feather from raw source to fashionable good.

Sourcing feathers

Non-native to Britain or Europe, ostrich feathers were acquired through global trade networks and were primarily sourced from two species of ostrich: the Arabian Ostrich of modern-day Syria and the Arabian Peninsula, which became extinct in the early twentieth century, and the African Ostrich of northern and western Africa supplied British consumers.[19] Ostrich feathers, especially those from the tail and wings, were desirable because their barbules, or individual filaments, are the same size and length on each side of the vane, creating a symmetrical shape. In addition, these barbules are unattached from each other, resulting in the ostrich's inability to fly, which creates a visually tantalizing, fluid material surface.[20] In 1810, Thomas Mortimer observed:

They are more prized as ornaments than those of any other bird; not only on account of their brilliant white hue, but their length, their suppleness, the beauty of their fringe, and the facility with which they may be cleaned, or stained of any colour, without the opposition of that tenacious oil so common in other species of feathers.[21]

These distinct qualities had positioned the ostrich feather as a privileged sartorial accent since antiquity, adorning the heads of gods in Egyptian hieroglyphics, and crossing the Mediterranean to feature in Roman, Assyrian and Babylonian art and material culture.[22] By the sixteenth century, ostrich feathers featured as a fashionable material for British and European dress and accessories, first for men and later for women.[23] They were frequently depicted in Tudor and Jacobean portraiture – in fans and as individual plumes held in the hand or on hats – adorning both male and female sitters.[24] Their painted popularity continued into the eighteenth century, appearing in the work of portraitists, including Joshua Reynolds, Thomas Gainsborough and George Romney.

The desirability for ostrich feathers in Europe, and later in North America, produced a cross-continental industry to supply growing demand. Prior to the establishment of ostrich farms in South Africa in the nineteenth century, where specific birds were bred for the quality of their feathers, ostriches were hunted on horseback for their meat as well as for their plumage.[25] Feathers, along with other raw materials like senna, ivory, gold dust and gum, were transported out of the interior regions of Africa in caravans across the trans-Saharan trade routes, then shipped across the Mediterranean from North African ports, like Tunis and Tripoli.[26] In a report on 1 February 1789, Robert Traill, the British Commerce Secretary, commented on the status of the slave trade in Tunis, to Lord Sydney, the British Consul General. He observed, 'The only articles of Commerce, which are brought by these Caravans, into this Kingdom are, Senna, Gold Dust, Ostrich Feathers & at times, a few Elephant's teeth.'[27] Traill reports that feathers were transported still attached to their skins, 'the Ostrich feathers are of little importance, as the quantity does not exceed one hundred unpluck'd Skins yearly, which are sold from five to six pounds sterl[ing] each skin', a minor value compared with that of enslaved people or senna.[28] These skins would later be tanned into leather. From the Barbary coast, feathers were funnelled through Leghorn, modern-day Livorno, or Venice to other Mediterranean ports.[29]

However, as maritime trade routes in the Atlantic expanded, the dominance of Mediterranean trade waned and ostrich feathers became one of the raw materials increasingly exported from the West Coast, a 'silent' commodity of the triangle trade.[30] The central artery system of the Atlantic world, the triangle trade was flourishing by the late eighteenth century. With ships benefitting from favourable winds to sail from Britain to the West Coast of Africa to the West Indies and American Colonies and back, a chain of British manufactured goods, enslaved persons and sugar, and coffee and tobacco was created.[31] Despite the considerable body of scholarship surrounding the triangle trade and its role in the Atlantic world, the exportation of African commodities, including ostrich feathers, into the British market remains largely overlooked.[32]

British merchants who focused their business specifically on material commodities rather than human cargo traded British-manufactured goods, like fabric, earthenware and muskets, for African raw materials, including ostrich feathers, ivory, gum, hides, wax and drugs.[33] Before the so-called colonial scramble, British merchant ships were not limited to nationally defined territories, enabling a wider scope of coastal trade. For example, the logbook of the British merchant ship, the *African Queen*, which traded in elephant teeth and gum, demonstrates the range of trading ports accessible to a merchant vessel in the 1790s and that British maritime trade routes of material commodities acted more as a line, rather than a triangle, bypassing the West Indies entirely.[34] It is possible that some ostrich feathers sailed across the Atlantic and back, completing the triangular pattern; however, British customs records affirm that ostrich feathers were imported primarily from Africa and other European countries.[35] The ledgers recording yearly imports into London by last port of call, rather than country of origin, from 1770 to 1780 detail that English merchant ships imported undressed ostrich feathers by the pound from Africa, Spain, Flanders, Italy, Turkey and Venice.[36] Following 1780, feathers were imported from Africa, France, Gibraltar and the Straights, Italy, Turkey and Ireland.[37] Some quantities of dressed, or already processed feathers, were imported from France to ports outside of London.[38] That ostrich feathers passed through Britain before reaching North America also suggests the significance of London, and its makers, as a necessary step in the feather's manufacture. In their undressed state, ostrich feathers required the skilled knowledgeable hands of the feather maker to finish them.

Making feathers

While newspaper advertisements indicate that customhouse sales of undressed ostrich feathers occurred in Dover, Liverpool, Weymouth and Southampton, the primary hub of the feather trade centred in London.[39] Unlike in Paris, where only one term, *plumassier*, was used to designate the tradesperson who worked and sold feathers, in London many names were employed: feather maker, feather manufacturer, feather dealer, feather merchant, feather dresser and feather worker, as well as *plumassier*. The tenor of these names – maker, worker, manufacturer – speaks to the physical labour that structured the trade and each person's place in the production of feathers. As Frieda Sorber has argued in her work on the French feather trade, 'for *plumassiers* feathers would be mere raw materials, requiring an artisan's care to transform them into useful objects or fashionable ornaments', highlighting the role of the feather maker to improve upon nature.[40] Employing men and women, the feather trade relied on the material literacy of its makers both in the skills of their hands and in their understanding of material, fashionable practice to transform grease-covered feathers into wearable plumes.

Ostrich feathers underwent a thorough process of washing, dying, shaping and styling. Unlike milliners, tailors or staymakers, whose trade had been described, at least briefly, in the numerous trade treatises and manuals of the century, an English account of the feather trade was not published in Britain until 1804 in *The Book of*

Trades, or Library of the Useful Art.[41] However, a systematic explanation of the material process was published in 1751, followed by five illustrative plates in 1771, in Denis Diderot and Jean-Baptiste le Rond d'Alembert's *Encyclopédie ou dictionaire raisonné des sciences, des arts et des metiers*. Here, the entry for the *plumassier panachier* addresses the multi-step procedure for cleaning, dying and shaping feathers.[42] First, feathers needed to be cleaned to de-grease them of the fatty, oily remnants left from the animal's skin. Subsequently white feathers could be dyed to various colours or whitened using Spanish clay or sulphur.[43] The white and black feathers of the male ostrich were the most desirable for their size and colour, while the female's brown, beige or grey-tipped feathers were less valuable as their colour was harder to change.

Although Diderot's textual description remains brief, his plates offer greater insight into the haptic skills of the feather maker. As Geraldine Sheridan has established, the illustrative plates of the *Encyclopédie* often reveal more of the maker, their context and their material process than do the textual entries.[44] Of the five plates, the first three feature the preparation of ostrich feathers (Figures 12.3, 12.4, 12.5).[45] In keeping with the order of plates across the *Encyclopédie*, a representation of the feather maker's workshop is depicted first, followed by details of the artisan's tools, methods, raw materials and products. Within each plate, individual figures are numbered and accompanied by a brief line of descriptive text. The plates depict feathers before and after they have been cleaned. In *Plate I,* garlands of degreased ostrich feathers are drying from the ceiling. Details of these tied bunches are portrayed in *Plate III*. 'Fig. 1' depicts a group of large ostrich feathers bundled together to be degreased, beneath which an individual feather is ready to be tied to the bunch with a loop of string. In the centre of the plate, 'Fig. 3' depicts a cluster of small ostrich feathers waiting to be degreased. To the right of each of these groupings, 'Figs. 2' and '4' portray two individual degreased ostrich feathers of different sizes, their symmetrical barbules clearly delineated for the viewer.

After feathers were treated, they were cut, curled, shaped and sewn together. *Plate II* portrays the gestures of these manoeuvres, a rare depiction of the haptic skills of makers caught in action. At a workbench, the top figure curls the barbules with a knife in his right hand. He stabilizes the base of the feather in his left hand, with his forearm resting on a box, which was used to store untreated feathers.[46] Beneath, a pair of hands uses a piece of glass to thin the central vane of the feather, while holding it against a board. Thinning the vane would have lightened the feather, aiding its upright buoyancy when worn.[47] Both of these techniques required learned skill and careful precision in order not to damage the fragile barbules. When advertising positions in the trade, feather makers specifically asked for good hands, in addition to good character.[48] For example, 'To Feather-Makers. WANTED some good HANDS to work at the OSTRICH and HACKLE business. Enquire at Lewis's Wholesale Manufactory, No. 374, Strand.'[49] In addition to working with a knife, the feather maker required adept skills with a needle. The final stages of the feather's production are depicted in the vignette of the workshop in *Plate I*. The female figure seated at the far corner of the workbench, labelled as figure 'b', is depicted in the process of sewing individual feathers together to form the plumes that her

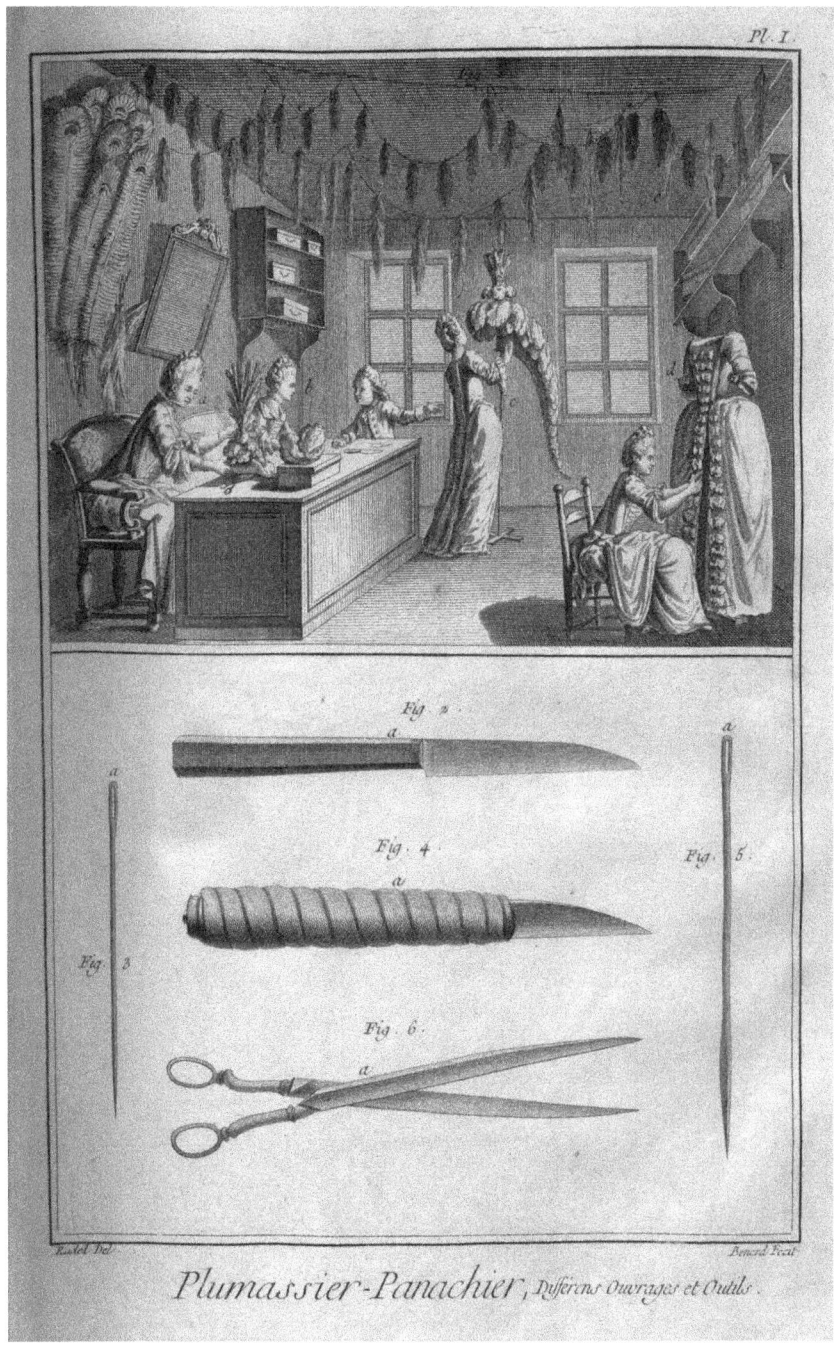

Figure 12.3 'Plumassier panachier', *Encyclopédie ou Dictionaire Raisonné des Sciences, des Arts et des Métiers*, Vol. 8, 1771, etching and engraving. Courtesy of the ARTFL Encyclopédia Project, University of Chicago.

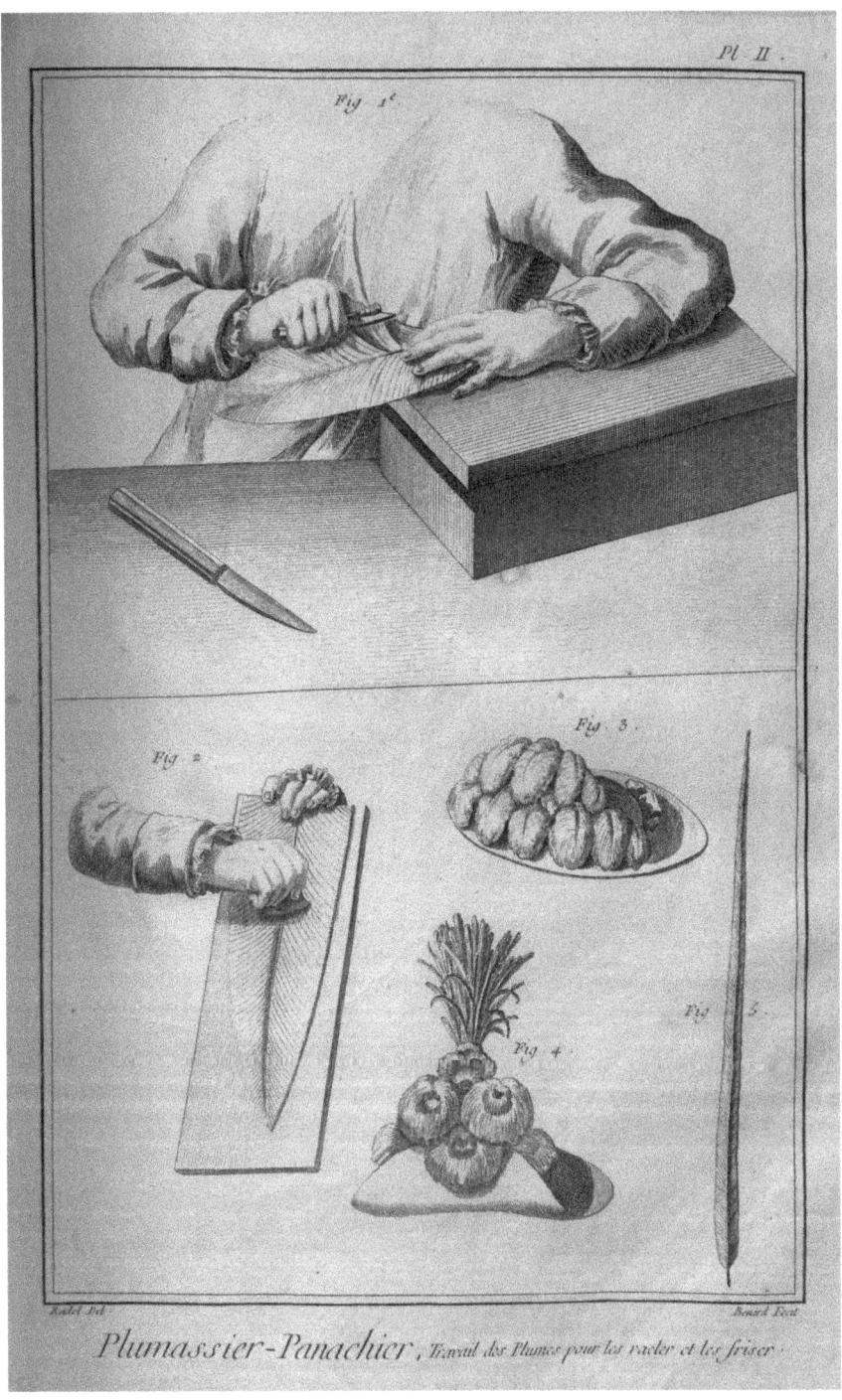

Figure 12.4 'Plumassier panachier', *Encyclopédie ou Dictionaire Raisonné des Sciences, des Arts et des Métiers*, Vol. 8, 1771, etching and engraving. Courtesy of the ARTFL Encyclopédia Project, University of Chicago.

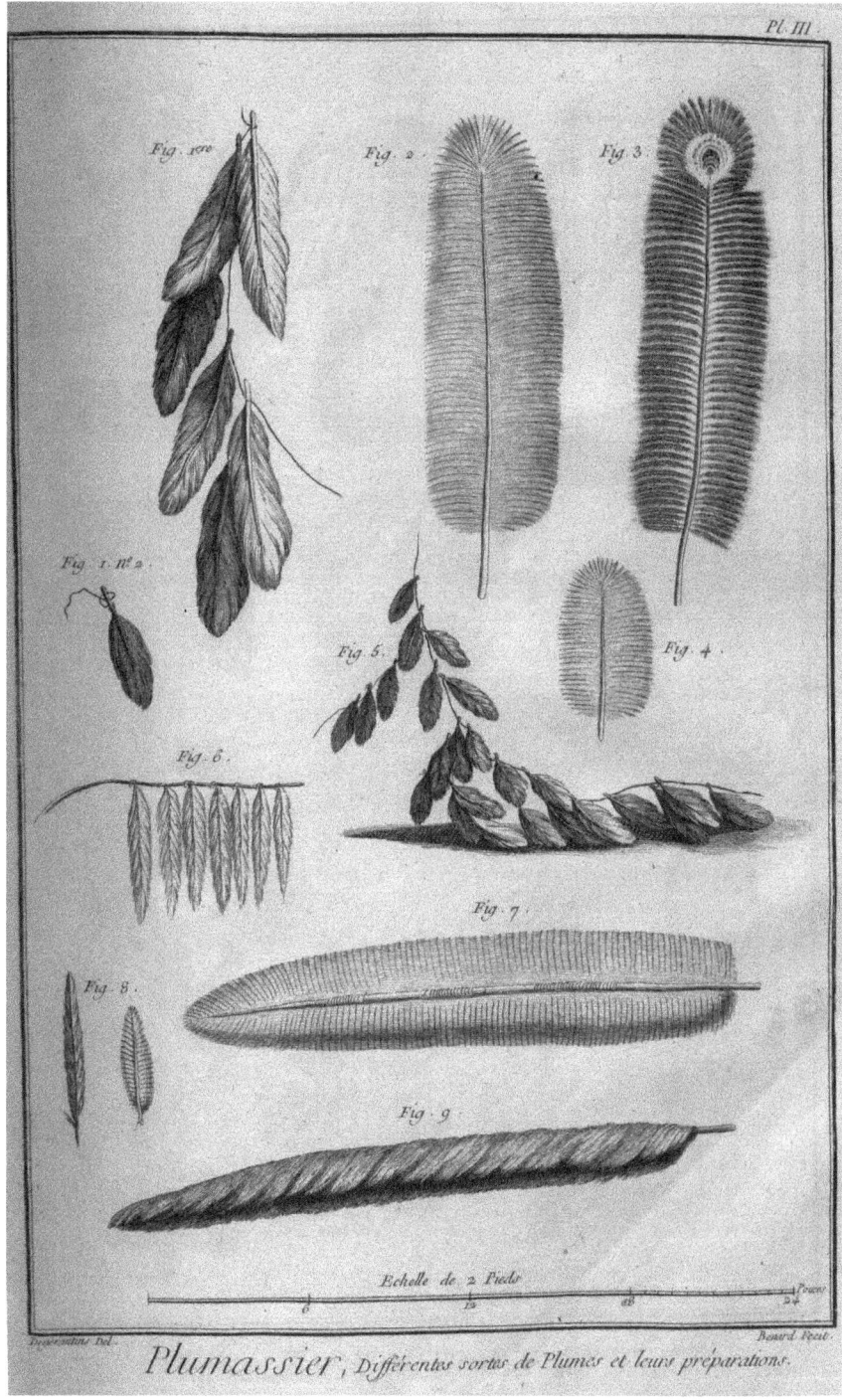

Figure 12.5 'Plumassier panachier', *Encyclopédie ou Dictionaire Raisonné des Sciences, des Arts et des Métiers*, Vol. 8, 1771, etching and engraving. Courtesy of the ARTFL Encyclopédia Project, University of Chicago.

fellow feather makers are using in their pieces. Individual feathers were matched by colour and size, and sewn along the fragile vanes to increase the density of barbules, giving greater body and fullness. This method is elaborated in *Gill's Techonological*, published in 1830:

> They pass a needle and thread between the fringe, all along one side of it, taking care that they do not entangle the fringe between the loops in thread. They secure each turn of the thread by a knot, and thus proceed, without cutting the thread, until they have reached the end of the feather.[50]

Like the delicacy required for shaping plumes, the precise fingerwork of looping and knotting the thread between each individual barbule, without tangling or damaging the fine filaments, necessitated meticulous, practised and materially proficient hands. Following their assemblage, a small piece of wire could be attached to the bottom of the quill.[51] Once finished, plumes were deployed in various fashions, as is portrayed by the three remaining feather makers in their workroom. Each applies feathers to a different mode: affixed as trimming to the front of a gown, as a horse's ceremonial headdress and as the plumes of a hat. Depictions of feathermaking demonstrate how the trade relied on multiple tools, skills and techniques and how at each stage makers had to apply their material literacy to intricate natural matter, while keeping in mind the feather's future application.

The business of feathers

The entry in Diderot's *Encyclopédie* offers access to understanding the material process, workspace and business of the feather maker. Lit by two large windows, the workspace is neatly presented with shelves for the boxes of undressed feathers lining both walls. A large workbench intended for smaller projects occupies one side of the room. The other side is left open for work on larger mounted projects, like the horse's headdress and gown. Although drying feathers hang on the wall and ceiling, this room is clearly intended as a space for making up feathered products, with the degreasing and dying occurring in a separate location. The gilded mirror and the upholstered bench upon which the two women sit, along with their finely kept appearances, all suggest that this room functioned also as a sales space.[52] Similar to the smartly presented furnishings, the women's dress, fashionable gowns and coiffeurs of the early 1770s, would have affirmed to customers the makers' needle skills and workmanship, projecting their up-to-date fashionable and material fluency.[53] Employed by British and French retailers alike, these marketing tactics had become an established practice by the end of the eighteenth century. Fashion retailers used their shop's space and inhabitants to display not only their wares, but an appealing environment to attract the British shopper.

Newspaper advertisements affirm the range of consumers who frequented feather makers' shops, despite their association as a narrow luxury fashion trade. Advertising to the top of society, firms like T. Hamilton and Mr. Carbery sought the patronage of courtly circles. In an advertisement 'For His Majesty's Birthday' in 1800, T. Hamilton, 'Artificial Florist, Fancy Trimming Manufactuer to Her Majesty and the Princess of Wales, &c. No. 66 St. James's Street, begs leave to inform the Nobility, Gentry and his Friends' that he now has for sale a variety of trimmings and 'sells every other Article for Court and Fancy dresses, and has always the greatest Choice of Ostrich and other fancy feathers'.[54]

Moreover, printed advertisements enable us to trace the growth and development of the British feather trade, as well as follow progress of individual makers, for example, the firm of Farmer and Taylor, Ostrich Feather Manufacturers. Beginning on 20 May 1779, the firm announced their removal from Dean Street in Soho to No. 90 Oxford Street, which would remain their business premises for the following decade.[55] Their early advertisements indicate that they began mainly as a military feather supplier, offering army officers 'a large assortment of soldiers feathers of all sorts and colours, and particularly scarlet hackel [*sic*] feathers that will not fade, which they are determined to sell at the very lowest prices'.[56] The presence of 'Feathers for the ladies' was only noted in the postscript.[57] The demand for military feathers alongside fashionable feathers is noted in the description and accompanying illustrative plate for a 'Feather Worker' in *The Book of Trades, or Library of the Useful Arts* (Figure 12.6).[58] The plate depicts a woman with short, cropped hair in a fashionable high-waisted gown, assembling a military hackle (a round grouping of feathers used in military dress), made of either cock or small ostrich feathers. The anonymous author cannily observes that 'within these few years, the trade for feathers, especially those worn by the military, has very greatly increased', no doubt aided by Britain's near constant state of military engagement during this period.[59]

Feather makers like Frances and Co. at No. 7 Princes St, Westminster, continued to produce feathers exclusively for the military, but other firms such as Farmer and Taylor chose to diversify their stock.[60] Over the 1780s, the firm would redirect its focus towards fashionable feathers and expand into the fur trade. Their changing stock and customer base are documented by advertisements, such as one from 1780 that 'begs leave to acquaint the Ladies, that they have a large assortment of ostrich feathers, of the most beautiful colours'.[61] Farmer and Taylor continued to advertise their wares into 1783, when they extended their offerings to include fur accessories, such as goat skin muffs, as well as 'a large quantity of the very finest White Ostrich Feathers that ever was in this country', imported from Leghorn.[62] However, following that advertisement Taylor's name stops appearing in conjunction with the business, and the firm continues solely under the name Mr C. Farmer. Throughout the latter half of the 1780s, Farmer progressively focused on fur accessories, which were increasingly in demand over the course of the decade. By the announcement of his retirement in 1790, Farmer was listing his stock as 'consisting of several thousands of Muffs, Tippets and Trimmings, and all other articles in the Fur Trade, of the very best quality and workmanship under prime cost; and likewise his great stock of white, black, coloured and fancy Ostrich Feathers, of the best qualities, considerably under prime cost'.[63] Nearly a year after Farmer first advertised his retirement, G. Sneath, 'successor to Mr. Farmer', took over Farmer's stock and shop at 90 Oxford Street. In Sneath's first advertisement, he not only marketed his wares, but also advertised for help: 'N.B. Wanted a number of Manufacturers of Feathers and Flowers; good hands will have constant employ, and

Figure 12.6 'Feather Worker' in *The Book of Trades, or Library of the Useful Arts,* Vol. 2, 1807, etching, Harvard University, Cambridge.

meet with encouragement; and a Young Man, with a good address, as Shopman.'[64] That new feather makers will 'meet with encouragement' is a small, but important phrase – denoting not only Sneath's role as employer, but also as mentor and teacher to those less experienced in the trade. In feather work, like other sartorial trades, the material literacy depicted in Diderot's prints was passed on through the mentorship of apprentice and junior makers. The language of Sneath's advertisement also speaks to the robust nature of the business when he took it over – a multi-person operation that employed makers and sellers as distinct roles. The first requiring good hands and the second a good address, the separate skills of these positions worked in tandem in Sneath's business, both fundamental facets to the successful manufacture and sale of his stock.

Beyond the feather maker

While those who advertised in the press represent only a small percentage of feather makers in London and Britain, the portrayal of their businesses sheds light not only on types of stock they produced and sold, but also on their intended consumers.[65] Those unable to shop in person were able to buy feathers from a distance. Consumers without access to a nearby feather maker's shop, like Lady Amelia Knollys in Winchester, employed the typical method of proxy shopping by correspondence, in Lady Amelia's case taking advantage of her brother Lord Wallingford's residence in London.[66] She writes in the autumn of 1784, 'I will be obliged to … you to buy me three White flat Ostrich feathers, 10 S, 6 D each, as I would have them long, and Handsome.'[67] Originally, she wrote 'three white feathers', but clarified for her brother that the feathers were to be flat and made of ostrich, identifying for him the physical characteristics of the feathers she desired, as well as the type of plumage so that he would be able to inspect the feathers on her behalf upon purchase. In addition to showing her understanding of the material quality and her material understanding as a consumer, her letter also reveals the cost, not an unsubstantial amount for sartorial accents. The choice to buy feathers remotely suggests either a lack of close retailers or the implied quality of a metropolitan supplier.

However, provincial feather makers were available elsewhere. Mrs Daniel in Bristol, for example, proprietor of the Fur and Ostrich Feather Warehouse at No 14, Clare St, informs the ladies that 'she has imported a large quantity of Ostrich Feathers, which she has manufactured in the most fashionable style, and will render them as cheap as any Warehouse in the Kingdom. Hatter and Milliners supplied on the best terms wholesale. NB. Feathers cleaned to appear as new.'[68] As a link in the chain of supply and manufacture, Daniel sold feathers to individual customers and at wholesale prices to hatters and milliners. As indicated by Hamilton and Farmer, who identified the scope of their professions as wider than feathers alone, ostrich feathers were also sold by hatters, furriers, haberdashers and milliners, who incorporated dressed feathers into their products or sold them individually. In particular, feathered ready-to-buy accents were fashioned by milliners. Robert Campbell, for instance, described the milliner as 'a Neat Needle-Woman in all its Branches, and a perfect Connoisseur in Dress and Fashion', who 'puts the Ladies Heads in as many different Shapes in one Month as there are different Appearances of the Moon.'[69] This combination of dexterous skill and fashionable understanding embodies the material literacy of the maker, which is realized in Laurie and Whittle's *Beauty and Fashion* (Figure 12.7), published on 24 January 1797. The print reworked an earlier image *The Pretty Milleners* (Figure 12.8), which was published on 7 November 1781 by Laurie and Whittle's former masters, R. Sayer and J. Bennet, and identified the sitters' trade in its title.[70] Published sixteen years later, *Beauty and Fashion* (Figure 12.7) depicts significant alterations in style from *The Pretty Milleners* (Figure 12.8). The later print (12.7) shows a pair of women sewing, one attaching an ostrich feather to a turban. The two women, who were originally portrayed wearing looped-style (polonaise) gowns with tightly fitting bodices, now sport the loose-fitting cut of the turn of the century with the waistline drawn under the bust. Their hairstyles have also evolved – from the large, rounded coiffeurs with sausage curls and gauzy caps of the early 1780s, to close-fitting hair topped with a turban and top hat.

In addition to these changes in fashion, the later print also updates the materials at hand. While the woman on the right of the print still sews a piece of striped fabric, faded by wear on the copper plate, the cap and bow that her companion was sewing in 1781 have been replaced by a turban and an ostrich feather in 1797. A finished plumed cap has also been added to the composition. The inclusion of the large ostrich feather corresponds with its reappearance in fashionable dress in the mid-1790s. While the ostrich feather was never was fully out of fashion from the 1770s onwards, by the 1780s it was smaller in proportion, at least relative to the wide style of hats and caps. Beginning in 1794, nearly twenty years after the first feather craze, large plumes again appeared to project from women's fashionable heads, saturating news and satirical prints alike. One anonymous commentator quipped, 'Such is the rage for wearing feathers, that the waiting-women now, instead of the usual phrase – *undressing* my lady, call it – *plucking her*.'[71] However, the plume in *Beauty and Fashion* (Figure 12.7) appears not atop a head, but in a hand. Unlike the characteristic trope of milliners in satirical prints, which emphasizes their skills with flirting rather than their skills with the needle, the 1781 and 1797 prints are unusual in that they portray these women solely as makers, drawing attention to their material knowledge.[72] In *Beauty and Fashion*, the print is centred around the pull of a needle and thread. The young woman, holding the needle in her left hand and the vane of the feather and fabric together in her right, is caught in the gesture of pulling the thread taught with her raised arm. The white of the carefully delineated feather and that of the thread are made clearly visible against dark patterned wallpaper. Framed by the women's bodies, the viewer's attention is drawn to this interrupted action. The repetitive, haptic motion of stitching with a needle and thread is portrayed as second nature, an inherent gesture not reliant on sight, but on physical, material memory. The upright gaze of the sewer, who looks directly across the picture space, not down at her work, appears to be engaged either in conversation or turned to some other attention than what she holds in her hands. The praxis of needlework is founded in a kinetic relationship between the fingers, needle and fabric.[73] While the title of the 1797 print erases the professional association of the sitters, instead attaching the potentially more-marketable generalizations of fashion and beauty, the

Figure 12.7 *Beauty and Fashion*, 24 January 1797, published by Laurie & Whittle, mezzotint, 35.4 × 25.2 cm. Courtesy of The Lewis Walpole Library, Yale University.

women's roles as makers remain central to the print's motion and subject. Similar to the feather makers in Figures 12.3 and 12.6, these women, professional or otherwise, are portrayed in relation to their ability to sew – demonstrating their haptic capability and material literacy, to which, by the 1790s, the ostrich feather has become a central part of their practice.

THE PRETTY MILLENERS.

Figure 12.8 *The Pretty Milleners*, 7 November 1781, published by R. Sayer and J. Bennet, mezzotint, 35.3 × 25.2 cm, © The Trustees of the British Museum.

Over the course of the late eighteenth century, the ostrich feather trade reached four continents. As well as London and provincial Britain, dressed feathers were exported and sold as far as India, the American colonies and later, the newly formed United States. Feathers were advertised for sale in Calcutta and the phrase, 'Just Imported from London', was a frequent headline in colonial newspapers.[74] For example, 'Ostrich feathers of different colours' were advertised on 5 December 1774 by Fisher Edwards in the *South Carolina Gazette*.[75] Just as the news of a giant pair of breeches topped with a plumed crown had reached colonial readers, ostrich feathers reached colonial wardrobes. Colonial importation of British-made ostrich feathers temporarily ceased during the war, as demonstrated by William Miller's 1782 advert for 'an elegant assortment of ostrich feathers', imported instead from Europe, in the *Pennsylvania Gazette*.[76] From being hunted in the arid deserts of Africa and the Arabian peninsula to being advertised for sale in British and colonial newspapers, the ostrich feather underwent a complex, multi-stage process to transform it from raw material to fashionable accessory.

Similar to the print *Beauty and Fashion*, which displayed the making of a feathered headdress, rather than the wearing of that headdress (in keeping with the majority of the feather's printed representation), this chapter has focused on the processes of the feather trade, rather than the wearer of its products. It has established the less-visible narrative of the ostrich feather, discerning how and where the feather arrived in Britain and in whose hands it was cleaned, cut and sewn to make it a wearable object. The journey of feathers and their manufacture shows not only the feather's extensive geographic footprint prior to the establishment of nineteenth-century ostrich feather farming, but also the specialized kinetic knowledge required to produce these fashionable plumes. The active material literacy of the feather maker highlights the material processes, business and targeted consumers of the feather trade, which have previously remained a shrouded branch of the sartorial trades. In conjunction with a cross-continental network of hunters, agents, merchants and retailors, the feather maker manufactured the material 'nodding plumes' of the 'CAPRICIOUS, airy, feather'd race', arming them with the fashionable accent that would define and dominate women's heads into the early twentieth century.[77]

Notes

1   Caitlin Blackwell, '"The Feather'd Fair in a Fright": The Emblem of the Feather in Graphic Satire of 1776', *Journal for Eighteenth-Century Studies* 36, no. 3 (2013): 353–76, 354.

2   Susan J. Vincent, *Hair: An Illustrated History* (London: Bloomsbury Visual Arts, 2018), 39–40, 66–7.

3   Richard Corson, *Fashions in Hair: The First Five Thousand Years* (London: Peter Owen, 1965), 333; Margaret Powell and Joseph Roach, 'Big Hair', *Eighteenth-Century Studies* 30, no. 1 (2004): 79–99, 88, 92.

4   Blackwell, 'The Feather'd Fair in a Fright', 355–58.

5   See Harriet Stroomberg, *High Heads: Spotprenten Over Haarmode in de Achttiende Eeuw* (Enschede: Rijksmuseum Twenthe, 2000).

6   *The Morning Post and Daily Advertiser, The London Chronicle or Universal Evening Post, Middlesex Journal,* and *Lloyds Evening Post,* 25 April 1775.

7   Ibid.

8   On observation in satirical print design, see Cindy McCreery, *The Satirical Gaze* (Oxford: Clarendon, 2004), 22–3; on the Darlys' inspiration and caricature, see Diana Donald, *The Age of Caricature: Satirical Prints in the Reign of George III* (London: Yale University Press, 1996), 3.

9   *The Morning Post and Daily Advertiser,* 25 April 1775.

10   Ibid.

11   Heather Carroll has highlighted these prints as a public articulation of the Duchess's reputation as the domineering figure within her marriage; see 'The Making and Breaking of Wedlock: Visualising Jane, Duchess of Gordon after Marriage', in *After Marriage in the Long Eighteenth Century: Literature, Law and Society,* ed. Jenny DiPlacidi and Karl Leydecker (Cham: Palgrave Macmillan, 2017), 129–58.

12   *The Virginia Gazette,* 24 June 1776.

13   Peter McNeil and Patrik Steorn have addressed the circulation of fashion through print culture transnationally between London and continental Europe, while Zara Anishanslin and Jenifer van Horn have explored the network of material exchange between London and British America. See Peter McNeil and Patrik Steorn, 'The Medium of Print and the Rise of Fashion in the West', *Konsthistorisk tidskrift/Journal of Art History* 82, no. 3 (2013): 135–56; Zara Anishanslin, *Portrait of a Woman in Silk: Hidden Histories of the British Atlantic World* (London: Yale University Press, 2016); Jennifer van Horn, *The Power of Objects in Eighteenth-Century British America* (Chapel Hill: University of North Carolina Press, 2017).

14   The concepts of object lifecycles and object-biographies were first introduced by Igor Kopytoff; see 'The Cultural Biography of Things: Commoditization as Process', in *The Social Life of Things,* ed. Arjun Appadurai (Cambridge: Cambridge University Press, 1986), 64–84.

15   On the challenges of artefacts' material survival, see Lou Taylor, *The Study of Dress History* (Manchester: Manchester University Press, 2002), 4–12.

16   Blackwell, 'The Feather'd Fair in a Fright', 354.

17   On the feather industry, see Sarah Abrevaya Stein, *Plumes: Ostrich Feathers, Jews, and a Lost World of Global Commerce* (London: Yale University Press, 2010); Ghislaine Lydon, *On Trans-Saharan Trails: Islamic Law, Trade Networks, and Cross-Cultural Exchange in Nineteenth-Century Western Africa* (Cambridge: Cambridge University Press, 2009), 141–3; Aomar Boum and Michael Bonine, 'The Elegant Plume: Ostrich Feathers, African Commercial Networks, and European Capitalism', in *The Southern Shores of the Mediterranean and Its Networks: Knowledge, Trade, Culture and People,* ed. Patricia M. E. Lorcin (Abington: Routledge, 2016), 27–41; Blackwell and Joseph Roach explore the connections between London fashion and Native American feather adornment in the 1710s and 1770s; see Blackwell, 368–73; Joseph Roach, *Cities of the Dead: Circum-Atlantic Performance* (New York: Columbia University Press, 1996), 119–31, 161–5.

18   On the French feather trade, see Frieda Sorber, 'Feathers, Nature Improved', in *Birds of Paradise: Plumes & Feathers in Fashion* (Tielt: Lannoo, 2014), 187–99; Jutta Wimmler, *The Sun King's Atlantic: Drugs, Demons and Dyestuffs in the Atlantic World, 1640–1730* (Leiden: Brill, 2017) 69–70; Wimmler, 'Material Exchange as

Cultural Exchange: The Example of West African Products in Late 17th and Early 18th-Century France', in *Cultural Exchange and Consumption Patterns in the Age of Enlightenment, Europe and the Atlantic World*, ed. Veronika Hyden-Hanscho, Renate Pieper and Werner Stangl (Bochum, Verlag Dr. Dieter Winkler, 2013), 136–8.

19   Stein notes that a smaller proportion of feathers came from the Middle East (*Plumes*, 86).

20   Sorber, 'Feathers, Nature Improved', 60.

21   Thomas Mortimer, *A General Dictionary of Commerce, Trade, and Manufacturers: Exhibiting Their Present State in Every Part of the World* (London, 1810, n.p).

22   Fiona Benson, 'A History of Ostrich Feathers', *The Bulletin of the Fan Circle International* 103 (2017): 23–4.

23   For a discussion of feathers shifting from a masculine to a feminine accessory in the sixteenth century, see Catherine Howey, 'The Vain, Exotic, and Erotic Feather: Dress, Gender, and Power in Sixteenth- and Seventeenth-Century England', in *Religion, Gender, and Culture in the Pre-Modern World*, ed. Alexandra Cuffell and Brian Britt (Basingstoke: Palgrave Macmillan, 2007), 211–40.

24   Howey identifies the portrait of Mary Howard, Duchess of Somerset and Richmond by Hans Holbein as the first portrayal of a feather worn by a British woman in 1532 ('The Vain, Exotic, and Erotic Feather', 211). The sixteenth-century Bristowe Hat includes four or five dyed green ostrich feathers sewn together to form the plume: HRP 3503710.

25   Stein, *Plumes*, 30–3.

26   John Wright, 'Sequins, Slaves & Senna: Tripoli's International Trade in 1767', *Africa: Rivista Trimestrale di Studi e Documentazione dell'Istituto Italiano per l'Africa e l'Oriente* 63, no. 2 (2008): 249–60; Stein, *Plums*, 88; Lydon, *On Trans-Saharan Trails*, 141.

27   FO 773, National Archives. For a discussion of this report in relation to the slave trade, see Ismael M. Montana, 'The Trans-Saharan Slave Trade in the Context of Tunisian Foreign Trade in the Western Mediterranean', in *The Southern Shores of the Mediterranean and Its Networks: Knowledge, Trade, Culture and People*, ed. Patricia M. E. Lorcin (London: Routledge, 2016), 27–41. My thanks to Kelly Fleming for directing me to this source.

28   FO 773, National Archives.

29   Michael Shrubb, *Feasting, Fowling, and Feathers: A History of the Exploitation of Wild Birds* (London: T &AD Poyser, 2013), 192; Stein, *Plumes*, 91.

30   Jutta Wimmler uses the term 'silently' to describe the understudied movement of raw commodities out of Africa in comparison to the more prevalently studied slave trade (*The Sun King's Atlantic*, 11). My thanks to Kazuo Kobayashi for his insights into the African commodity trade.

31   For the development of Atlantic history, see Bernard Bailyn, *Atlantic History: Concepts and Contours* (Cambridge: Harvard University Press, 2005), 1–56.

32   In Wimmler's work on French-African trade, she establishes that Atlantic ports in addition to the Barbary Coast resulted in larger supplies of feathers, which along with other perishable goods, were shipped directly to France, see 'Material Exchange', 136; *The Sun King's Atlantic*, 5, 29, 69–70. For a discussion of British shipping patterns, see Kenneth Morgan, *Bristol and the Atlantic Trade in the Eighteenth Century* (Cambridge: Cambridge University Press, 1993), 55–88.

33   For example, the logbook for the ship *Hector* provides an itemized account of goods purchased in Britain intended to be used 'to purchase Negroes, Ivory &c' (National

Maritime Museum, AML/Y/1). Customs records also demonstrate the range of manufactured goods imported and exported to and from Africa (CUST 3; National Archives).

34   The *African Queen* made a round trip from Bristol to the West Coast lasting from 28 February to 4 December 1790. Log M/64, National Maritime Museum.

35   In 1787, a small quantity of feathers was imported from Jamaica, but this was an exception (CUST 17/10, National Archives).

36   CUST 3/70-80, National Archives.

37   CUST 5/1A; CUST 17/1-22, National Archives. Following 1772, the format of the ledgers shifts from listing by country to listing by commodity. See Marion Johnson's discussion of this source material in relation to the slave trade in *Anglo-African Trade in the Eighteenth* (Leiden: Centre for the History of European Expansion, 1990).

38   CUST 3/70-80, National Archives.

39   *World*, 13 August 1789; *World*, 26 August 1789; *World*, 23 May 1791; *World*, 27 April 1792; *Sun*, 14 February 1798.

40   Sorber, 'Feathers, Nature Improved', 189.

41   For example, Campbell, *The London Tradesman*.

42   Sorber, 'Feathers, Nature Improved', 189. D'Alembert was co-editor until 1759.'Plumassier', in *Encyclopédie ou Dictionnaire raisonné des sciences, des arts et des métiers*, Vol. 12, ed. Denis Diderot and Jean le Rond d'Alembert (Paris, 1751), 798. Translated by Marine Kiesell.

43   Sorber, 'Feathers, Nature Improved', 190.

44   Geraldine Sheridan, *Louder than Words: Ways of Seeing Women Workers in Eighteenth-Century France* (Lubbock: Texas Tech University Press, 2009), 4.

45   The final two plates portray the finished products of other feathers, such as heron and egret.

46   Sheridan identifies the hands in this plate as male, due to the placement of cuffs at the wrists – one of the many examples in the *Encyclopédie* where the disembodied hands are male, while the fully embodied figures in the vignette are female (*Louder than Words*, 226-7).

47   Thomas Gill, *Gill's Technological & Microscopic Repository; or, Discoveries and Improvements in the Useful Arts, Being a Continuation of His Recent Technical Repository* (London, 1830), VI: 251.

48   *Morning Post and Daily Advertiser*, 20 October 1796.

49   *Oracle and Daily Advertiser*, 12 November 1799.

50   The entry entitled 'On the Art of the Plumassier' in *Gill's Technological,* which was an English translation of the French entry in *Dictionnaire Technologique,* provides a much more detailed, multi-page description of the step-by-step process than Diderot's brief paragraph (248–54, 251). For the original French source, see Tome Seizième, *Dictionnaire Technologique, ou Nouveau Dictionnaire Universal des Arts et Métiers, et de L'Economie Industrielle et Commerciale* (Paris, 1829), XVI, 331–7.

51   *The Book of Trades, or Library of the Useful Arts* (London, 1804), III: 56.

52   On the development of shopping and shoppers, see Jon Stobart, Andrew Hann and Victoria Morgan, *Spaces of Consumption: Leisure and Shopping in the English Town, c. 1680-1830* (Abingdon: Routledge, 2007); Kowaleski-Wallace, *Consuming Subjects*; Berry, 'Polite Consumption: Shopping in Eighteenth-Century England', 375–94.

53   Sheridan, *Louder than Words*, 89.

54   *True Briton*, 2 June 1800.

55   *Morning Post and Daily Advertiser,* 20 May 1779.

56 *Morning Post and Daily Advertiser*, 22 June 1779.

57 Ibid.

58 *Book of Trades*, 51–56.

59 Ibid., 51. An extant military hackle survives in the collection of the Metropolitan Museum of Art, 2009.300.2683a-e.

60 *World*, 11 July 1793.

61 *Gazetteer and New Daily Advertiser*, 3 June 1780.

62 *Morning Chronicle and London Advertiser*, 21 January 1783.

63 *World*, 6 November 1790.

64 *World*, 11 November 1791.

65 The Sun Insurance Records provide extensive data of London feather makers, many of whom are not advertised in the press (MS 119363, London Metropolitan Archives).

66 See Lambert, "'Sent from Town'", 66–82.

67 1M44/75/13, Hampshire Archives and Local Studies.

68 *Felix Farley's Bristol Journal*, 10 May 1788.

69 Campbell, *The London Tradesman*, 207.

70 Former apprentices, Laurie and Whittle, took over Sayer's business upon his death in 1784.

71 *Morning Chronicle*, 31 December 1795.

72 For example, see *Obediah Tempting the Pretty Milliner*, publ. by R. Sayer, 10 January 1788, British Museum; *Old q-iz the Old Goat of Piccadilly*, publ. by R. Dighton, 25 February 1779, British Museum. On flirtatious milliners, see Smith, *Women, Work, and Clothes in the Eighteenth-Century Novel*, 162; Batchelor, *Dress, Distress and Desire*, 52.

73 On this point, see Maureen Daly Goggin, 'Introduction: Threading Women', in *Women and the Material Culture of Needlework and Textiles, 1750-1950*, ed. Maureen Daly Goggin and Beth Fowkes (Farnham: Ashgate, 2009), 1–12, 4. See also Crystal B. Lake in this volume.

74 *Calcutta Chronicle and General Advertiser*, 3 April, 31 July and 28 August 1788.

75 *South Carolina Gazette*, 5 December 1774.

76 *Pennsylvania Gazette*, 9 October 1782.

77 The phrase 'nodding plumes' was first attributed to the plumed helmets of ancient Rome and Greece; however, as this poem suggests, by the end of the eighteenth century, the helmeted warrior had been replaced by the feathered female. *Weekly Miscellany*, 28 April 1739; *London Chronicle, or Universal Evening Post*, 6–8 April 1775; republished in *Public Advertiser*, 17 March 1788.

13

Tomahawks and scalping knives: Manufacturing savagery in Britain

Robbie Richardson

In the 3 January 1759 issue of the London *Public Advertiser*, an advertisement titled 'Just arrived from AMERICA, And to be seen at the New-York Coffee-house in Sweeting's Alley', promised visitors an audience with 'A Famous Mohawk Indian Warrior'. The advertisement goes on to declare this warrior to be a hero who captured a French general at the Battle of Lake George, but his identity, if indeed he was a Mohawk, remains unknown today.[1] The genuine traits of his dress, in particular, come into prominent play as the announcement continues: he is 'dressed in the same Manner with the native Indians when they go to War, his Face and Body painted, with the Scalping-knife, Tom-ax [*sic*], and all other Implements of War that are used by the Indians in Battle. A Sight worthy the Curiosity of every True-Briton'. This combination of martial dress and objects offered newspaper readers a 'worthy' focus of their curiosity and an opportunity to exercise a kind of patriotic material literacy in viewing such a heroic Indian ally. But Samuel Johnson responded differently to this description and, after quoting it at length, declares in *The Idler* no. 40 that it 'conveys rather horror than terror' since '[a]n *Indian*, dressed as he goes to war, may bring company together; but if he carries the scalping knife and tom ax, there are many true Britons that will never be persuaded to see him but through a grate' (226).[2] Johnson's rereading of the spectacle turns the curious scene staged by the advertisement into one of 'horror', underscoring for spectators willing to brave the sight of the 'Indian Warrior' the savage practices embodied by these objects, the 'scalping knife and tom ax'. Yet by this time, and indeed for a century prior, tomahawks and scalping knives such as the ones brandished in this London coffeehouse in 1759 were exclusively of European manufacture. Similar to Britons in the eighteenth century, North American Indigenous people too had undergone their own 'consumer revolution', incorporating European commodities into their lives as luxury goods, practical instruments and ceremonial objects.[3] But in the British imagination, particularly beginning in the 1750s, the tomahawk and scalping knife became a trope easily deployed to evoke Indian cruelty – synecdoches for savage warfare – that omitted any understanding of their transcultural provenance.

This chapter will trace the material and textual history of tomahawks and scalping knives, and the practices associated with them, in the British imagination. Their varied conditions of production and the ways they were interpreted by British audiences, often incorrectly as foreign-manufactured items, articulate how material literacy could fail or fall short. These objects form an important part of the material culture of colonialism, a key site in understanding both the historical processes and the imaginative work of empire. As Laura Peers writes, 'in many ways, the empire existed materially, not simply as networks of people and politics but as the things they worked with, traded, made, gave to each other, sold, looted, brought home, commented on, and consumed: an "empire of goods" as much as of political structures.'⁴ These weapons in particular bear a rich symbolic function in transatlantic Anglo American culture; as Timothy Shannon argues, '[n]o other artifact associated with the European-Indian encounter has contributed as much to the racist stereotyping of Indians as the tomahawk', which stands 'as symbolic shorthand for the Indian warrior's primitive bloodlust and his providential extermination at the hands of a superior civilization'.⁵ Scalping, which has also vastly contributed to racist depictions of Indians, was represented with singular horror and outrage by British people, and accounts of it filled the pages of captivity narratives and newspapers in the mid-eighteenth century. This 'inhuman practice' involved the removal of an enemy's scalp as a trophy, and was in British texts a crucial distinction between British and Indigenous cultures, but as with the tomahawk, it was in fact produced by intercultural contact and spread by the very forces of modernity it supposedly transgressed.

In tracing the history of material exchange between cultures, this chapter will borrow from anthropology's insights into the dynamics of gift exchange and acculturation.⁶ More specifically, the notion of 'entanglement' is key to understanding the negotiations in meaning embodied in these weapons. Nicholas Thomas describes 'entangled objects' in the context of the relationship between Pacific cultures and Europeans, showing how forms of exchange can (re)contextualize objects in various ways. What is given by one culture is not the same thing that is received by the other; as Thomas provocatively suggests, 'objects are not what they were made to be but what they become'.⁷ Thus, the apparent certainty of the object claimed in some approaches to material culture, its fixed signification guaranteed by its materiality, is undermined by the mutability and 'promiscuity of objects'. Material literacy in the case of entangled, ethnographic objects is therefore a subjective experience that can be coloured by other discursive forces and contexts.

Tomahawks and scalping knives are particularly useful objects through which to understand the fraught position that North American Indigenous peoples occupied in the eighteenth-century British imaginary. Variously admired, despised and pitied, the manner in which Indians were represented frequently reflected ongoing anxieties and negotiations in Britain over new cultural phenomena often associated with the modern.⁸ I begin by looking at tomahawks as entangled objects, from the origins of the word to the emergence of the British-manufactured pipe tomahawk later in the eighteenth century. The rise in the British steel industry allowed for a greater output in trade objects to the colonies, leading to higher levels of consumption of British goods among Indigenous nations.⁹ Next I turn to the scalping knife, which, like the tomahawk, was a fraught commodity. In looking at the discourse around the practice of scalping in the eighteenth century, I will additionally show how savagery itself is tied to modernity in ways that go beyond its functioning as an abject other. Throughout

I will consider the sites of encounter with these objects and how they were read and reinscribed with meaning. While they were initially traded as unornamented tools, Indigenous people decorated them in a variety of ways; tomahawks were often given carved wooden handles, or were inlaid with wampum, valuable beads made from shells that sometimes served as currency in the colonies, while knives were typically given leather cases decorated in beads or quillwork and worn around the neck or waist. These cases are among the more common eighteenth-century North American objects in museum collections today, but they were not produced by Native peoples until the widespread introduction of British knives in the 1740s.[10] They appeared in early British collections, such as in the Leverian Museum, as examples of Indigenous material culture, as did the British-made knives. The concept of 'transculturation', often used for interactions between peoples, is helpful in looking at such material culture; in the case of the Chinese objects that were so popular to eighteenth-century Britain, a transcultural lens can help make sense of 'the simultaneous staging and disavowal of difference' which animates them.[11] In the case of the objects born out of European and Indigenous contact, but read as Indian, transculturation helps to illuminate the failure of material literacy to understand unfamiliar objects that are 'out-of-place' or hybrid.[12]

Tomahawks

As a word, 'tomahawk' entered English in the early seventeenth century, through relations between the Jamestown settlers in Virginia and local Powhatan people. Words similar to 'tomahawk' appear in various Algonquian languages, and in Virginian Algonquin dialects it refers to an instrument for cutting.[13] The first appearance in an English text is in John Smith's *A Map of Virginia with a Description of the Countrey, the Commodities, People, Government and Religion* (1612), in which Smith provides a short glossary of Indian words and includes 'tomahacks', translated as 'axes', and 'tockahawks', translated as 'pickaxes' (3).[14] And while this early usage corresponds to the subsequent definition of a tomahawk as a steel hatchet, its meaning was unstable throughout the seventeenth century. In Thomas Cecil's *A Relation of Maryland* (1635), he writes that the Indians 'use in warres, a short club of a cubite long, which they call a *Tomahawk*'.[15] William Wood, in *New England's Prospect* (1635), describes it as an Indian executioner's club.[16] In the first few decades of the seventeenth century, in short, the word shifts from a cutting weapon, to a hatchet, to seemingly any type of Indigenous striking weapon. This uncertainty would cloud the legibility of actual objects as they were encountered by non-Indigenous people.

The first tomahawks to appear in Britain and be named as such by British collectors similarly conflated hatchet weapons with war clubs, and *Musaeum Tradescantianum* (1656), the catalogue of the father and son Tradescants whose collection would become the Ashmolean Museum at Oxford, includes a listing for 'Tamahacks, 6 sorts'.[17] There is no further description in the book, but these items still survive at Oxford and are in fact a variety of ball-headed clubs and paddle-shaped clubs from North and South America.[18] These are presumably among 'all the rarities at Oxford' that Moll would have seen during her performance of gentility and attempt to 'look like quality for a week' in Daniel Defoe's *The Fortunes and Misfortunes of the Famous Moll Flanders* (1722).[19] Other collectors would speculate about the aesthetic designs of these weapons, reading their surfaces and suggesting that their owners had inscribed them with religious significance. Such efforts at interpretation represent an attempt at a new kind of collecting, a way of rendering ethnographic objects legible to both fellow collectors and potential visitors to the collections.[20] In Nehemiah Grew's catalogue of the Royal Society, *Musaeum Regalis Societatis* (1681), he lists a 'TAMAUHAUKE' as a carved '*Brasilian* Fighting-CLUB'.[21] In his description he elaborates by speculating that it contains 'a rude Representation of some one of their *Idols*, whose help they expect' (367). Ralph Thoresby's description of his collection in Leeds, *Ducatus Leodiensis* (1715), contains a

> *Tomahaw*, or fighting Club from *North-Carolina*: it is a yellowish Wood like *Box*, above two Foot long, tapering from a little more than an Inch broad at the Handle, to three Inches at the other End, where it terminates in a Knob or Ball eight Inches round: Upon one side is drawn an odd Figure supposed to represent one of the Idols whose assistance they implore; upon the Bowing at the End is a Lizard nine Inches long, cut out of the same Piece of Wood, artificially enough, considering its being wrought with Flints by the Native *Indians*.[22]

Thoresby further notes that a similar weapon appears in the hands of one of the Four Indian Kings in a portrait painted during their visit to London in 1710. Another 'curious *Tamahauke*' in Thoresby's collection is described as 'inlaid with seven Rows of white Studs perforated, that seem to be the best Sort of *Wampampeage*, and Brass Annulets' (473). Despite its 'curious' appearance, Thoresby asserts that this is 'a desperate Weapon, being armed with a blewish Marble or Flint, a Foot long, and sharp at both Ends' (473). While it does not match the iconic shape of later tomahawks, it is nonetheless invested with the cruelty for which they would later become shorthand. For these collectors, savage weapons constitute curious objects of knowledge, whose ferocity is contained by the dispassionate eye. Of course, as Thomas suggests, such curiosity undermines the supposed disinterest of the collector.[23]

For the Indigenous people who initially crafted these weapons, the clubs formed part of a broader system of literacy and communication frequently disregarded by colonialist culture.[24] The specific clubs in these collections no longer survive, so it is hard to know their precise origins or meanings. But many of the other surviving war clubs in museums are inscribed with similar carvings. Numerous European sources of the eighteenth century describe the Iroquoian (and perhaps also Algonquin) practice of leaving war clubs beside the bodies of their victims, to communicate to others their prowess or power.[25] These 'calling cards' were carved with several types of information, including the identity of the owner, their clan affiliations and an account of their exploits. The Moravian missionary John Heckewelder explained that,

> when they have made their stroke, they leave a war club near the body of the person murdered, and make off as quick as possible. This war club is purposely left

that the enemy may know to what nation the act is to be ascribed, and that they may not wreak their vengeance on an innocent tribe.[26]

Thus this 'weapon-mediated communication' between nations provided a system that could be read by all Woodland nations.[27] This meaning, of course, is lost, or re-contextualized, in the museum encounter, and the singularities of each club become representative weapons, a process of reinterpretation that overwrites their individual meanings.

Yet early visitors to collections such as the Ashmolean and Thoresby's museum in Leeds did not necessarily seek the information from these objects that the catalogues attempted to describe or that the Iroquois warriors communicated. While the Enlightenment museum ostensibly aimed for a kind of static classification of representative objects, preserved in a hermetically sealed environment, early museums carried with them the legacy of curiosity cabinets.[28] Principles of wonder and rarity were more important than knowledge preserved in glass cases. In addition, the birth of the public museum with the Ashmolean in 1683 meant that gentlemanly collections were now open to the scrutiny of non-elite people, who brought different kinds of epistemologies to the visitor experience. While sight was a key interpretive lens, part of the appeal for visitors was also to have 'an intimate physical encounter with rare and curious objects'.[29] Handling objects in museums, including paintings, was a common practice throughout the eighteenth century, and while it doubtlessly hindered the conservation of materials, this tactile access shows the importance of touch to the understanding and interpretation of objects.[30] This, in turn, suggests the perceived importance of material literacy to the museum experience: objects could become more legible to visitors through sensory experience. Thus, German visitor Zacharias Conrad von Uffenbach observed upon his visit to the Ashmolean,

> It is surprising that things can be preserved even as well as they are, since the people impetuously handle everything in the usual English fashion and, as I mentioned before, even the women are allowed up here for sixpence; they run here and there, grabbing at everything and taking no rebuff from the *Sub-Custos*.[31]

Early visitors discuss the weight and feel of objects, as when Celia Fiennes, visiting in 1694, notes that 'there is a Cane which looks like a solid heavy thing but if you take it in your hands its as light as a feather'.[32] Von Uffenbach himself, despite his criticism of female visitors such as Fiennes, notes the hair of a stuffed reindeer is 'almost as stif as horse-hair'.[33] There are not, unfortunately, accounts of visitors handling the Indian war clubs, but we can certainly assume they did, feeling their weight and balance, and using their material literacy to engage with these entangled objects.

In Thoresby's description of one of his tomahawks, he references a painting of one of the Four Indian Kings to provide a clearer picture of his object. This portrait (Figure 13.1), by John Verelst, is of Etow Oh Koam, a Mahican who joined the three Mohawks on their embassy to London in 1710.[34] He is indeed shown holding a ball-headed war club, raised in his right hand. He also wears a European sword, perhaps suggesting his alliance with the British. The turtle behind him marks his clan affiliation. But curiously,

Figure 13.1 Etow Oh Koam, King of the River Nation, 1755. Courtesy of the Yale Center for British Art.

there is a steel tomahawk on the lower right side of his portrait, next to his feet, and a similar tomahawk appears in three of Verelst's four portraits of the other Indian Kings. Its position likely refers to the Indigenous practice of 'burying the tomahawk/hatchet'.[35] In this painting, however, the tomahawk is clearly of European provenance, with a sharpened steel head rather than the stone head of pre-contact hatchets. The painting was made widely available as a print and it is unlikely that British people understood the transformation of a British-made commodity into an Indian object or could read its significance.

Indigenous people had, in fact, been using European hatchets for at least a century by this point, and earlier iron weapons from Vikings had circulated across Native America hundreds of years before that. Throughout the seventeenth century, there was a wide variety of axes made for trade with Indians, mostly originating in Spain and the Netherlands, but by the century's end the majority of the axes traded in North America were made in Sheffield and Birmingham.[36] These became important in both the fur trade, as payment for animal furs used in high fashion, particularly the beaver pelts used for felt hats, and in treaty negotiations, where as many as 300 axes could be given out in one treaty meeting.[37] Sir William Johnson estimated the Northern Indian Department would need 10,000 such hatchets in the year 1765 alone.[38] Domestically, the British steel industry was in its ascendency in the beginning of the eighteenth century; while Britain's population grew by 80 per cent during the course of the century, its production and consumption of iron grew at a much higher rate due to domestic consumption and export.[39] Steel would become a huge success story for Georgian Britain, and, as Maxine Berg observes, a 'material culture based in the processing and working of iron and the making of fine steel provided a powerful symbol of national authority'.[40] Chris Evans and Alun Withey go so far as to describe British steel products as 'enlightened goods', products of refined industrial technology and global systems of commerce.[41]

It is therefore peculiar that some of these 'enlightened goods', produced by such a vast and internationally integrated industry, should come to be defined as markers of distinctly non-European savagery. Indeed, as the century progressed, British people on both sides of the Atlantic almost solely associated the tomahawk's physical embodiment with the steel hatchet.[42] At the same time, the word itself became associated with the brutal practices of Indian warfare. And by the mid-eighteenth century, these same mass-produced weapons were brought back to Britain and collected in museums and private collections. Their re-emergence as foreign objects of curiosity conveys the failure of Britons to properly read their provenance. The ultimate embodiment of this complex entanglement is the pipe tomahawk, a weapon that combined the hatchet with the calumet, or 'peace pipe', which was often used by Indigenous people in ceremonies and treaty negotiations. Manufactured in Britain but catering to the consumer desires of Indians, pipe tomahawks were highly prized and expensive objects among eastern Native peoples that manifested the creative potential of the contact zone. And yet these too would become representative of savage cruelty; as Shannon notes, the Seven Years' War 'linked the pipe tomahawk to racial violence in the European imagination, associating it with a kind of warfare that included scalping, mutilation, and the murder of noncombatants'.[43] It is the same weapon that Peter Williamson holds in the frontispiece (Figure 13.2) to his captivity narrative, which is copied from the

Engraved for the Grand Magazine, June 1759.

Mr Peter Williamson *in the Dress of a* Delaware Indian, *with his* Tomohawk, Scalping knife, &c.

Figure 13.2 Frontispiece engraving depicting Mr. Peter Williamson in the Dress of a Delaware Indian, 1759. Courtesy of the John Carter Brown Library at Brown University.

depiction of the Mohawk who appeared in London in 1759; Samuel Johnson rightly notes that 'true Britons' would cower from such weapons, not recognizing their own commodities transformed into symbols of savage violence.

Scalping

As a word and a verb, 'scalping' by Indians began to appear with regularity in English texts in the late seventeenth century, primarily through the New England Puritan captivity accounts of Cotton Mather and translations of French missionary Louis Hennepin which were printed in London in the 1690s.[44] And while these Puritan accounts never quite had the same impact in Britain as in the colonies, they nevertheless introduced settler paranoia over supposed Indian depravity to many readers. It is also worth noting that they depicted colonists scalping too, as in the infamous story of captive Hannah Dustan, a Puritan woman captured by Indians in Massachusetts during King William's War in 1697 who subsequently killed and scalped a family of ten Abenakis in their sleep, including six children.[45] However, more often scalping, and the threat of scalping, performed as a powerful anxiety in representations of colonial warfare.[46] In British texts of the eighteenth century, it became a potent symbol of savagery, and was perceived as a uniquely Indian practice that was testament to their cruelty and disregard of European conventions of warfare. But the scalping knife, like the pipe tomahawk, is an object that represents and reveals the cross-cultural collaborations that underwrite a supposedly savage act. Steel blades made in British and colonial mills make plain the implication of Europeans in the scalping economy. This in turn reveals how potent symbolic acts, however distasteful to British sensibilities, could become commodified and made more brutal by the same forces of modernity ostensibly meant to spread peace and liberty, and prevent such acts of savagery.

After the well-publicized embassy of the Four Indian Kings to Queen Anne in 1710, depicted not only in the portraiture discussed above but also in poems, ballads and periodicals such as *The Spectator*, many Britons were left with a sense that their Iroquois allies were perhaps not the naked savages which the press and colonists' accounts had suggested but rather a society not so utterly different from themselves.[47] Yet Daniel Defoe complained that the Mohawks, the subjects of these same so-called kings, 'were always esteem'd as the most Desperate, and most cruel of the Natives of *North-America*; and it was a particular Barbarity singular to them, that when they took any Prisoners, either of the *English* or other Natives, they always *Scalp'd them*'.[48] Despite Defoe's horror and anger over the reception of the Indian kings into the English court in 1710, particularly by rival Tories, British newspapers were already reporting by this time that English colonists themselves were participating in the scalping economy. Indeed, in this same year, an item in the London *Evening Post* reports that Colonel Nicholson told the French Governor of Quebec that if 'he should incourage [sic] the Indians barbarously to Scalp the English Prisoners, he would use the same Severity upon the French Inhabitants in Port-Royal'; thus this 'particular Barbarity', as Defoe describes it, was already not singular to the Mohawks. And Nicholson, then governor of Maryland, was so pleased by Hannah Dustan's scalping of an Indian family that he rewarded her above and beyond the bounty she claimed from her ten Abenaki scalps.[49] Thus scalping, and the knives used in the practice, often carried competing material meanings; the legibility of scalping knives was subject to reinterpretation depending on whose hands wielded the weapon.

Newspaper reports played a key role in reshaping the cultural legibility of scalping. On 10 October 1722, the London *Daily Journal* optimistically reported that war with the Indians in New England was improving since the introduction of a bounty on their scalps by the colonial government, and thus should soon be concluded. The *Post Boy* on the following day explained that anyone could claim a reward on the scalp of a male Indian aged twelve and above, but warned 'if any produce any Scalp not being the Scalp of an Enemy or Rebel Indian, with an Intent to deceive, is liable to suffer three Months Imprisonment, and forfeit double the Sum which would otherwise have accrued to him for an Enemy or Indian Rebel'.[50] Throughout the 1720s, there are frequent stories in the press of British soldiers claiming their reward for scalps. In 1722, for example, the *Evening Post* reports a Captain Robinson arrived in Boston with 'two prizes taken from the Indians; also with two Indian Scalps and a canoe', and in 1725 another soldier claimed £1000 for ten Indian scalps.[51] British authorities would burn or bury scalps upon their receipt so they could not be claimed twice. Clearly the British must have understood their own intervention into the practice, commodifying an existing cultural activity rather than suppressing it, and, at this stage at least, they seem relatively untroubled by it.[52] But what did the British think scalping meant in its original context? Did their introduction of bounties simply magnify its purpose, or transform it entirely? As is often the case, it is difficult to ascertain since there was frequently little effort to understand non-European cultural acts outside of bare description or moral judgement. But in James Adair's *The History of the American Indians,* published in London in 1775, he writes that the custom known by 'the hateful name of scalping' was first practised when Indians were far from home and unable to cut off the heads of those they killed in war and carry them hundreds of miles without putrefaction, so removed them as 'speaking trophies of honour'.[53] He explains how they stretched them on hoops to dry and then painted them; one such scalp was in the British Museum by the 1760s, perhaps earlier, and is still in their possession. It appears in numerous city guides to London in the latter half of the century, simply listed alongside other Indigenous material as 'an Indian scalp' (significantly, not as an 'Indian's scalp').[54]

Similar to tomahawks, scalping – and collectible, physical scalps – was reinterpreted by antiquarians eager to render them legible objects to British readers. In 1782, antiquarian and former colonial governor Thomas Pownall tried in part to integrate Indians into his broader stadial history, *A Treatise on the Study of Antiquities as the Commentary to Historical Learning.* He attempts to 'restore and re-edify history from the ruins amidst which it lies'.[55] Pownall traces the universal origins of writing across a shared developmental history, seeing in North American Indians the earliest stages of communication. He notes that 'every act of communication … is pledged by some token' among Indians, and with the hatchet and kettle expressing war, bands standing for prisoners, and scalps as enemies killed.[56] Thus, a scalp for Pownall is like a hieroglyph, a precursor to the written word.

James Axtell has suggested that Indigenous scalping represented 'the transference of power and identity into the victor's hand', and this seems to be understood in Britain by the middle of the eighteenth century, certainly by James Adair. But this ethnographic analysis has avoided British acts of scalping and their reasoning for engaging in the activity. Often the reason seemed to stem from vengeance, the terrifying capacity for which Britons almost always attribute to those they colonized. According to Cadwallader Colden's *The History of the Five Indian Nations* (1747), Indians 'greatly sully ... those noble Virtues' by their thirst for revenge.[57] But in British colonial discourse, those who scalp shall be scalped; thus, General Wolfe declared before the 1759 siege of Quebec, where he famously died: 'The general strictly forbids the inhuman practice of scalping, except when the enemy are Indians, or Canadians dressed like Indians.'[58] Those who cross-dress as Indians forfeit the right to civility in warfare; the importance of dress in this instance cannot be overstated, and leaves the ambiguities of hybrid frontier fashions up for potentially fatal interpretation.[59]

This material distinction between Indians and Europeans is often described in texts as being represented by the scalping knife itself, which, alongside the tomahawk, became the most potent symbol of Indian savagery beginning in the 1750s.[60] The spread of scalping was facilitated by steel blades, which replaced older implements of stone and flint, and these blades were predominantly made in Britain. As one author wrote in 1759, 'this cruel custom' was not 'introduced among the *Americans*, til the *Europeans* provided them with instruments proper for the purpose, and promised them a reward for every scalp'.[61] Commodities and commerce, products of the modern world, produced such apparent degeneracy. Sheffield steelworkers called their knives 'tormentors' in reference to their use for scalping on colonial frontiers.[62] But it is curious that the focus on scalping, beginning in the 1760s, went from the act of scalping itself to a simple evocation of the scalping knife, a trade object. In Francis Parsons's 1762 portrait of Cherokee leader Cunne Shotte or Standing Turkey in London (Figure 13.3), the subject's fierce grip on the scalping knife is a menacing reminder of his savage heart, but, again, his knife is an object of European manufacture. His shirt, blanket and ornaments such as his silver gorget, the armour hanging from his neck, are also trade items; indeed, his gorget is marked with the initials of George III and his medallions depict the monarch and Queen Charlotte.[63] His ostensibly fierce visual vocabulary reveals more nuance than appears at first glance, yet the failure to comprehend such subtle visual nods to the crown and British trade led contemporaries to interpret the image as a 'terrifying' portrait which 'signal(led) an indisputable notion of savagery'.[64]

In addition, so-called scalping knives were readily available for purchase in Britain; they frequently appear in auctions in the pages of newspapers, as in a 4 January 1758 advertisement in the *Public Advertiser* for a sale at Garraway's Coffeehouse in Exchange Alley, or in one from the *Gazetteer and London Daily Advertiser* for a sale in Plymouth that includes forty-one scalping knives.[65] The 1771 sales catalogue for the weapons collection of Daniel Campbell includes a scalping knife among the category of weapons of the world.[66] And scalping knives also began to appear in public collections, as in the

Figure 13.3 Cunne Shote, the Indian chief, a great warrior of the Cherokee Nation, 1762, after Frances Parsons. Courtesy of the John Carter Brown Library at Brown University.

Hibernian Museum at Dublin and William Bullock's Museum at Liverpool. Antiquarian Richard Greene's Museum of Curiosities at Lichfield contained an '*Indian* Scalping knife' by 1773, well before his friend Samuel Johnson visited it with James Boswell in 1776.[67] Irish baronet Sir John Caldwell had himself painted wearing various Indigenous objects that he collected from several Native groups during his time as a soldier in North America, including a European-manufactured pipe tomahawk and scalping knife.[68] This desire to collect and display savage weaponry as a way to access authentic articulations of primitive subjectivity depended upon the effacement of the material origins and manufacturing behind such weapons. They are only legible as Indian objects, and as such material literacy fails to convey and communicate their entangled origins.

Despite the frequency of scalping appearing in writing on North America, there are relatively few descriptions of the act itself. In Tobias Smollett's *History of England* (1762), however, we find a more detailed explanation of scalping:

> The operation of scalping, which, to the shame of both nations, was encouraged both by French and English, the savages performed in this manner: The hapless victim being disabled or disarmed, the Indian, with a sharp knife, provided and worn for the purpose, makes a circular incision to the bone round the upper part of the head, and tears off the scalp with his fingers … The Indian strings the scalps he has procured, to be produced as a testimony of his prowess, and receives a premium for each from the nation under whose banners he has been enlisted.[69]

Smollett acknowledges the entangled nature of scalping, a product of shameful intercultural systems of meaning through the encouragement of competing European powers and their exchange of efficient tools for the task. His character of Scotsman Lieutenant Obadiah Lismahago in *The Expedition of Humphry Clinker* (1771) is likely the most famous victim of scalping in eighteenth-century literature, a rare survivor of a practice that typically targeted dead victims. Indeed, the scalp of a living person seemed to defeat the purpose of such a trophy. In Peter Williamson's *French and Indian Cruelty* (1759), for instance, an Irish soldier passes out drunk and 'An Indian skulking that way for prey … found him, and made free with his scalp, which he plucked and carried off'. His commanding officers tell him he has been scalped and will die, to which he cries 'are you joking me?', but he survives and is perfectly healed, although Williamson insists 'no [other] instance of the like was ever known'.[70]

Smollett revisits the outcomes of live scalping via Lismahago's first appearance in *Humphry Clinker*. Lismahago's scalped head is nearly the first thing we learn about him, when his head is exposed by a fall that causes his hat and periwig to fall off. Witnesses see 'a head-piece of various colors, patched and plastered in a woful condition'.[71] The women scream, while 'certain plebeians' laugh, thinking 'that the captain had got either a scald head, or a broken head, both equally opprobrious' (155). Lismahago later explains that his disgraced cranium

> is neither the effects of disease, nor of drunkenness, but an honest scar received in the service of my country. He then gave us to understand, that, having been wounded at Ticonderoga, in America, a party of Indians rifled him, scalped him, broke his skull with the blow of a tomahawk, and left him for dead on the field of battle; but that, afterwards found with signs of life, he had been cured by the

French hospital, though the loss of substance could not be repaired; so that the skull was left naked in several places, and these he covered in patches. (157–58)

Much, of course, has already been said on Lismahago's captivity episode and transculturation, but his having been scalped, presumably by a steel knife, and battered with a pipe tomahawk particularly emphasizes Tara Ghoshal Wallace's point that the torture of Lismahago later in the text is 'a monstrous reenactment of European incursions into American territories [in which] Indians use the invader's weapons to penetrate and destroy them'.[72] But I wonder too, what is the 'loss of substance' that could not be replaced in the French hospital? Christopher Flynn suggests it is connected with English civility, and since the French were allied with Indians they could not replace it.[73] Yet if we bear in mind Smollett's own understanding of scalping as itself a practice spread by both the French and the English, it seems this lost, irreplaceable substance is more likely the pretence to civility; that is, Lismahago's woefully patchwork cranium, described more as a thing than the seat of a person's soul, reveals the perverse commodifying logic of colonialism. Neither periwig nor hat can hide it, although this colonial logic is re-written as savage practice. His description in the text as a 'curiosity' and a 'high-flavored dish' emphasizes his transformation into commodity, and the patches on his head conceal the wounds brought about by a British commodity circulated in the colonies. Such concealment, however, remains a transitory act at best, as demonstrated by the initial fall that exposes Lismahago's patched scalp. Later Lismahago attempts to fill in the explanation of his appearance, a textual act to describe a part that has been already interpreted and judged as material evidence of savagery. His efforts to elaborate upon his patches stand as a complex reminder of how the material signs of colonialism are subject to interpretation and reinterpretation. His 'loss of substance could not be repaired' but nonetheless he wields his material literacy to cover such losses with patches and to justify his colonial experiences to his British audience.

Scalping in eighteenth-century Britain was simultaneously understood as a uniquely savage act, and as a product of colonial incursion and European trade. As such it provided a profound indictment of the instrumentalizing forces of Enlightenment, or those forces that transform people into things. In a famous image (Figure 13.4) from Carver's *Travels through the Interior Parts of America* (1789), the Indian warrior entering his wigwam with a scalp is depicted with various aspects of racial difference from the period to mark his profound otherness, typical of how scalping was evoked by this time. His hair and skin colour could just as easily be evocative of other non-European groups. And yet, the viewer is implicated, being presented with these scalps inside the Indian's home. And the Indian's posture is strangely familiar: indeed, it is in imitation of one of the most dearly loved statues of the eighteenth century from Classical European tradition, the Apollo Belvedere. It is possible that this similarity allies the Indian with ancient Europeans and marks him with the contradictions of the 'noble savage'. However, his European-made tomahawk and his knife-sheath for a steel blade appear to paradoxically signify his fearsome savagery. The failure of material literacy to recognize such entanglements helped produce harmful stereotypes and concealed the complexities of the material culture of empire. At the same time, of

An Indian Warrior

Entering his Wigwam with a Scalp.

Figure 13.4 An Indian Warrior, Entering his Wigwam with a Scalp, 1789. Courtesy of the John Carter Brown Library at Brown University.

course, this failure on the part of British writers, collectors and consumers to recognize their own materials, transforming them instead into markers of barbarism, served the interests of empire. It was ultimately a useful tool of colonialism that successfully absolved British manufacturers and allowed them to continue exploiting the market of the colonial frontier.

Notes

1   Timothy Shannon has argued that this man could have been Scottish performer Peter Williamson, who had made claims of being an Indian captive and performed in Indian dress across Britain. See Shannon, *Indian Captive, Indian King: Peter Williamson in America and Britain* (London: Harvard University Press, 2018), 145–6.

2   The term 'tom-ax' is listed in the *OED* as a variation of 'tomahawk', but I believe this initial usage, and Johnson's repetition of it, is an error.

3   James Axtell, *Natives and Newcomers: The Cultural Origins of North America* (New York: Oxford University Press, 2001), 104–20.

4   Laura Peers, 'Material Culture, Identity, and Colonial Society in the Canadian Fur Trade', in *Women and Things, 1750-1950: Gendered Material Strategies*, ed. Maureen Daly Goggin and Beth Fowkes Tobin (Farnham: Ashgate, 2009), 55–74, 55.

5   Timothy Shannon, 'Queequeg's *Tomahawk*: A Cultural Biography, 1750-1900', *Ethnohistory* 52, no. 3 (2005): 589–633, 590.

6   For a useful overview of anthropological views of objects and exchange, see Nicholas Thomas, *Entangled Objects: Exchange, Material Culture, and Colonialism in the Pacific* (London: Harvard University Press, 1991), 7–34.

7   Ibid., 4.

8   For a comprehensive discussion of modernity and colonialism, see my book *The Savage and Modern Self: North American Indians in Eighteenth-Century British Literature and Culture* (Toronto: University of Toronto Press, 2018).

9   For a broader consideration of the Indigenous consumption of European goods, see Craig N. Cipolla ed., *Foreign Objects: Rethinking Indigenous Consumption in American Archaeology* (Tucson: University of Arizona Press, 2017).

10   Christian F. Feest, 'Quilled Knife Cases from Northeastern North America', *Anthropology, History, and American Indians: Essays in Honor of William Curtis Sturtevant*, ed. I. Goddard and W. L. Merrill (Smithsonian Contributions to Anthropology 44: Washington, DC, 2002), 263–78.

11   Anna Grasskamp and Monica Juneja, 'Introduction', in *EurAsian Matters: China, Europe, and the Transcultural Object, 1600-1800*, ed. Anna Grasskamp and Monica Juneja (Heidelberg: Springer, 2018), 8.

12   Ibid.

13   William R. Gerard, 'The Term Tomahawk', *American Anthropologist* New Series 10, no. 2 (1908): 277–80, 277. The Powhatan word *tamahaac* is from the Algonquin root *temah-*, or 'to cut off by tool'.

14   William Strachey, secretary of the Virginia Company, also referred to the 'Tahahauke' as a hatchet. Quoted in Daniel K. Richter, 'Tsenacommacah and the Atlantic World', in *The Atlantic World and Virginia, 1550-1624*, ed. Peter Mancall (Chapel Hill: University of North Carolina Press, 2012), 30.

15   Thomas Cecil, *A Relation of Maryland* (London, 1635), 30.

16 William Wood, *New England's Prospect* (London, 1635), 80.

17 John Tradescant, *Musaeum Tradescantianum: or, A Collection of Rarities Preserved at South-Lambeth* (London, 1656), 46.

18 For illustrations and a discussion of these objects, see Arthur MacGregor, *Tradescant's Rarities: Essays on the Foundation of the Ashmolean Museum, 1683* (Oxford: Oxford University Press, 1983), 110–20.

19 Daniel Defoe, *The Fortunes and Misfortunes of the Famous Moll Flanders* (London, 1722), 59.

20 See Isabel Yaya, 'Wonders of America: The Curiosity Cabinet as a Site of Representation and Knowledge', *Journal of the History of Collections* 20, no. 2 (2008): 173–88.

21 Nehemiah Grew, *Musaeum Regalis Societatis, or a Catalogue & Description of the Natural and Artificial Rarities Belonging to the Royal Society* (London, 1681), 367.

22 Ralph Thoresby, *Ducatus Leodiensis* (London, 1715), 472. Hereafter cited in parenthesis in the text.

23 Thomas, *Entangled Objects*, 126–9.

24 As Heidi Bohaker asks, 'Are people who did not create archives of alphabetic writing inherently less knowable than those who did?'; See Bohaker, 'Indigenous Histories and Archival Media in the Early Modern Great Lakes', in *Colonial Mediascapes: Sensory Worlds of the Early Americas*, ed. Matt Cohen and Jeffrey Glover (Lincoln: University of Nebraska Press, 2014), 99–137.

25 Nikolaus Stolle, 'Recorded Honours/Signs of Conquest: On the Meaning of War Clubs in Indigenous Eastern North America before 1850', in *Weapons, Culture and the Anthropology Museum*, ed. Tom Crowley and Andrew Mills (Cambridge: Cambridge Scholars Publishing, 2018), 69–71.

26 Ibid., 73–4.

27 See Scott Meachum, '"Markes upon Their Clubhamers": Interpreting Pictography on Eastern War Clubs', in *Three Centuries of Woodlands Indian Art*, ed. J. C. H. King and Christian F. Feest (Vienna: ZKF Publishers, 2007), 68.

28 For a discussion of the impact of Indigenous materials from the Americas on early modern collecting, see Silvia Spitta, *Misplaced Objects: Migrating Collections and Recollections in Europe and the Americas* (Austin: University of Texas Press, 2009) and Yaya, 'Wonders of America', 173–88.

29 Constance Classen, 'Museum Manners: The Sensory Life of the Early Museum', *Journal of Social History* 40, no. 4 (2007): 895–914, 897.

30 Ibid., 900–1.

31 Quoted in Martin Welch, 'The Ashmolean Described by Its Earliest Visitors', in MacGregor, *Tradescant's Rarities*, 59–69, 62.

32 Quoted in Classen, 'Museum Manners', 896.

33 Ibid.

34 For an overview and discussion of this visit, see Eric Hinderaker, 'The "Four Indian Kings" and the Imaginative Construction of the First British Empire', *The William and Mary Quarterly* 53, no. 3 (1996): 487–526.

35 As Robert Beverley explains in *The History and Present State of Virginia* (London, 1705), the Indians 'use formal Embassies for treating, and very ceremonious ways in concluding of Peace, or else some other memorable Action, such as burying a *Tomahawk*, and raising an heap of Stones thereon' (27).

36 Charles A. Heavrin, *The Axe and Man: The History of Man's Early Technology as Exemplified by His Axe* (Mendham: Astragal Press, 1988), 99.

37   On pelts and fashion, see Ann M. Carlos and Frank D. Lewis, *Commerce by a Frozen Sea: Native Americans and the European Fur Trade* (Philadelphia: University of Pennsylvania Press, 2010), 15–35.

38   Axes and hatchets are synonymous; those traded to Indians were often just the heads of these tools. Harold Peterson, *American Indian Tomahawks* (New York: Museum of the American Indian, 1971), 13.

39   Alan Birch, *Economic History of the British Iron and Steel Industry* (New York: Routledge, 1967), 15.

40   Berg, *Luxury and Pleasure in Eighteenth-Century Britain*, 155.

41   Chris Evans and Alun Withey, 'An Enlightenment in Steel?: Innovation in the Steel Trades of Eighteenth-Century Britain', *Technology and Culture* 53, no. 3 (2012): 533–60. The raw materials of British steel had to be imported in vast quantities from Sweden and Russia throughout the century.

42   Peterson, *American Indian Tomahawk*, 4.

43   Shannon, 'Queequeg's Tomahawk', 600–3.

44   James Axtell and William C. Sturtevant, 'The Unkindest Cut, or Who Invented Scalping', *The William and Mary Quarterly* 37, no. 3 (1980): 451–72, 462. King Phillips War in 1675 saw an increase in the practice and its corresponding description spread into the lexicon. Its first pictorial depiction is in a 1591 engraving by Theodor de Bry.

45   Dustan has many monuments scattered throughout the eastern seaboard of America. See Barbara Cutter, 'The Female Indian Killer Memorialized: Hannah Duston and the Nineteenth-Century Feminization of American Violence', *Journal of Women's History* 20, no. 2 (2008): 10–33.

46   For more on the complex history of scalping, see Axtell and Sturtevant, 451–72. It is likely that scalping in some form existed in North America prior to colonization, but certain that Europeans encouraged its spread through the introduction of bounties while practising it themselves.

47   See Hinderaker, 'The "Four Indian Kings"'.

48   Quoted in Alden Vaughan, *Transatlantic Encounters: American Indians in Britain, 1500–1776* (Cambridge: Cambridge University Press, 2006), 129.

49   This account is in Cotton Mather, *Magnalia Christi Americana* (London, 1702), 91.

50   *Post Boy*, 11–13 October 1722.

51   *Evening Post*, 11–13 December 1722. The account of ten scalps is in *Daily Courant*, 8 May 1725.

52   The same pattern can be seen in the case of wampum, which was transformed into currency through European exchange. See George Snyderman, 'The Functions of Wampum', *The Proceedings of the American Philosophical Society* 98, no. 6 (1954): 469–94.

53   James Adair, *The History of the American Indians* (London, 1775), 147.

54   See, for example, *A Companion to Every Place of Curiosity and Entertainment in and about London and Westminster* (London, 1767), 103.

55   Thomas Pownall, *A Treatise on the Study of Antiquities as the Commentary to Historical Learning* (London, 1782), 52.

56   Ibid., 180, 186–8. The phrases 'take up the hatchet' and 'bury the hatchet' reference Indigenous practices around declaring war and peace, as does 'boil the war kettle'. 'Bands' refers to the straps with which prisoners were tied. These objects and expressions frequently appear in European texts from the period, although their meanings were typically much more complex in Indigenous societies.

57   Cadwallader Colden, *The History of the Five Indian Nations* (London, 1747), vi.

58   Quoted in Simon Harrison, *Dark Trophies: Hunting and the Enemy Body in Modern War* (New York: Berghahn Books, 2012), 42.

59   On the hybrid 'Indian fashion', see Timothy J. Shannon, 'Dressing for Success on the Mohawk Frontier: Hendrick, William Johnson, and the Indian Fashion', *The William and Mary Quarterly* 53, no. 1 (1996): 13–42.

60   See Richardson, *The Savage and Modern Self*, 171.

61   *The Modern Part of an Universal History*, Vol. XLIV (London, 1759), 9.

62   Dionysius Lardner, *The Cabinet Cyclopedia*, Vol. II (London: Longman et al., 1833), 15.

63   These were gifts from George III; see Susan C. Power, *Art of the Cherokee: Prehistory to the Present* (Athens: University of Georgia Press, 2007), 73–4.

64   Kate Fullager, *The Savage Visit* (Los Angeles: University of California Press, 2012), 103.

65   *Gazeteer and London Daily Advertiser*, 23 January 1759. Both of these sales are for the plunder of captured French ships.

66   *A Catalogue of the Curious and Entire Collection of Arms of Various Nations, of the Late Daniel Campbell of Deargachy, Esq* (London, 1771), 4.

67   Richard Greene, *A Descriptive Catalogue of the Rarities, in Mr. Greene's Museum at Lichfield* (Lichfield, 1773), 15.

68   See Beth Fowkes Tobin, 'Wampum Belts and Tomahawks on an Irish Estate: Constructing an Imperial Identity in the Late Eighteenth Century', *Biography* 33, no. 4 (2010): 680–713.

69   Tobias Smollett, *History of England* (London, 1762), 184.

70   Peter Williamson, *French and Indian Cruelty* (London, 1759), 75–6.

71   Tobias Smollett, *The Expedition of Humphry Clinker*, 2nd ed. (London, 1771), 155. Hereafter cited in the text.

72   Tara Ghoshal Wallace, *Imperial Characters: Home and Periphery in Eighteenth-Century Literature* (Lewisburg: Bucknell University Press, 2010), 100.

73   Christopher Flynn, *Americans in British Literature, 1770–1832: A Breed Apart* (Burlington: Ashgate Publishing, 2008).

14

The lady vanishes: Madame Tussaud's self-portrait and material legacies

Laura Engel

In 1802, Marie Tussaud travelled from France with her small child and crates of wax sculptures to London where she established a wax works show that became enormously popular and still survives today as a leading tourist attraction and multi-million pound enterprise.[1] From the outset, the concept of the wax museum worked as a place where people could interact with 'living portraits'; Madame Tussaud dressed her figures in real clothes and hunted down historically accurate props to evoke the idea of authenticity.[2] Tussaud exercised her literacy in multiple materials to create her figures: textiles, hair, accessories, wax and the human form. Spectators, however, were invited to suspend their own material literacy, as wax replaced flesh and rigid stillness replaced corporeal movement. Like an embodied magic-lantern show, waxworks offered audiences a chance to witness recreations of famous figures from the past and present. Patrons could interact with life-size sculptures of kings and queens along with leading actors, actresses, aristocrats, politicians, criminals and even representations of Tussaud herself.

Wax is an astonishingly supple medium that is both extremely durable and fragile. Although a few of the original figures displayed at Madame Tussauds have survived (the most famous is a sculpture of sleeping beauty modelled after Madame du Barry, which appears to breathe on its own), most of the statues Tussaud carried across the sea to Britain have now been lost.[3] Similarly, in modern wax museums, new sculptures of celebrities and notable people continuously replace those that have become yesterday's news. Often the original objects are melted down and destroyed. Most of the history and material legacies of wax figures, particularly those from the late eighteenth and nineteenth centuries, are preserved now only in archival photographs. These images do not exist in a particular, physical archive, but tend to be scattered randomly on-line. Considering these photographs of wax sculptures provides compelling ways of theorizing the concept of material legacies – the echoes and traces of material objects and their makers across genres and time periods.

Kamilla Elliott's concept of 'picture identification' is useful in thinking about the relationship between Madame Tussaud, material literacy and material legacies. Elliott explains:

Picture identification is a cultural *use* of portraiture; an intersemiotic practice that most commonly matches an embodied presented face to a named, represented face to verify social identity ... Today we match embodied faces to the named photographs of passports, driving licenses, national identity cards, employee and student IDs and other membership cards, and we tag photographed faces with proper names on social networking sites to indicate identity. Between 1764 and 1835, faces were matched to painted, sculpted, wax, engraved, printed, drawn, written and spoken images of faces to produce social identification.[4]

Madame Tussaud participated in facilitating the material literacy of picture identification by sculpting celebrities and notable people in wax and putting them on display. She engendered a mode of embodied social interaction with famous faces and bodies that remains a hallmark of contemporary celebrity culture. As an artist, Madame Tussaud identified herself with particular images and then produced her own self-representation. Her wax self-portrait is an archive of her own material literacy, not only in her blending of specific sculptural techniques and sartorial details, but also in her representation of her image as a recognizable celebrated figure based on other well-known portraits of famous women.

In this chapter I examine an archival photograph (perhaps once a souvenir postcard) of a youthful self-portrait of Madame Tussaud dated to 1784 (Figure 14.1), at a point when she was still in France designing wax works shows with her uncle and artistic mentor, the wax modeller Phillipe Curtius.[5] Tussaud's version of herself in wax must have been amongst the statues that she brought with her to London at the turn of the century. The image of a wax figure of a young Madame Tussaud is phenomenally spooky, both because it takes a second to realize that the figure is made of wax (something that happens with all photographs of wax figures) and also because it represents several impossibilities at once – photography did not exist as a medium in 1784, Madame Tussaud could not have been photographed in her youth and the image could not have been produced when the figure was originally made. The photograph was taken later on (in the late nineteenth century or early twentieth century), possibly when the figure was on display. It appears to be seated on a velvet chair, placed against a wall of rich brocaded fabric. The iconography of the image resembles formal aristocratic portraits as well as the compositions of early photographs and *cartes de visite* postcards. The portrait contains a clash of modalities (it is a photograph which captures a present real moment) of something not real, made from wax, but that looks real because of the medium of photography. The portrait may also be a glimpse at an earlier display space and exhibition practices; it is difficult to tell if she is standing or sitting, but she seems posed as if she had just emerged from a painting. The image is an archive of an archive.

The photograph documents precise and specific sartorial details: Tussaud's dress of multi-layered fabrics with a variety of textures (lace, muslin, quilting), along with her choker necklace, matching bracelet and fashionable hair with hanging ringlets. Her dress elements and accessories mark a particular moment in time. She also wears a flower in the centre of her bosom, a symbol of fertility and sexuality, and a contrast of the live with the dead. Although her figure appears lifelike, Madame Tussaud's expression is vacant and empty. She stares directly at the viewer without any particular

MME. TUSSAUD,
AT THE AGE OF 24 (1784).

COPYRIGHT.

Figure 14.1 Marie Tussaud, French artist and Wax Sculptor, English Photographer, twentieth century, Private Collection © Look and Learn, Elgar Collection, Bridgeman Images.

emotion, almost as if she were caught in a specific moment unaware that someone was creating her image. She appears to be holding a tool for sculpting in her hand: the only clue in the portrait that this is an image of the artist herself.

Tussaud's early self-image exists now only in photographs, but its visual presence in the contemporary realm invites us to recreate the possibilities of past performances, as well as the echoes and resonances of material traces across media and time periods. In doing so we may begin to chart a narrative about Madame Tussaud as an artist, a story that runs counter to the established tale of her as a savvy business woman and the face of an international brand. Taking as a starting point this volume's theme of material literacies (which I extend to include the idea of material legacies) I want to consider Madame Tussaud's role as an original creator of material objects. Focussing in particular on the mystery of her early self-portrait in wax captured in an archival photograph from the early twentieth century, this chapter explores how and why Madame Tussaud's youthful self-representation has been ignored and elided in academic and popular studies of her life and career. Why doesn't anyone write about this sculpture, which is now easily accessible online and licensed by several major photo banks? This striking image of Tussaud is also the cover photo for the paperback edition of Pamela Pilbeam's comprehensive study *Madame Tussaud and The History of Waxworks*, but the image is not discussed in the book itself, nor is it mentioned in Kate Berridge's recent biography of Tussaud, a book that emphasizes Tussaud's significance in understanding the history of celebrity and the connections between famous people and image making.[6] In contrast, most scholars and biographers who write about Tussaud mention the self-portrait she made in 1842, an image of her as a bespectacled old woman, which has come to represent Tussaud's longevity and her material presence in the wax gallery.[7] Copies of this figure and other representations of Tussaud in action (sculpting and making death masks in the chamber of horrors) are still present in contemporary exhibitions at Madame Tussauds around the world. The younger version of Tussaud has vanished.

The history of Tussaud's early wax sculpture of herself can only be pieced together by considering the possibility of its relationship to other materials, and to legacies of objects and their makers. Thus, analysing Tussaud's self-portrait foregrounds several important aspects of the concept of material legacies. Tussaud's sculpture emphasizes the idea that material legacies are multi-modal and multi-generic. Echoes of Tussaud's figure can be found in sculpture, portraiture, photographs, tapestries and costumes. The existence of Tussaud's wax figure in an archival photograph underscores the connection between material legacies and the disruption of linear time. Tussaud's fashions may very well mark her temporally, but the competing material and visual forms ultimately undermine the dating of her dress and accessories. Various bodies (the artist, the subject and the spectator) interact in various encounters at different moments in this photograph of a wax object dressed in late eighteenth-century clothes. Taking into account how material legacies may disrupt linear concepts of time invites us to imagine who is looking at what across time periods.

Written documents, such as Tussaud's exhibition catalogues and her memoirs, provide additional ways of thinking about the significance of her early self-portrait. Tussaud's London exhibition catalogue from 1823 includes a description of her self-portrait followed by an entry for sculptures of her two children. In making portraits of her own children and including them in her exhibition, Tussaud followed in the footsteps of her famous contemporary Elisabeth Vigée Le Brun, who painted several portraits of herself with her beloved daughter Julie. Material legacies for women are often specifically related to maternal legacies (Chloe Wigston Smith notes a similar material-maternal legacy in the case of Hannah Roberston and her daughters). When eighteenth-century female artists portray themselves with their children, they legitimize and feminize the act of creation. In addition, Tussaud's visual imitations of Vigée Le Brun through the sartorial details of her self-image highlight the connection between material legacies and self-fashioning. Tussaud's dress, accessories and expression mirror images of Vigée Le Brun and her well-known portraits of Marie Antoinette, creating iconographic links between seemingly disparate figures, materials and genres. Female artists and women in the public sphere – particularly in the high arts (sculptors and painters) – have often been labelled Sapphic.[8] There is something particularly threatening about women creating artistic objects on their own. Female sculptors reimagine the patriarchal Pygmalion myth: the vision of a male artist crafting an ideal female figure that then comes to life. A woman creating her own three-dimensional image or crafting the bodies of other women disrupts this original fantasy of male genius.[9] Female material legacies are thus also in many ways, Sapphic legacies. Finally, the ephemerality of Tussaud's wax self-creation underscores the relationship between material legacies and archives. The photograph of the sculpture provides evidence of something that once existed in the archive and has now disappeared. The photo itself represents another kind of archive, extending the legacy of evidence of artistic performances and material practices forward into new realms.

Material legacies and the disruption of time by wax and women artists

The specific properties of silhouettes and waxworks as media are analogous to how women artists have appeared and disappeared in the archives, as well as the ways in which women's embodied experiences are mediated through material representations that both preserve and obscure their presence.[10] As Roberta Panzanelli explains, 'The history of wax is a history of disappearance – transformed, softened, liquefied, and sometimes lost forever. The vast majority of wax objects have disappeared from view, either because they have perished or fallen out of fashion, or because they have undergone semiotic changes that make them no longer meaningful.'[11] Wax sculptures, as material objects, are often disassociated from their creators. While the artists in museums are still privileged as individual authors and creators, in modern wax museums teams of experts assemble wax figures using both updated digital image technology and older methods of portrait painting, where the subject comes and 'sits' for his or her portrait. Celebrities are not 'paid' to do sittings for the wax museum, but do so willingly because they see it as a certain marker of their status as a celebrity icon, a material testament to the longevity of their image.[12] In the wax-sculpting process, the artist's role as an individual creator thus disappears and is overshadowed by the celebrity figure that seems to emerge from the ether as an almost living being. No one knows who the sculptors are at Madame Tussauds, a fact that shifts the emphasis of the objects themselves from the process to the product.

Waxwork has long been associated with female artists and artisans. As Marjan Sterckx reminds us, 'Women's close association with modelling, especially wax modelling, dates back at least to the Middle Ages, when nuns made candles, wax flowers and small statues of saints and the Virgin Mary for convents and private chapels. This tradition continued through the seventeenth century.'[13] The connections among women artists, craftworks and wax have led to a near dismissal of the medium as a serious art form. Art historian Luke Syson remarks, 'The standing of waxwork makers was not enhanced by the fact that from the late seventeenth century onward, many were women. Beyond Madame Tussaud, now remembered as more a show woman or brand than significant artist, many other women operated successfully in this arena and a handful achieved a certain celebrity in their lifetimes.'[14]

Taking Madame Tussaud seriously as an artist invites viewers to negotiate the ways in which wax, like portraiture, mimics faces and figures of the past. Wax looks just like skin, so it is an ideal material for representations of the human body. Panzanelli suggests that spectators are immediately drawn to wax figures because the medium's 'ambiguities create a sense of instability as viewers try to reconcile its many binary oppositions: warm and cold, supple and solid, life and death, ephemeral and permanent, amorphous and polymorphous.'[15] While some wax models are made from life, others are sculpted from moulds, and all are carefully enhanced and manipulated after the initial impression into something that resembles a realistic image.[16] This artistic process is largely undocumented by practitioners, which has reinforced the idea that wax figures are authentic copies of original beings. In this way, 'Wax retains "the memory" of the impressed form time and again.'[17] The idea of wax remembering – of material literally holding onto traces of embodiment – anticipates performance theorist Rebecca Schneider's important question: 'How can we account not only for the way differing media cite and incite each other but for the ways that the meaning of one form *takes place* in the response of another?'[18] Waxworks seem to enact what Schneider describes as 'temporal inter-(in)animation by which times touch, conversations take place inter-temporally, and the live lags or drags or *stills*'.[19] When the spectator confronts a wax sculpture, she is immediately drawn into the uncanny presence of an object that appears to be life-like, but is in reality just a collection of materials. This is as true today as it was in the earliest wax exhibitions.[20] Oddly, despite tremendous advances in contemporary image-making technology, spectators' reactions to wax museums have not changed. The virtual age cannot account for the desire to touch and be in the presence of famous people and to imagine that these figures might suddenly come alive.

Material and maternal legacies

In crafting her self-portrait Madame Tussaud had to navigate the gendered expectations and parameters for female professionals operating in the public sphere in the early nineteenth century. She borrowed specific techniques from other female artists, most notably her contemporary, the well-known portrait painter Vigée Le Brun to present herself as a sympathetic figure. Vigée Le Brun (the Queen's designated court painter) was known for her dazzling self-portraits with her daughter and for attempting to rescue Marie Antoinette's damaged reputation by representing her alongside her children. While Vigée Le Brun became one of the first women to be accepted into the *Académie royale*, on the suggestion of the Queen, she has only recently been given her due as an extraordinary artist, with a comprehensive exhibition at the Metropolitan Museum of Art in 2017.[21]

Vigée Le Brun and Tussaud shared similar backgrounds. They came from humble origins, they spent periods of time in apprenticeship and at court, they escaped the revolution and travelled widely promoting their careers in London and elsewhere in Europe, and they were by all accounts devoted mothers. As Mary D. Sheriff has shown, Vigée Le Brun's status as a legitimate artist was continuously under attack because of her gender.[22] Tussaud managed to escape this kind of public censure largely because she worked in a craft-related genre and, as several chapters in this collection note, craft was often viewed as an appropriately domestic activity for women and girls. Somehow the anxiety over Vigée Le Brun's access to male and female figures (particularly nude bodies) did not translate to Tussaud's process of wax sculpting, which similarly involved intimate proximity to human flesh. Tussaud did not achieve the same artistic legitimacy as Vigée Le Brun and French portraitist Adelaide Labille-Guiard.[23] She nevertheless engaged in similar self-fashioning strategies to promote her work and her public image. She made self-portraits in wax, wrote her own catalogue descriptions and scripted an exaggerated dramatic version of her life in her memoirs, published a few years after Vigée Le Brun's *Souvenirs* (1835).[24]

The best evidence we have of the existence of Tussaud's early wax self-portrait is a description from her London exhibition catalogue from 1823.[25] Each item in the catalogue has a label along with a brief sentence about the modelling process for the image. Item number 51 is listed as 'The Artist: Taken by herself'. This descriptor 'taken by herself' is significant. In the 1823 catalogue many of the figures are 'taken from life' by the artist (George IV in 1821, The Duke of York in 1812, Queen Charlotte in 1809) or in the case of the 'Duke of Wellington' taken from 'a bust executed by a celebrated artist in 1812'.[26] The booklet thus provides an archival record of when Tussaud made these figures as well as a glimpse of the scene of the object's creation. For Tussaud's self-portrait the caption reads:

> Madame Tussaud is a native of Bern in Switzerland. At the age of six years, she was sent to Paris to be placed under the care of her Uncle M. Courcis, Artist to Louis XVIth, by whom she was instructed in the fine arts, of which he was an eminent professor. Madame Tussaud had the honor of instructing Madame Elizabeth to draw and model, and was employed by that amiable Princess until October, 1789. In 1802 she left France, and since that period has exhibited her Collection of Model Figures in the principal cities and towns of Great Britain and Ireland, and is proud to say, she has received the most flattering testimonials of public approbation.[27]

In her entry Tussaud makes it clear that she has been associated with the French court, entrusted to be an instructor to the King's beloved sister Elizabeth, and that after 1789 she made a name for herself exhibiting 'her collection of model figures' in England and Ireland to great acclaim.[28]

The following entries in Tussaud's catalogue (figures 52 and 53) are 'The Son and Daughter of the Artist'. For this caption, Tussaud carefully stages the inclusion of her relations for visitors to the exhibition:

> Madame Tussaud hopes it will not be considered that the introduction of her own family portraits is improper. It may be naturally supposed that she would model those persons with whom she is connected by the ties of nature and affection, and as she flatters herself the likenesses are good (of which every visitor of actual observation is able to judge) she trusts that the placing of them in the exhibition will be pardoned. They are only brought forward as specimens, to shew by living comparisons, her skill in the art of modeling.[29]

Here Tussaud weaves together a verbal portrait of herself as a 'natural creator', modelling the lovely creatures she gave life to, while at the same time mounting a defence of herself as an excellent artist and skilled craftswoman. She uses language associated with art criticism and theory (likeness, observation, judgement, exhibition) as well as the scientific term 'specimen'. Her children, she explains, are only on display to showcase her remarkable talents. At the same time, the wax sculptures of her son and daughter work to position them as her most significant creations. However, in an eerie twist, although Madame Tussaud had two sons who eventually took over her thriving wax empire, Madame Tussaud's daughter did not survive infancy. The wax figure of the girl is an imagined replica of what her female child might have looked like. Tussaud's portrait of her daughter is a haunting memorial gesture. Although the figures of Tussaud's children now appear to be lost, archival photographs do document Tussaud's wax figure of Marie Antoinette, flanked by sculptures of her two children. The arrangement of the group evokes Vigée Le Brun's portrait of Marie Antoinette with her children (1787). Madame Tussaud's representation of her deceased daughter may have been partially a self-fashioning strategy designed to align her public persona with images of other famous women and their offspring. Vigée Le Brun also made a point of representing her own daughter Julie in portraits throughout her career. In doing so she provided a simultaneous visual record of Julie's development into a young woman as well as her own maturation as an artist.[30] These echoes of women artists, their children and objects past and present evoke how material legacies are intertwined with maternal legacies, whether on canvas or in wax.

Madame Tussaud as an artist

Madame Tussaud refers to herself as an artist at several points in her memoir *Memoirs and Reminiscences of France* (1838). In her introduction she declares,

> At that period modeling in wax was much in vogue, in which representations of flowers, fruit, and other subjects were most beautifully executed; and to such

a perfection had Madame Tussaud arrived at giving character and accuracy to her portraits, that, whilst still very young, to her was confided the task of taking casts from the heads of Voltaire, Rousseau, Franklin, Mirabeau, and the principal characters of that period, who most patiently submitted themselves to the hands of the fair artist.[31]

Later she remarks, 'So much did the taste for resemblances in wax prevail during the reign of Louis the Sixteenth, that he, the Queen, all the members of the royal family, and most of the eminent characters of the day, submitted to Madame Tussaud, whilst she took models of them' (58). While calling herself 'the fair artist' and praising her own loveliness and accomplishments, Tussaud showcases her own power, particularly in the process of creating her work. Making a wax figure from a live sitter involves a particularly embodied process where the subject must trust the artist while she places clay over their face (straws in the nostrils were used to help with breathing) to make a preliminary cast. Later hot wax would be poured over that mould to capture the original shape of the person's features.[32] The artist then added details, colour and hair to the figure in order to complete the piece. Tussaud's physical proximity to famous people and her remarkable reproductions of them provide evidence of her varied and significant artistic performances.

Much of Tussaud's *Memoirs* are descriptions of her time at court and her relationships with royalty and aristocrats.[33] In her initial depiction of Marie Antoinette, for instance, she describes the Queen through the words of Vigée Le Brun: 'Her complexion was so extremely fair that Madame Le Brun, the celebrated portrait painter of that period, observed, when taking the picture of the Queen, that it was impossible for the art of colouring to render justice to the exquisite delicacy and transparency of her skin' (38). It seems significant that Tussaud presents her thoughts about the Queen through Vigée Le Brun's ideas about the difficulty of painting the Queen's portrait. Vigée Le Brun worried that she would not be able to capture Marie Antoinette's beauty, particularly the transparency of her skin. Ironically showcasing this kind of 'transparency' is something that only a medium like wax could accomplish. In this moment, Tussaud acknowledges her awareness of Vigée Le Brun and her status as an artist, but she also hones in on an instance of instability in Vigée Le Brun's creative process, a process that in many ways mirrors her own. Challenging traditional generic hierarchies that privilege oil painting above wax works, Tussaud also suggests here that the flexibility of wax has the potential to create a more life-like image than two-dimensional portraiture.

In *The Exceptional Woman: Elizabeth Vigée-Lebrun and the Cultural Politics of Art*, Sheriff writes extensively about the gender politics surrounding Vigée Le Brun's career and works, noting particularly how she was considered an anomaly as a female artist who attempted to be both a portrait and a history painter.[34] According to Sheriff, 'No real being could actually fill the category of woman history painter.'[35] Portraiture, on the other hand, was considered a more appropriate genre for a female artist; as Sheriff explains: 'In theory, portraiture did not tax the mental powers of imagination and judgment, central components of reason; it required only manual skill and the ability to copy or mimic nature … Mimicry was allowed only within certain limits; it was not "natural" for women to mimic men in most circumstances.'[36]

Sheriff understands Vigée Le Brun's self-portraits as not merely displays of likeness or mimicry but as interwoven, layered series of codes and conventions that perform and produce the idea of a portrait. As she argues,

> If, as Butler argues, identity is created by performances accumulated over time, we must imagine that in representing her identity Vigée- Lebrun condensed her real and fantasied performances into a single image and synthesized them onto the surface of her fictive body. The layering or imbrication of signifiers suggests various temporal moments or actions condensed into one pictorial space.[37]

The same codes that were at work for Vigée Le Brun were likely in play for Tussaud. Wax sculpture is mimicry to an excessive degree, but it is also a form of portraiture. In curating and staging her exhibitions of wax figures, Tussaud was able to highlight the relationship of her own figure to other sculptures in her London show. Tussaud's decision to make a self-portrait in 1784 the year after Vigée Le Brun and Labille-Guiard displayed their own self-portraits at the *Académie royale* seems particularly strategic. Tussaud's version of herself from 1784 echoes Vigée Le Brun's 1783 self-portrait (Figure 14.2) in a number of ways, including the position of the body, the facial expression, the placement of the artist's tools (paint brushes and paints for Vigée Le Brun and sculpting tool for Tussaud), and her sartorial self-presentation as a fashionable young woman. Tussaud and Vigée Le Brun imagine themselves in costumes and accessories that mimic trends of the moment as worn by the most famous woman of the day, Queen Marie Antoinette. While Vigée Le Brun's image pays homage to Rubens portrait of his wife Susanna (1622–5), particularly in her use of vivid colour and characteristic *chapeau de paille*, Sheriff also points out that Vigée Le Brun's choice of costume, the alluring chemise, connects her to her own scandalous depiction of Marie Antoinette as well as other important aristocratic female figures: 'In associating her self-portrait with her portrayals of the Queen and her circle, Vigée-Lebrun suggests not only her attachment to Marie Antoinette, but also an intimacy among women.'[38] While we do not know if the costume in the archival photograph of Tussaud's self-portrait is exactly what was worn by the original figure, she appears to also be dressed in a version of a chemise dress, with a necklace, that mirrors a necklace from an early portrait of Marie Antoinette by François Drouais.

A lithograph of a young Tussaud from 1778 (later used as the frontispiece for her memoirs published in 1838) seems to be the only image of Tussaud, herself, that resembles her early self-portrait in wax.[39] Tussaud's hairstyle, headdress and costume (adorned with a rose in the middle of the bosom) as well as the shape of the face and expression echo her wax rendition of herself. In the lithograph Tussaud's neck is bare; however, curiously, Tussaud's wax figure wears a necklace or choker that appears to be part diamond and part velvet ribbon. A portrait of Marie Antoinette by François-Hubert Drouais painted in 1773 (Figure 14.3) depicts the young Queen wearing a very similar necklace.[40] Marie Antoinette's hairstyle, oval face and expression also echo Marie Tussaud's self-representation. Archival photographs of wax figures of Marie Antoinette from the early twentieth century show her wearing a version of this necklace

Figure 14.2 Self Portrait (oil on canvas), Elisabeth Vigée Le Brun, Elisabeth Louise (1755–1842), The Pushkin State Museum of Fine Arts, Moscow © Bridgeman Images.

but there is a velvet ribbon where the diamond bow should be, exactly like the necklace on Madame Tussaud's self-portrait.[41]

The possibility that Tussaud may have modelled her own portrait after the Queen suggests several significant questions about material legacies and self-fashioning. Sculpting one's own flesh has an embodied element that goes beyond painting one's portrait. Shaping one's image from life as a kind of hybrid of oneself and the Queen of

Figure 14.3 Portrait of Marie Antoinette (1755–1793), archduchess of Austria and Queen consort of Louis XVI (1754–1793), King of France, painting by Francois Hubert Drouais (1727–1775). Courtesy of De Agostini Picture Library, J. L. Charmet, Bridgeman Images.

France is at the same time a wish fulfilment activity (perhaps for beauty, rank and fame pre-revolution), a self-promotion strategy and an imagined merging of two personas into one figure. As I have suggested, Tussaud's self-portrait can also be compared to

Vigée Le Brun's image of herself, a painting that also directly references portraits of the Queen. What does it mean for Tussaud to have inserted her figure as 'the artist', in the form of a sculpture that self-consciously seems to mimic images of a well-known portrait painter and a Queen into the exhibition installation itself? And how do the performative visual versions of the 'Maries' work together to create a fantasy of the ideal female portrait?

Material legacies and the Sapphic

In her now-famous essay, 'Marie Antoinette Obsession', Terry Castle wonders: 'What is it about Marie Antoinette – and Marie Antoinette alone – that she should become so extraordinarily present?'[42] Considering Marie Antoinette as a primary example of the apparitional lesbian, Castle ponders the effect of conjuring her presence on specific individuals: 'One cannot help but feeling at the end perhaps that there is something bizarrely liberating, if not revolutionary, about the transmogrification of Marie Antoinette into lesbian heroine … in the act of conjuring up her ghost, they were also, I think, conjuring something new into being – a poetics of possibility.'[43] Castle is referring here to her analysis of several cases of Marie Antoinette ghost sightings plus the vast variety of representations of Marie Antoinette as a lesbian icon in the early half of the twentieth century. Susan S. Lanser in her ground-breaking study, *The Sexuality of History*, echoes Castle in her assertion that representations of the Sapphic across genres signal the possibility of what she calls 'imaginative agency'.[44] Although there is no distinct evidence that Tussaud engaged in intimate relationships with women, her position as a single mother who left her husband behind in France and made her way alone in London establishes her as someone who 'evaded or exploited heterosexual economies'.[45] Accusations of unnatural masculinity levelled at Vigée Le Brun associated both her persona and the subject matter of her paintings with the Sapphic.[46]

If Marie Tussaud's self-portrait is an amalgam of herself, Vigée Le Brun and Marie Antoinette – the resulting composite figure represents a kind of fantasy of identity (as most self-portraits are) – created through adopting and synthesizing various images that convey the complexities of positioning oneself as a woman artist and creator during this period. In synthesizing these varied faces, Tussaud transforms the painted portrait, which captures the sitter in space and time, into wax sculpture, a medium that appears to bring the sitter back to life. In making herself (literally and metaphorically) Tussaud engages in what Lanser characterizes as 'resistant practices that lie both within and outside the seemingly normative'.[47] For Sheriff these 'resistant practices' are epitomized by the practices and presence of eighteenth-century women artists:

> When Butler explains how the isolated willful act performed outside of normative boundaries does not actually subvert the law because that act cannot be registered or named while repeatedly performed, she inadvertently describes the position of the eighteenth century exceptional woman whose isolated freedoms did

not challenge the constraints under which other women lived, but who could potentially become a real 'threat' to the law if joined by to many other exceptional women, all acting improperly.⁴⁸

Tussaud's channelling of Vigée Le Brun and Marie Antoinette through self-representation invokes the Sapphic because of its intimate connection to visual trajectories of desire, the need to make and create art, that is coded as female. That Tussaud's identity as an artist has vaporized into the ether is connected to a revised version of her life that does not include the idea of her as a maker or creator of an artistic vision – in terms of both the art objects she made herself and the staging or installation of them in the gallery. The status of wax works as a lesser genre than painting and sculpture also contributed to the invisibility of Tussaud's artistic practice. In wax sculpture the identity of the artist is displaced even when the artist and the sitter are the same. Tussaud's legacy as an original creator is a threatening challenge to dominant discourses about artistic genius, influence, inheritance and hierarchy. The history of Tussaud's self-portrait represents a material legacy of women recreating other women through objects.

Waxworks of historical figures create an opportunity for spectators to confront the presence of bodies from the past. While very few of Tussaud's original wax sculptures exist, her waxworks remain visible through photographs and her catalogue descriptions. Tussaud's arresting mix of celebrity, sculpture and flesh has produced a significant material afterlife, in both popular culture and contemporary art, that reminds us of the unstable material boundaries between different media and between flesh and wax. In 2018 contemporary artist Hiroshi Sugimoto captured an eerie moment of presence across time in his photograph of the wax figure of the iconic actress, Norma Shearer, in her most famous role as Marie Antoinette. In this work, the persona of the actress Shearer is conjured through her portrayal of the doomed Queen in one of the most lavish film productions of the early twentieth century. The sculpture is placed in front of an ornate dressing table with a large mirror, framed theatrically by drapery, echoing the many layers of refracted, reflected and ghostly performances operating in this installation. The reproduction of a life-size figurine of Norma Shearer as Marie Antoinette in the Madame Tussaud's exhibition space is connected to what David Roman terms 'genealogies of performance': repetitions of performances across time and genre that create a complex trajectory of representations. As Spencer Bokat-Lindell observes of Sugimoto's work:

> The imposition of photographic distance has a kind of embalming effect on Sugimoto's subjects, rendered somehow more lifelike in the act of preservation. 'Photographs,' Susan Sontag once wrote, 'are a way of imprisoning reality.' But in *Portraits*, Sugimoto uses his camera to opposite effect, creating counterfeit realities that give history back to the dead: 'However fake the subject,' he writes, 'once photographed, it's as good as real.'⁴⁹

As Sugimoto suggests, photography offers tangible proof of an object's existence in the world, even as it muddies the lines between authenticity and performance. Thus, the presence of the photograph of Madame Tussaud's early self-portrait in the digital realm presents 'evidence' of its realness, documenting Tussaud's past performances beyond the afterlife of the original form. The material artwork, however, no longer exists, in keeping with the ephemeral history it represents. Looking at Madame Tussaud's self-portrait as evidence of her artistry suggests alternate ways of thinking about presence in the archives. By considering the objects themselves as containing traces of past performances, we become more attuned to who and what is looking back at us. Such attention demands a material literacy that looks across time and artistic form.

Recently Sean Silver has suggested that when 'we are interested in things coming into being, in recapturing the tacit dimension of knowledge, or the openness of persons to things, we should remain attentive to materials while they are still vulnerable, while they are malleable, porous, deformable: the properties exemplified by wax, fabric, silk, leather, and clay'.[50] Tussaud's figures are hybrid works that combine wax with textiles, wigs, ribbons, necklaces, hair and paint. Their composition and iconography echo images in diverse material forms (particularly painting and sculpture) by other artists. The particular vulnerabilities of wax, its dynamic and malleable properties, reinforce the complex material legacies of women artists, and their varied relations to traditions of high art. Madame Tussaud's early self-portrait connects us to these material legacies, threads that operate within and between media and across history, in the porous boundaries that tie the living to the dead.

Notes

1   Kate Berridge, *Madame Tussaud: A Life in Wax* (New York: William Morrow, 2006), 1.
2   Uta Kornmeier, 'Almost Alive: The Spectacle of Verisimilitude in Madame Tussaud's Waxworks', in *Ephemeral Bodies: Wax Sculpture and the Human Figure*, ed. Roberta Panzanelli (Los Angeles: Getty Research Institute, 2008), 67–82, 74.
3   'Sleeping Beauty' was featured most recently in the exhibition 'Like Life: Sculpture, Color, and the Body' at the Metropolitan Museum of Art in 2018.
4   Kamilla Elliot, *Portraiture and British Gothic Fiction: The Rise of Picture Identification* (Baltimore: Johns Hopkins University Press, 2012), 2.
5   There is no concrete evidence to support this date. However, in 1784 Madame Tussaud would have been twenty-three years old, which approximates the assumed age of the wax figure.
6   See Pamela Pilbeam, *Madame Tussaud and the History of Waxworks* (London: Hambledon and London, 2003) and Kate Berridge, *Madame Tussaud*.
7   For more on Madame Tussaud's later self-portrait, see my chapter on 'Women Artists, Silhouettes and Waxworks', in *Women, Performance and the Material of Memory* (London: Palgrave, 2019).
8   See analyses of the life and career of the sculptress Anne Damer in Clara Tuite's '"Comedy, Too Fatal Emblem": Anne Damer and Occult Theatricality', *Eighteenth-Century Fiction* 27, no. 3–4 (2015): 557–96; Andrew Efenbein 'Lesbian Aestheticism on the Eighteenth-Century Stage', *Eighteenth-Century Life* 25, no. 1 (Winter 2001): 1–16.

9 For more on women sculptresses reversing the Pygmalion myth, see Alison Yarrington, 'A Female Pygmalion: Anne Seymour Damer, Allan Cunningham and the Writing of a Woman Sculptor's Life', *Sculpture Journal* 1 (1997): 32–44, 34–5; Marjan Sterckx, 'The Invisible "Sculpteuse": Sculptures by Women in Nineteenth-Century Urban Public Space-London, Paris, Brussels', *Nineteenth-Century Art Worldwide* 7, no. 8 (Autumn 2008): http://www.19thc-artworldwide.org/autumn08/38-autumn08/autumn08article/90-the-invisible-sculpteuse-sculptures-by-women-in-the-nineteenth-century-urban-public-spacelondon-paris-brussels, accessed 22 May 2020.

10 For a fuller discussion of these appearances and disappearances, see Laura Engel, *Women, Performance, and the Material of Memory: The Archival Tourist, 1780–1915* (Basingstoke: Palgrave Macmillan, 2019).

11 Roberta Panzanelli, 'The Body in Wax, the Body of Wax', in *Ephemeral Bodies: Wax Sculpture and the Human Figure*, ed. Roberta Panzanelli (Los Angeles: Getty Research Institute, 2008), 1–12.

12 Billy Heller, 'The Whole Ball of Wax: Step Right Up to the Stars at Madame Tussaud's', *New York Post*, 24 February 2001.

13 Marjan Sterckx, 'Pride and Prejudice: Eighteenth-Century Women Sculptors and Their Material Practices', in *Women and Material Culture, 1660–1830*, ed. Jennie Batchelor and Cora Kaplan (Basingstoke: Palgrave Macmillan, 2007), 86–102, 93.

14 Luke Syson, 'Polychrome and Its Discontents', in *Like Life: Sculpture, Color, and the Body*, ed. Luke Syson, Sheena Wagstaff, Emerson Bowyer and Brinda Kumar (London: The Metropolitan Museum of Art, Yale University Press, 2018), 14–41, 28.

15 Panzanelli, 'The Body in Wax', 1.

16 Ibid., 2.

17 Ibid., 3.

18 Rebecca Schneider, *Performing Remains: Art and War in Times of Theatrical Reenactment* (London: Routledge, 2011), 168.

19 Ibid.

20 Ibid.

21 For more on Vigée Le Brun's life and career, as well as a lavishly illustrated catalogue of her works, see Joseph Baillio, Katharine Baetjer and Paul Lang, *Vigée Le Brun* (London: The Metropolitan Museum of Art, Yale University Press, 2016).

22 Mary D. Sheriff, *The Exceptional Woman: Elisabeth Vigée-Lebrun and the Cultural Politics of Art* (Chicago: University of Chicago Press, 1996).

23 For more on Adelaide Labille-Guiard, her life, work and relationship to Vigée Le Brun, see Laura Auricchio's *Adelaide Labille-Guiard: Artist in the Age of Revolution* and Catherine R. Montfort, 'Portraits of Self: Adelaide Labille-Guaird and Elisabeth Vigée Lebrun, Women Artists of the Eighteenth Century', *Pacific Coast Philology* 40, no. 1 (2005): 1–18.

24 Elisabeth Vigée Le Brun, *The Memoirs of Elisabeth Vigée Le Brun*, trans. Sian Evans (London: Camden Press, 1989).

25 Michelle Bloom provides this caption for an image of the older figure of Madame Tussaud: 'Madame Tussaud's self-portrait in wax. The head mould for the figure was made by Tussaud herself in 1842. Courtesy of the Madame Tussaud Archives, London'; see *Waxworks, a Cultural Obsession* (9).

26 *Biographical and Descriptive Sketches of the Whole Length Composition Figures, and Other Works of Art, Forming the Unrivalled Exhibition of Madame Tussaud* (Bristol, 1823), 7.

27 Ibid., 31.

28 Tussaud's assessment of herself as a significant artist and public figure is noticeably absent from later exhibition catalogues. For example, in the 1876 catalogue Madame Tussaud is listed as figure number 84 but there is no mention of her role as 'the artist'.

29 *Biographical and Descriptive Sketches*, 31.

30 In *The Second Sex*, Simone de Beauvoir famously criticized Vigée Le Brun for making herself the subject of so many of her portraits and for never wearying of 'putting her smiling maternity on her canvases'. See Simone de Beauvoir *The Second Sex* (1952; New York: Vintage Books, 1974), 786. Sheriff has critiqued this dismissive account of Vigée Le Brun's work arguing that it misses the point of the artist's carefully constructed portrayals of herself and her daughter. See *The Exceptional Woman*, 43–6.

31 Madame Marie Tussaud, *Memoirs and Reminiscences of France, Forming an Abridged History of the French Revolution* (London, 1838), 17. Hereafter cited parenthetically in the text.

32 For more about Tussaud's sculpting process, see Berridge, *Madame Tussaud*, 55.

33 Pilbeam and Berridge have both suggested that Tussaud's memoir is probably not entirely accurate. They disagree, however, on whether or not Tussaud was actually present immediately after the execution of Marie Antoinette, in a recent BBC documentary, *Marie Tussaud: A Legend in Wax*, which aired on Thursday, 23 February 2017. According to a transcript of the film on the blog *Rodama*, Pilbeam argues for the veracity of the story, adding that 'somebody had to make them. And who was going to make them if Marie didn't?', while Berridge quips, 'This was all part of a very elaborate self-propaganda of suffering and hardship' (*Madame Tussaud*, 2017).

34 According to Sheriff, the *femme homme* 'mimicked the virile position by performing the man's intellectual or political function in the social order. She was a mixed creature, having the sex organs of a woman but acquiring or imitating the social characteristics proper to those with male sex organs' (ibid., 183).

35 Sheriff, *The Exceptional Woman*, 191.

36 Ibid., 191.

37 Ibid., 204.

38 Ibid., 213.

39 On the frontispiece, see Pilbeam, *Madame Tussaud*, 33.

40 This portrait of Marie Antoinette by François Hubert Drouais is one of several versions of the original portrait by Drouais made for Louis XV of France in 1772 now at the Musée Conde, Chantilly. This portrait uses the same pose and face, but appears to be mounted on a fabric background. According to the description of another version of the painting from the V & A Museum, London: 'Despite the scepticism of the critics at the time about this series of portraits, which lack of vitality and likeness, the paintings achieved quite a great success and 529–1882 was probably chosen as the model for the tapestry made by the Cozette father and son in the Gobelins Manufactory (Chambre de Commerce et d'Industrie, Bordeaux), which is inscribed on left with "Drouais pit en 1773" and, on the right, with "Cozette ext en 1774."' In fact, the tapestry reproduces the exact same garments and jewels as the V&A painting. The tapestry was commissioned by the financier and court banker Nicolas Beaujon (see letter dated 5 October 1774, to d' Angivillers, *Directeur des Bâtiments*): https://vads.ac.uk/large.php?uid=189740. It is possible that this version of the painting from 1773 is from an image of the tapestry described above, accessed 22 May 2020.

41 The replacement of the diamond necklace with a velvet ribbon could be something that occurred over time, or perhaps it was a way to detract attention from Marie Antoinette's reputation for excessive spending and her involvement in the diamond necklace affair. For more on representations of Marie Antoinette, see Dena Goodman's *Marie Antoinette: Writings on the Body of a Queen* (New York and London: Routledge, 2003), particularly Sara Maza's essay 'The Diamond Necklace Affair Revisited (1785–1786): The Case of the Missing Queen', 73–98. For more on British women writers' reactions to Marie Antoinette, see Katharine Binhammer, 'Marie Antoinette Was "One of Us": British Accounts of the Martyred Wicked Queen', *The Eighteenth Century* 44, no. 2–3 (2003): 233–55.

42 Terry Castle, *The Apparitional Lesbian: Female Homosexuality and Modern Culture* (New York: Columbia University Press, 2003), 117.

43 Ibid., 148–9.

44 Susan S. Lanser, *The Sexuality of History: Modernity and the Sapphic, 1565–1830* (Chicago and London: The University of Chicago Press, 2014), 11.

45 Ibid., 7.

46 Sheriff, *The Exceptional Woman*, 117.

47 Lanser, *The Sexuality of History*, 7.

48 Sheriff, *The Exceptional Woman*, 195.

49 Spencer Bokat-Lindell, 'Hiroshi Sugimoto's Portraits Bring the Dead Back to Life', *The Paris Review*, 28 February 2018.

50 Sean Silver, 'Afterward: What Do We Mean by Material?' *Eighteenth-Century Fiction* 31, no. 1 (2018): 213–30, 228.

15

Learning to craft

Beth Fowkes Tobin

Craft, as Richard Sennett argues, arises out of face-to-face interactions sustained over extended periods of time in the workshop space, where neophytes can watch and imitate expert craftsmen, receive instruction and correction, and learn by doing. This kind of learning, what educational psychologists call 'situated cognition', occurs when the learner is embedded within a community of practice, such as a workshop, where the embodied practices of the artisan are visible, and learning is based on observation and imitation.[1] However, those without access to the formal training of an apprenticeship with its 'regular exposure to processes and procedures' or the near proximity to a workshop, did manage to learn to craft.[2] As Ariane Fennetaux and Chloe Wigston Smith have argued, women who were denied access to workshop spaces did acquire skills and material competences from other women across generations within households and in schools. This chapter looks beyond the household and the school to explore how the embodied techniques of the skilled craftsperson were transferred in sites of informal learning, specifically in shops that sold craft materials and provided expert advice on crafting and in exhibition spaces that displayed crafted objects. Although not designed as sites of formal instruction, shops and exhibits offered their clients opportunities to interact with materials and to study the results of embodied techniques.[3] Another site of learning outside the workshop and the school explored in this chapter are printed manuals designed to instruct readers in the secrets of decorative arts and crafts, manuals that provided readers with imagined communities of practice. My examples come from the overlapping realms of decorative arts and natural history, sometimes called nature fancywork, including shellwork, taxidermy and flower drawings: subjects that I have encountered in my study of natural history collecting practices.[4] With the transfer of embodied techniques of craftwork as my focus, this chapter begins by examining informal learning in commercial spaces where bodies and things interacted in face-to-face, short-term encounters. We then turn our attention to the question of whether printed books could promote the development of material literacy. Were books effective teachers of craft techniques when words that described embodied practices had to substitute for the experience of watching bodies interact with materiality?

Shops as sites of informal learning

For the past two decades, historians of science have argued that to understand more fully how scientific knowledge was produced and exchanged in the early modern period, historians needed to examine more closely the commercial sphere where long-distance merchants and local tradesmen were instrumental in the production of knowledge about the natural world.[5] More recently in eighteenth-century studies the role of commerce in generating and disseminating natural knowledge has been explored by James Delbourgo in his study of Hans Sloane and by Sarah Easterby-Smith, who has studied the role of commercial nurserymen in London and plantsmen with their long-distance trade in plants as brokers of botanical knowledge.[6] Those of us interested in the intersection of natural history and decorative arts could take a lesson from these historians of science to examine commercial sites, and in particular, shops as spaces that functioned as sites of exchange, not just of money for objects, but as sites where knowledge transfers occurred. As Serena Dyer and Kate Smith have argued, shops were sites of knowledge production, specifically where customers acquired material literacy, learning what to purchase and how to use those purchased items.[7] Focussing on shopping as a form of learning, I examine two London shops that advertised themselves as offering instruction to their customers: George Humphrey's Shell Warehouse on St. Martin's Lane, where it did a brisk business in shelling shells in the 1760s and 1770s, and Hannah Robertson's shop that supplied customers with materials to do fancywork in the 1780s.[8]

In 1762 George Humphrey opened his first shop on St Martin's Lane, calling it the Shell Warehouse, where 'shells, corals, fossils and other Natural History Curiosities' could be purchased for 'collections, Grottos, or for making artificial Flowers'. This shop, next to John's Coffeehouse and 'opposite Cecil-Court', had 'the greatest variety of … [shells] in England'.[9] The Shell Warehouse was run as a family business in the 1760s and 1770s until Humphrey senior died in 1776. His son George inherited the shop, eventually moving it to Leicester Square in the 1780s, where he became known as 'the chief commercial naturalist in this country'.[10] He acquired shells from the Pacific by purchasing them from officers and sailors on board Captain Cook's voyages, and he sold them to customers, who, like the Duchess of Portland, could afford such exotic and rare molluscs.

The Humphreys were a fascinating family of artworkers and retailers. William Humphrey, the younger George's brother, was an illustrator, engraver and printseller, who opened a print shop on Gerrard Street in 1774, and eventually began dealing in oil paintings, becoming an expert in old English portraits. Their sisters, Elizabeth and Hannah, were prize-winning shellworkers, who exhibited at the Free Society of Artists several times in the 1760s. Hannah, however, went on to open Humphrey's Print Shop, employing and eventually living with caricaturist James Gillray, becoming the main purveyor of his prints, and Elizabeth married Jacob Forster, a mineral dealer, and ran their mineral shop in London not far from the Shell Warehouse.[11]

Tracking the Shell Warehouse in the advertisement section of the London newspapers in the 1760s and 1770s reveals how complex this retail space was in terms of the range of activities it supported and supplied. Shoppers could purchase shells, corals and fossils, and 'Ladies may be furnished with the Implements used in making Shell-Work' (Figure 15.1).[12] Humphrey's sisters were featured in the shop's advertising as willing to teach 'Ladies the said Art at Home or Abroad'.[13] Elizabeth and Hannah Humphrey taught customers how to do shellwork within the warehouse space as well as going to their clients' homes to assist with their shell projects. Other shell-related services were also advertised. Humphrey offered to recommend professional shellworkers for large projects: 'any Gentleman or Lady inclined to build a Grotto' may enquire at the Shell Warehouse for the name of a person 'qualified for such an undertaking, and may be furnished … with all Materials necessary for such a Work'.[14] This strategy of offering instruction and brokering expertise was, of course, designed to boost sales in shells. Teaching women to do shellwork would develop a steady stream of customers in need of shells and supplies. Advising their more elite customers on how to go about getting a grotto built in their gardens was also a way to support their sales at the shop. This blend of instruction and proffered expertise transformed Humphrey's store from a site of mere commercial transactions into a more complicated social space. Combining the transfer of goods with the transmission of knowledge and the acquisition of material literacy made Humphrey's shop into a productive space, where customers learned to make new things from the materials provided by the shopkeepers. One could argue that retailers have always offered instruction to their customers, teaching them what to desire and, in some cases, how to use the goods they sold, but the quality and

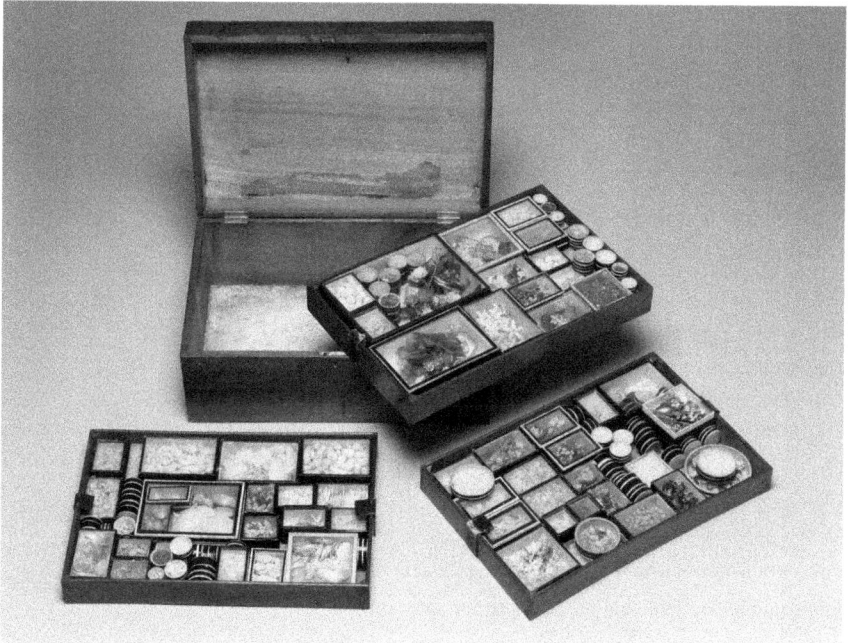

Figure 15.1 Box of shells, *c.* 1800, given in memory of Mrs Joan Griffith, W.5:1 to 4–2010 © Victoria and Albert Museum, London (see Colour Plate 8).

kind of instruction offered by the Humphreys at their shop went beyond the usual commercial practices to provide their customers with technical expertise and natural history knowledge, transforming the shop space into a site where the transmission of knowledge occurred through the rubric of craftwork.

The Humphrey sisters' activities as instructors in craftwork were part of a larger commercial operation focused on the sale and acquisition of shells. Craft in the Shell Warehouse was subordinate to and dependent on its commercial goals and was unlike the Robertsons' shop with its exclusive focus on the materials needed for decorative arts and crafts. According to Robertson, the shop she ran with her daughter Anna was the first of its kind to make craftwork and craft technique its core mission. Normally, crafting necessitated visits to multiple shops to gather supplies, including stationer's shops and toyshops, natural history warehouses, and apothecaries' and druggists' shops. To make a decorative watercolour drawing of a flower, for instance, a person would have to buy pigments and ink at a colour shop, gum arabic at a druggist's or apothecary's shop, and her paper at a stationer's shop. The Robertsons also offered their customers instruction in the use of the materials they sold. Robertson wrote in her brief autobiography: 'Our shop was crowded with nobility, and we were also employed in teaching many of the first families.' She boasted that her daughter was responsible for newly ornamenting Windsor Castle with 'elegant works of fancy'.[15] Like the Humphrey sisters, Anna Robertson exhibited her crafts at the Free Society of Artists, perhaps as a way to advertise her skills in paper crafting and to drum up business as a teacher of paperworking techniques, including filigree, gum flowers and cut paper flowers.

Anna Robertson no doubt learned to craft from her mother who, as Chloe Wigston Smith has explained, was the author of very popular recipe book and craft manual, *The Young Ladies School of Arts*, which underwent several editions in the space of forty years. It contains recipes for making pickles, preserves, medicine, cosmetics and the chemical compounds needed for many art practices, including lacquerwork, china painting, gilding and shellwork. Robertson's preface states that the purpose of her book is to show how 'Ladies as have time to spare, may do a great many pretty things at a small expence, for their own pleasure, or the ornament of their houses, and in case of misfortunes happening in life, which is not uncommon, their knowledge of such things will be of real use to them.'[16] Robertson also sets off these decorative art practices from needlework, of which she says, 'it is well known how small a value is set on women's work, so that the cleverest needle can scarcely earn subsistence'. She goes on to promote 'a knowledge of the various branches [of the Arts] mentioned in this Treatise', and contends that 'any girl capable of painting, japanning, gum-flowers, pongs, &c. will always find employment among fashionable people, and especially in towns of trade and commerce, by which they may earn a tolerable living'.[17] Robertson's audience was young ladies of leisure and impoverished gentlewomen in need of an income. Robertson herself was just such an impoverished gentlewoman who turned to her material literacy in the decorative arts to make a living to support her children. Robertson had married an unscrupulous spendthrift who eventually died, leaving her to make her way in the world on her own. She taught decorative arts in Edinburgh and then York, teaching her students the requisite decorative and domestic skills that they were expected to master as part of their education and preparation for their roles as wives in the middling and upper classes. She joined her daughter in London to assist with running their shop that supplied crafters with materials, advice and instruction.

When Humphrey and Robertson advertised their shops as spaces where their customers could learn how to do shellwork and other crafts and offered their daughters as experts who could introduce clients to crafting techniques, tools and materials, they transformed the socialscape of their shops, adding an educational dimension to the commercial transactions that structured their activities. In these shops customers purchased materials for craftwork and in the process acquired making skills and material literacy. This blending of shopping with the transmission of embodied knowledge underscores the complexity of social interactions within retail spaces and points to a long tradition of shops being sites of informal instruction, especially within the field of decorative craftwork.

The manipulation of materiality

What exactly went on in these shops when customers were offered instruction? How can we know with any specificity what techniques the Robertsons and the Humphrey sisters taught their customers? Robertson's very popular recipe book offers us glimpses of the bodily techniques and the materials needed to produce decorative pieces of craftwork. For instance, Robertson's manual provides detailed directions on how to clean shells. She recommends using a combination of aqua fortis (nitric acid), emery files and boars' hair brushes to remove the shell's epidermis, its dull or crusty coat. Once this is removed, the shell must be 'rubbed with linen cloth impregnated with common soap', and 'if after this the shell, when dry, appears not to have so good a polish as was desired, it must be rubbed over with a solution of gum arabic; and this will add greatly to its gloss without doing it any sort of injury'.[18] The process was sometimes lengthy, involving 'repeated dippings' in nitric acid, a practice which Robertson warns was dangerous. 'In the effort of work, the operator must always have the caution to wear gloves, otherwise, the least touch of the aquafortis will burn the fingers and turn them yellow; and often, if it is not regarded, will eat off the skin and nails.'[19] Some mollusk shells, particularly sea snail shells, provided shellworkers with special challenges, involving the removal with acid of the crusty outside shell, while protecting the delicate inside of the shell from the corrosive effects of the acid:

> The limpet, auris marina, the helmet shells, and several other species are of this kind, and must have this sort of management; but as the design is to shew the hidden beauties under the crust, and not to destroy the natural beauty and polish of the inside of the shell, the method of using the aquafortis must be this, a long piece of wax must be provided, and one end of it made perfectly to cover the whole mouth of the shell; … and the mouth being stopped by the wax, the liquor cannot get into the inside to spoil it; … When the repeated dippings into the aquafortis shew that the coat is sufficiently eaten away, then the shell is to be

wrought carefully with fine emery and a brush; and when it is polished as high as can be by this means it must be wiped clean, and rubbed over with gum-water or the white of an egg.[20]

Robertson's text demonstrates how well she knew her materials, which in this case was a limpet's shell and how its surface would react to the corrosive powers of nitric acid. Her manual is filled with such knowledge, describing how to handle many different kinds of shells and to get them looking their best. The tools and ingredients mentioned above include nitric acid, stiff brushes made of boars' hair, wax, emery files, gloves and gum arabic, and in another section she describes how to make flowers out of shells, suggesting that pale dull shells can be made beautiful by painting them with a Brasilwood-infused gum arabic wash. She also offers a recipe for making cement out of stone flour, sulphur, white rosin and beeswax: 'incorporate all together over a gentle fire, and then knead it with your hands in warm water'.[21] Robertson's words convey what a customer might have witnessed first-hand in these shell shops: the actions of skilled practitioners who were guided by knowledge they had gathered through experience.

Although Robertson's recipe book offers compelling and detailed descriptions of working with shells, the capacity of textual descriptions such as hers to teach material literacy and the embodied techniques of craft production has been questioned by museum curators and art historians, who privilege the object's materiality over verbal descriptions of that materiality. Miriam Clavir has argued that for conservators 'the objects contain knowledge in their very fabric and should be preserved intact so that this knowledge can be learned either now or in the future'.[22] Echoing this sentiment, an eighteenth-century drawing manual registers ambivalence about teaching drawing through printed words: 'it is not Words but Matter that should be consulted, and … Volumes are not always the most instructive'.[23] Malcolm Baker has argued that 'the relationship between making and materials' and 'texts as evidence for practices of making is indeed problematic', since it is extremely difficult to ascertain how a written text translates into practice, noting the 'difficulty of using words to suggest or define such processes'.[24] His concern is supported by Ian Hankey, a glassmaker who, in trying to reproduce Venetian glassmaking techniques, has discovered that the ability to make such intricate glasswork was dependent on 'the material itself as well as the art of the maker'. For Hankey, tacit skill is based on 'a feeling for the material, a feeling that cannot be fully articulated'.[25] Two concerns about textual transmission of embodied knowledge are voiced here. One is the failure of language to adequately capture the tactility of bodily encounters with materiality, and the other concern is methodological, questioning historians' use of 'texts as evidence for practices of making'. Indeed, many scholars have assumed that the skill and knowledge required of crafting were transferrable in textual formats, and the plethora of recipe books and household manuals, such as Robertson's, has been interpreted as evidence of the popularity crafting in the latter half of the eighteenth century. However, the relationship between the printed text and the embodied practice is quite complicated and one that bears scrutiny.

Heeding Baker's concern about the tenuous relationship between written recipes and embodied practice, we should consider the possibility that a recipe for varnish is not a reliable indicator that a reader of that recipe made that varnish. To read back from a recipe to how it was used requires the positing of a docile and obedient reading subject, and we know from current scholarship on how books were used in the early modern period that readers used printed books in all sorts of ways that had little to do with dutifully reading printed words on a page and then putting those words into action. Books provided the raw materials for other forms of archiving and communication, as in grangerizing, the cutting out the letters to form new texts, or turning a book into a repository for one's thoughts in the shape of annotations or notes, or even using the book as a receptacle to store important items, including locks of hair and love letters. Although we must not assume that a book's recipe for varnish is the same as that varnish having been made or used, we can assume that the recipe book was purchased with the idea that someone, including friends or family members, might consult such a book. The recipe book could be a witness to an aspiration to engage in craftwork rather than operating as an index of craft practices performed in real time and space, just as those cookbooks on your shelf that have never been used testify, despite their disuse, to your interest in cooking and food.

But what about Baker's and Hankey's casting of doubt on the ability of words to convey bodily techniques and language's failure to adequately describe the material object? To deal with, or at least acknowledge the validity of these objections, and yet proceed with an examination of the textual transmission of bodily practices, I return briefly to Robertson's recipe book as an example of how bodies and materials can be figured in a text. 'Founded on many years of experience', her recipes are attuned to the needs of the uninitiated reader, providing a level of detail that helps readers to visualize complex processes as varied as those involved in shellwork, taxidermy or lacquerwork.[26] Robertson's expertise as a teacher in informal and formal sites of instruction no doubt had an impact on her ability to foster her readers' material literacy by providing much more than the usual list of ingredients and materials needed to make something.

Although Robertson's book is ultimately a recipe book for domestic pursuits involving cookery and decorative arts, it also includes a section on taxidermy, perhaps the first description of taxidermy in an English book. Taxidermy in this era was not limited to the male-coded realm of the hunt. Preserving and posing the lovely feathered bodies of dead birds were considered by many women to be a decorative art practice. However, women who wanted to do taxidermy had to ask those who possessed the right to shoot game birds (landed gentlemen) for their help in procuring these items for their often elaborate decorative pieces that featured taxidermied birds perched on a branch. Robertson makes taxidermy approachable by drawing upon techniques for stuffing and preserving birds not that far removed from those used in cooking and needlework:

When you get a bird fresh killed, open the venter from the lower part of the breast-bone, down to the anus, with a pair of fine pointed scissors; take out all the intestines, liver, stomach, &c. fill up the cavity with the following mixture of salt and spice; and then bring the lips of the wound together, and stitch it up, so as to prevent the stuffing from falling out.

After stuffing the bird's gullet with this mixture, the next step is to:

Open the head near the root of the tongue, with the scissors, and after having turned them around three of four times to destroy the brain, fill this cavity likewise with the mixture. You may leave the wings and thighs in the natural state, for the salt will soon penetrate into these parts; the bird must then be hung up for about two days by the legs; then it must be placed in a frame to dry, in the same attitude we usually see it when alive on the plain, or on a tree. In this frame it must be held up by two threads, the one passing from the anus to the lower part of the back, and the other through the eyes; the ends of these threads are to brace up the bird in its natural attitude, and fastened to the beam of the frame above: let the feet be fixed down with pins or small nails. In this situation, it must remain for a month or more, until the bird is perfectly dry, which will be known by its stiffness; then it may be taken out of the frame and placed according to fancy, and it will require no other support than a pin through each foot, fastened into the box or bracket it stands on; the eyes must be supplied with proportionable glass beads, fixed in with strong gum-water. Common salt one ounce, allum powdered four ounces, pepper ground two ounces, all mixt together.[27]

The bird's body is present in Robertson's description – its head, stomach, liver, breast-bone and anus are called into the text – and the taxidermist's hands are implicitly there with the twisting of scissors in the bird's head to scramble and remove its brain and in the stitching up of the bird's gut that has been stuffed with salt and spice.

Robertson's sections on taxidermy and shellwork convey quite vividly the body's engagement with materiality. Equally detailed is her description of how to paint decorative motifs on glass and silk, but when it comes to making the paint, Robertson's text provides just three recipes and a single page on how to turn commercial pigments into oil paints and watercolours for decorative uses.[28] Robertson's mere listing of pigments and comparative lack of detail when it comes to making paint is not uncommon in art manuals and colour treatises, which often supply little detail when it comes to the body's engagement with these materials. The few art manuals and colour recipe books that included descriptions of bodies interacting with materials merit our attention because they appealed to the senses, in particular sight, touch and smell, making them potentially more instructive than the usual mere listing of ingredients. As we shall see, an inexperienced practitioner of decorative arts could have learnt from a few choice books how to prepare the materials she needed to colour her drawings and prints of flowers or to paint images of birds and butterflies on glass.

Making colour

Before colouring a drawing in watercolours, paints had to be made and paper or vellum had to be prepared to receive these colours. These activities required material literacy, depending on skill and knowledge as well as hours of grinding, sifting, mixing, straining, squeezing and waiting for fermentation to take place. Recipes for watercolours appear in a variety of texts, including household manuals and guides to the decorative arts, drawing and painting manuals such as *The Delights of Flower-Painting* (1756), and guides to the mechanical arts such as Robert Dossie's *The Handmaid to the Arts* (1758). In general, two kinds of recipes for watercolour paint appear in these books: recipes for making pigments from flowers and other plant-based materials, and recipes for dealing with professionally made pigments, often inorganic in origin and usually bought in druggist shops or colour shops; these latter recipes focused on how to work with these pigments to turn them into pliable, stable watercolours. Not until the mid-1780s were professionally produced watercolours widely available when the William and Thomas Reeves introduced sets of ready-made watercolour cakes in what is now their characteristic lozenge shape and sold them in wooden boxes complete with compartments for brushes, water, palettes and other painting paraphernalia. Before Reeve's watercolours and those of competitors appeared widely for sale, most watercolourists, amateur and professional alike, made their own colours from plants and from commercially available pigments such as carmine, white lack, Prussian blue, vermilion and verdigrises. Because ready-made watercolours like Reeve's were expensive, watercolourists continued to make their own paints well into the nineteenth century, and if one lived out of the reach of pigment purveyors, plant-based colours continued to be an important resource for amateur artists.[29]

The typical recipe for making watercolour paint using commercial pigments makes little mention of bodily techniques other than that implied by the verb to grind, which presumably meant using a muller against a slab of stone, implements rarely named in these texts. For instance, to make red paint from 'Lake', which consists of 'the grosser Particles or the Lees of Carmine' or the granular residue left over from the process of making Carmine, is quite difficult. As *The Delights of Flower-Painting* (1756) tells us, lake 'requires a great deal of Grinding, and to be strong gummed'.[30] The word 'grinding' is shorthand for a complex process that involves the hand holding the stem of the muller and placing the rounded base on a stone, ceramic or glass slab, the arm moving in a circular motion with the wrist rigid, and the shoulder bent over the moving hand, putting weight behind the arm's circular motion and pressing the muller against the stone, the pressure being applied depending on the degree of resistance in the substance being ground. Grinding also suggests the texture of pigments as does the mention of pigments' solubility in gum water, a mixture of powdered gum arabic and water. The degree and rates at which pigments dissolve in liquid are sometimes noted: sap-green is 'of so fine a gummy Substance, that it will dilate in Water, and, without any farther Trouble, becomes useful', and gamboge, a yellow pigment, 'dissolves the Minute the Water touches it, therefore wants neither grinding nor gumming'.[31] Behind these comments on the grain and solubility of pigments is the chemistry of watercolours that involves making a solution of finely ground pigments suspended in gum water that adheres to paper or vellum, and once dried does not flake off because it is made pliable by adding sugar or honey to the solution.[32]

Recipes for how to make watercolours from commercial pigments also included advice on how to purchase the right pigments. Teaching what qualities to look for in a product, a form of material literacy, *The Delights of Flower-Painting* urges its readers when buying sap-green 'see that it looks very black and bright', and cautions its readers to make sure they buy 'good' Indian-Ink: 'try it before you buy it, unless you can depend upon the Honesty of the Vender'. Saff-flower, used to make red, 'is scarce, and still scarcer to be got good; it is sometimes found in the Colour Shops, at other may find it at the Druggists'.[33] Introducing doubt into the quality of pigments sold in colour shops may have been a strategy to persuade readers to shop for pigments at this book's printseller's shop, as announced at the bottom of the title page: 'Printed for D. Voisin, Printseller, in *Middle-Row, Holborn* Where may be had all manner of Colours ready prepared, Variety of Drawing-books, Views, Japanning, Prints, &c.' *The Delights of Flower-Painting* is not the only art manual to offer advice on how to shop for pigments. *Arts Companion or a New Assistant for the Ingenious* (1749) addresses its readers as beginners who need to be told where to acquire pigments and oil-based paints: 'At most Colour-shops of note in *London*, we may meet with Colours of several sorts ground in Oil, and tied up in little Bladders to be sold at Three pence, a Groat, or Six pence a piece, according as they are more or less valuable' (Figure 15.2).[34] Another manual offers advice on where to buy a good lead pencil, saying that 'Tis a great Pleasure to a Draughtsman to work with a good Pencil, and as great a Plague to have a bad one', which can be 'hard, gritty, and full of Knots'.[35] This manual goes on to provide very specific information about where to purchase individual colours. Carmine, for instance, should be bought at 'Mr. *Goupee's*, the great Fan-Painter in *King-street, Covent-Garden*', and 'at the Colour-shop, the Sign of the *Bell*, against *Arundel-street* in the *Strand*'. 'The Best Lake I have met with, is at the great Colour-shop at the White-hart in Long-acre, near James's-street, Covent-Garden, ready prepared in Shells of Water-colours.'[36] These texts position themselves as instructing their readers how to purchase the best materials for their art projects, operating under the assumption that their readers reside in or near London or other cities where painting supplies could be purchased.

Some texts, on the other hand, position themselves as offering advice to readers who live in the countryside and cannot easily acquire these commercial pigments and paints. Recipes for plant-based colour, as opposed to those that deal with commercially available pigments, can be quite detailed in their depiction of complex chemical processes, and it is in these recipes that the body and the materials it manipulates are palpably present. Several recipes in *The Art of Painting in Miniature* (Figure 15.3) for making the colour green from flowering plants appeal to the senses, in particular the sense of smell:

> Take Leaves of *Iris:* Mince or chop them very small, and put them into a Glass or *Delft* Vessel; or, which is better, into a Brass or Copper Pot, with powder'd *Allum* and Quick-lime. Let the whole rot together for ten or twelve Days; and then being rotten, press it into Shells; for that the blue Colour may become green, it is better that the Flowers putrefy. The *Green* is darker and more lively, when only the Leaves are bruised, and pressed immediately, without letting them rot, after having strewed powdered *Allum* upon them.

The potentially foul-smelling juice of rotting buds and leaves squeezed into bags of cloth or goatshair and then poured into empty clam and scallop shells produces a

Figure 15.2 Trade card of Dorothy Mercier, print seller and stationer at the Golden Ball, Windmill Street, London, engraving and etching, *c.* 1761, E.291–1967 © Victoria and Albert Museum, London.

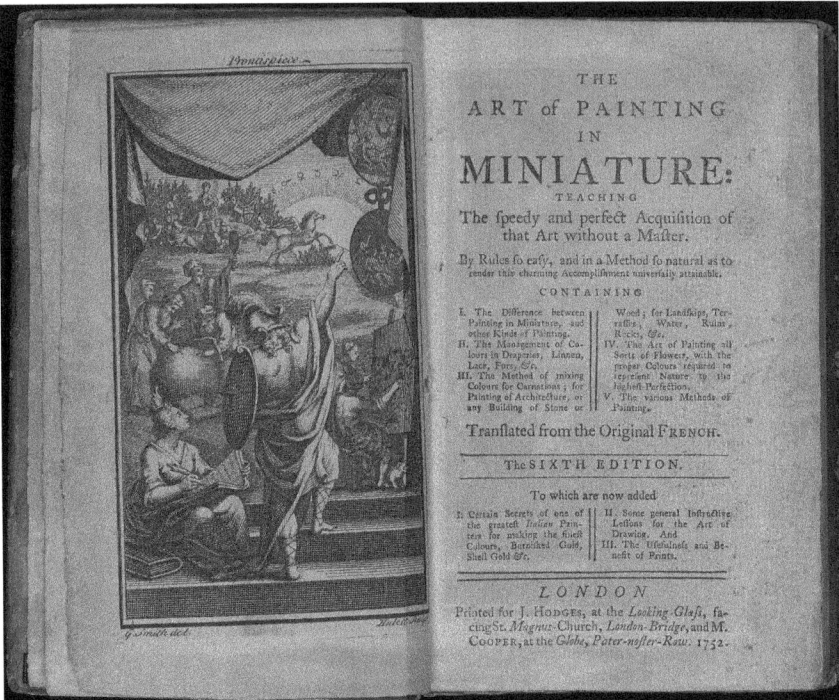

Figure 15.3 Title page, *The Art of Painting in Miniature: Teaching the Speedy and Perfect Acquisition of that Art without a Master* (London, 1752). Courtesy of Beinecke Rare Book and Manuscript Library, Yale University.

lovely green pigment that might be 'mouldy and Glewy' if not set out in the sun to evaporate properly.[37] Needless to say, fermentation, the chemical process involved in turning plant matter into pigments and dyes, could be a smelly business. Indigo, for instance, most famously stank so much that the pots in which the indigo plants were rotting were kept at a distance from residences of the rich and powerful.[38]

Another art manual, *The Art of Drawing, and Painting in Water-Colours* (1731 and 1735), contains pastoral descriptions of flower-gathering excursions in preparation for making plant-based pigments, in this case, blue:

> The chief of the Ingredients it is composed of may be easily had during four of the Summer Months, that is, the Cyanus or blue Cornbottle-flower, which abounds in almost every Corn-field; Children may gather it, without hurting any Thing, about the Skirts or Verges of the Corn-field. This Flower has two Blues in it, one of a pale Colour in the larger outward Leaves, and the other a deeper Blue, which lies in the middle of the Flower; ... but the deep Blue of the Middle produces much the best Colour, as one may try, by rubbing it while it is fresh, so hard upon a piece of good

writing Paper, as to press out the Juice, and it will yield an excellent Colour, which
will not fade, as the Experience of two or three Years has shewed me.[39]

This beautiful blue pigment, not a product of a chemical reaction, is literally squeezed
out of this plant by rubbing and pressing the centre of the flower on paper. The
anonymous author of this manual credits Robert Boyle with this recipe, although
Boyle's own description does not include children gathering these flowers.[40]

Most art manuals and colour recipe books are, in Sarah Lowengard's words,
'synthetic' texts.[41] Cobbled together from a variety of printed texts, many of these books
are the products of hacks and grub-street encyclopediasts, not based on the experience
of the author as expert practitioner. Even one of the most respected art manuals of the
period, Constant de Massoul's *A Treatise on the Art of Painting, and the Composition
of Colours* (1797), was 'culled from … others who have discussed this subject', as a
hostile reviewer in the *Monthly Review* wrote. Arguing that the treatise was 'a work of
deception', the reviewer pointed out that de Massoul, a French colourman whose shop
was located on Bond Street, presented information and recipes as his own, which were
actually taken from published treatises on colour.[42] Lowengard, having traced some
of this plagiarized material, is quite blunt in her summation: 'His disingenuousness
– or laziness – was extreme, as much of his information appears to have been culled
from the *Encyclopédie* and the *Encyclopédie méthodique* and not the original works of
his source-authors.'[43] Compared with de Massoul's and countless other books filled
with cribbed recipes, a small number of art manuals and colour recipe books stand
out as based on experience. For instance Robertson's recipe book was, by and large,
based on her experiences as a craftswoman, teacher and decorative arts shop-keeper,
especially those sections on shellwork and taxidermy, and this experience shows in her
verbal descriptions of artisanal practices. This air of authenticity can be found in the
anonymous *The Art of Drawing, and Painting in Water-colours*, in *The Art of Painting
in Miniature*, and a few other books, and arises out of the way in the body is called into
a description and the amount of attention paid to tactility of materials, detailing their
affordances and constraints.

For example, George Edwards, a prolific natural history illustrator and author,
offers advice on how to tell if a sheet of paper is sufficiently gummed and ready for
the application of watercolour, saying the paper 'ought to be neither over nor under-
gummed'. The practice of laying down a wash of a gum arabic solution was a technique
to ensure the paint's adhesion to the paper. As we have seen in several colour recipes,.
gum arabic was often added to the paint as a way to deal with this problem, but too
much gum arabic could leave the dried paint looking too shiny, and so, this was an
alternative to heavily gummed watercolour paint. In describing how to apply a gum
arabic wash, Edwards explains the reasons for applying the gum wash properly. Paper
that is

too much gummed, or sized [seized] in the making, is so hard and close it will not
take in the colours at all; and what is laid on at first, one is apt to wash off again in
the second shadowing, and so on, which is very inconvenient. An under-gummed

paper hath a contrary inconveniency; for the colours are apt to run through it, and spread beyond your design on the outline. A proper paper may be chosen by touching it with your tongue; an ungummed paper will stick very strongly to the tip of the tongue when touched: an over-gummed paper will hardly stick at all, by which a proper medium may be found, that only sticks a little to the tongue.[44]

Offering his readers this tidbit of advice – using their tongues to test the degree of gumminess – Edwards turns the tongue into a tool by which to measure the degree of adhesiveness. John Jane, purportedly the author of *The Delights of Flower-Painting*, also worries over gumminess and greasiness of paints and paper: 'if your Vellum or Paper be free of Grease, you may make use of it only well gummed.' Sap-green, which can be a 'gummy substance', also can turn greasy if the watercolourist uses her lips to taper the paintbrush's bristles: 'Be careful how you apply your Pencils [paint brushes] to Lips, for besides that some of the Colours, especially Sap-Green, is not very palatable, your Pencil is not so fit for Use, and will make your Colours work greasy.'[45] Tongues and lips as tools, and the tacky, sticky, greasy textures of paper and paints call attention to the embodied and material practices of decorative artwork.

Art manuals such as these may have been somewhat effective in transmitting material literacy and the tacit, embodied knowledge of artisanal practice, but were probably never completely sufficient as a form of instruction. Manuals and recipe books were most likely supplemental, acting in concert with multiple sources of informal instruction, like what was on offer at shops such as the Robertsons and the Humphreys. In addition to shops where customers engaged in face-to-face interactions with experts who distributed advice along with materials for craftwork, exhibition spaces offered their viewers sight of the objects they wanted to make. Seeing these artefacts was a form of instruction, as today's curators argue: 'objects contain knowledge in their very fabric'. The Free Society of Artists held exhibitions that featured the usual art museum fare of paintings, drawings and watercolours combined with examples of decorative craftwork. This is where Anna Robertson and the Humphrey sisters exhibited their pieces and countless other women displayed their craftwork, ranging from butterflies made from shells, flowers 'cut in cork', and bouquets 'modelled in coloured wax' to landscapes made with seaweed or 'wrought with needle, in human hair', and a 'likeness of a gentleman, done with his own hair'.[46] Other kinds of exhibitions, short-term, informal and often held in shops, were advertised in the London newspapers. An exhibit of 'a curious Collection of Fancy-Work', including 'raised Filligree, such as never has before been done by any one', was held at Cook's, No. 13, in Holborn's Read-Lion Street. The exhibitor, Mrs. Cook, charged admission – one shilling – and assured 'ladies it is her own work and not what has been purchased from others', and listed some of her artefacts on view: 'an elegant Wafer-Work superior to any in London; Cloth-Work in variety of figures entirely new; some elegant raised ditto, in Fruit and Flowers; Paper-Work in Flowers and Landscapes', and the list goes on ending with embroidery. She also advertised her services as a teacher – 'Ladies taught at home or abroad' – and the materials needed to make filigree: 'All sorts of Foundations for Filligree, by the maker, on reasonable terms. Exceeding good Coloured and Gold Paper, warranted even and good edges.'[47] A broker of expertise and materials, Mrs Cook must have thought that displaying her work for paying customers was the best way to catch the attention of women eager to develop their literacy as paper fancy-workers.

Although Baker cautions us that the burgeoning publication of craft and art manuals does not necessarily reflect an increase in women engaged in craft projects, printsellers and shopkeepers seem to have believed that there was a market for texts dispensing craft knowledge. In what were rapidly changing times, mothers, aunts and grandmothers were no longer perceived as sufficiently expert sources of knowledge about the domestic arts and crafts that needed to be mastered as a sign of elevated social rank and domestic bourgeois femininity. Traditional ways of transmitting knowledge within the family might have been perceived as outdated and the skills taught within the household as no longer applicable to the latest fashion trends in the decorative arts. To keep up with changing tastes, amateur craftworkers could supplement what they had learnt at home and school by reading books, visiting exhibits and spending time in shops. The exhibit and the shop provided young women and other amateurs the opportunity to learn craft techniques from experts like Hannah and Anna Robertson, Elizabeth and Hannah Humphrey, Mrs Cook and countless other nameless women whose skilled embodied practices remain elusive and their creations ephemeral. Traces, however, of these practices are to be found in the print record, primarily in manuals eager to convey the joy of making decorative objects, testifying to the presence of these overlooked communities of practice, however fleeting or imagined, where the customer and the reader could learn to craft and experience the pleasures of making.

Educational psychologists have argued that watching someone make something is key to learning a craft. The question I have addressed in this chapter is how do people learn to make things if they did not have access to workshop environments where watching and doing are the prime learning tools, and more specifically, how did young women learn to make decorative objects if knowledge transfer systems within households could not keep up with latest fashion trends in the decorative arts. I offer the shop and the decorative arts manual as sites where novices could have learnt to craft and to make the materials needed for decorative practices. All sorts of shops, including natural history dealers' warehouses, stationery stores and apothecary shops, offered advice on using their products and even in some cases, offered formal instruction in material literacy, as in George Humphrey's shell warehouse where his daughters taught customers how to make shell sculptures. Manuals and recipe books positioned themselves as delivering information on how to make things; however, their success in doing so was constrained by how effectively they conveyed embodied knowledge. A list of ingredients was not enough to teach someone how to assemble something. The body had to be called into the text with sensory language and attention to haptic and bodily processes if a manual were to aid in the learning process. However helpful manuals might have been in providing guidance to makers of decorative objects, they were ultimately supplemental and could not completely replace the experience of hands-on learning and face-to-face encounters within communities of practice with their visual and tactile economies of production.

Notes

1   See Richard Sennett's *The Craftsman*, which attempts to recover craft as intellectual engagement with material, arguing that 'making is thinking', and that there is a 'dialogue between concrete practices and thinking' and that knowledge is 'gained in the hand through touch and movement' (7–10). Learning a craft emerges within 'communities of practice', a term educational psychologists use to describe apprenticeship and learning from observation and physical engagement rather than from precept. See, for instance, John Seely Brown, Allan Collins and Paul Duguid, 'Situated Cognition and the Cultures of Learning', *Educational Researcher* 18, no. 1 (1989): 32–42.

2   Sarah Lowengard, *The Creation of Colour in Eighteenth-Century Europe* (New York: Columbia University Press, 2006); http://www.gutenberg-e.org/lowengard/, accessed 22 May 2020.

3   Key texts in the literature on bodily technique are Pierre Bourdieu, 'Bodily Knowledge', *Pascalian Meditations* (Palo Alto: Stanford University Press, 2000) and Marcel Mauss, 'Techniques of the Body', *Economy and Society* 2, no. 1 (1973): 70–88.

4   For the overlapping realms of decorative arts and natural history, see Tobin, *The Duchess's Shells* and Tobin, 'Bluestockings and the Culture of Natural History', in *Bluestockings Now: The Evolution of a Social Role*, ed. Deborah Heller (Abingdon: Ashgate, 2015), 55–70.

5   Landmarks in this approach are Harold J. Cook's *Matters of Exchange: Commerce, Medicine, and Science in the Dutch Golden Age* (London: Yale University Press, 2007) and Paula Findlen and Pamela H. Smith's edited volume, *Merchants and Marvels: Commerce, Science, and Art in Early Modern Europe* (London: Routledge, 2002).

6   James Delbourgo, *Collecting the World: Hans Sloane and the Origins of the British Museum* (Cambridge: Harvard University Press, 2017); Sarah Easterby-Smith, *Cultivating Commerce: Cultures of Botany in Britain and France, 1760–1815* (Cambridge: Cambridge University Press, 2017).

7   See Dyer, 'Shopping and the Senses', 694–703; Smith, *Material Goods, Moving Hands*.

8   The scholarship on eighteenth-century shopping is beyond the scope of this chapter; however, the following have helped shape my approach to shopping as a complex social practice: Stobart, Hann and Morgan, *Spaces of Consumption*; Walsh, 'Shop Design and the Display of Goods in Eighteenth-Century London', 157–76; and Jon Stobart, 'The Shopping Streets of Provincial England, 1650–1840', in *The Landscape of Consumption: Shopping Streets and Cultures in Western Europe, 1600–1900*, ed. Jan Hein Furnée and Clé Lesger (London: Palgrave Macmillan, 2014), 16–36. See also the essays by Serena Dyer, Jon Stobart and Sarah Howard in this volume.

9   *Lloyd's Evening Post and British Chronicle*, Friday 26 March 1762.

10   William Swainson, *Taxidermy: Bibliography and Biography* (London: Longman, 1840), 219.

11   Timothy Clayton, 'William Humphrey', *Oxford Dictionary of National Biography*.

12   *Lloyd's Evening Post and British Chronicle*, Friday 26 March 1762.

13   *Gazetteer and New Daily Advertiser*, 18 January 1765.

14   *Lloyd's Evening Post and British Chronicle*, 7 March 1763.

15   *The Life of Mrs. Robertson, Grand-daughter of Charles II, Written by Herself* (Derby, 1791), 40.

16   Hannah Robertson, *The Young Ladies School of Arts* (York, 1784), viii–ix.

17 Ibid.

18 Robertson, *School of Arts* (York, 1777), 10.

19 Ibid.

20 Ibid.

21 Ibid.

22 Miriam Clavir, *Preserving What Is Valued: Museums, Conservation and First Nations* (Vancouver, BC: University of British Columbia Press, 2002), quoted in Mary M. Brooks, 'Decay, Conservation, and the Making of Meaning through Museum Objects', in *Ways of Making and Knowing: The Material Culture of Empirical Knowledge*, ed. Pamela Smith, Amy Meyers and Harold J. Cook (Ann Arbor: University of Michigan Press, 2014), 377–404, 389.

23 *The Delights of Flower Painting* (London, 1756), 8.

24 Malcolm Baker, 'Epilogue: Making and Knowing, Then and Now', in *Ways of Making and Knowing*, 405–13, 408, 412.

25 Harold J. Cook, Pamela H. Smith and Amy R. W. Meyers, 'Introduction', in *Ways of Making and Knowing*, 8. See also Hilary Davidson's chapter on practice as research.

26 Robertson, *School of Arts* (Edinburgh, 1766), xvii.

27 Robertson, *School of Arts* (1784), 46–47. Taxidermy appears in the 1777 edition, but not the 1764 or 1766 editions. See my essay 'Women, Decorative Arts, and Taxidermy', 311–30.

28 Robertson's three recipes are for carmine, sap-green and 'colours extracted from flowers' (1784, 51). Her description of painting on silk is on page 58 (1784).

29 Ann Bermingham's definitive study *Learning to Draw* locates drawing's place as potentially a high and low art form within the class and gender dynamics of the period. Her discussion of the commercialization of amateur art practised mostly by bourgeois women provides an overview of Rudolph Ackermann's contribution to the commodification of amateur art practices. Ackermann's activities post-date most of the literature examined in this chapter.

30 *The Delights of Flower-Painting* (London, 1756), 11.

31 Ibid., 13–14.

32 A recipe for watercolours that does not contain sugar, honey or some form of glucose suggests to me that the author is not a practising watercolourist. Adding glucose to make the paint stable and flexible once dried does, however, introduce problems since flies are attracted to the sugar and have been known to eat the paint off of watercolour pictures. Another ingredient, coloquinth, is necessary if glucose is included in the recipe; it is repellent to flies.

33 *Flower-Painting*, 10–13.

34 *Arts Companion or a New Assistant for the Ingenious* (London and Dublin, 1749), 12.

35 *The Art of Drawing, and Painting in Water-colours* (London, 1735), 8–9.

36 *Art of Drawing*, 39–40.

37 *The Art of Painting in Miniature*, 6th ed. (London, 1752), 100–1. The author of the anonymously published treatise has traditionally been identified as most probably the Frenchman Claude Boutet, although Sarah Lowengard casts some doubt on this attribution. See Lowengard, *Creation of Colour*, chapter 11, 'Artists and Colourmen', http://www.gutenberg-e.org/lowengard/C_Chap11.html#txt4, accessed 22 May 2020. Pouring the prepared solution of watercolour paint into empty, clean shells was the standard way of storing, transporting and using watercolours before the rise of commercially produced watercolour cakes made by Reeves and other colourmen.

38 See Andrea Feeser's *Red, White, and Black Make Blue: Indigo in the Fabric of Colonial South Carolina Life* (Athens: University of Georgia Press, 2013) for the history of indigo production in South Carolina where enslaved Africans made indigo and had to live closer to these indigo works than their owners.

39 *The Art of Drawing*, 48.

40 Boyle's version elides the body's engagement with these flowers by using the passive voice: 'when they are freshly gather'd, they will afford a Juice, which when newly express'd, (for in some cases 'twill soon enough degenerate) affords a very deep and pleasant Blew.' See *Experiments and Considerations Touching Colours* (1660), http://www.gutenberg.org/files/14504/14504-h/14504-h.htm, accessed 22 May 2020.

41 Lowengard, 'Artists and Colourmen', in *Creation of Colour*, http://www.gutenberg-e.org/lowengard/C_Chap11.html#txt4, accessed 22 May 2020.

42 *Monthly Review*, 2d. series, 29 (1798): 108–9; quoted in Lowengard, 'Artists and Colourmen', http://www.gutenberg-e.org/lowengard/C_Chap11.html#txt4, accessed 22 May 2020.

43 Lowengard, 'Artists and Colourmen'.

44 George Edwards, *Essays on Natural History* (London, 1770).

45 *The Delights of Flower-Painting*, 12, 8.

46 Algernon Graves, *The Society of Artists of Great Britain, 1760–1791 and The Free Society of Artists, 1761–1783; A Complete Dictionary of Contributors and Their Work from the Foundation of the Societies to 1791* (London: George Bell and Sons, 1907), 242, 6, 77, 13, 141, 74

47 *The Times*, 16 April 1787, issue 726, pg. 1. © Times Newspapers Limited. Gale Document Number: CS16909456.

Select Bibliography

Adburgham, Alison. *Women in Print, Writing Women and Women's Magazines from the Restoration to the Accession of Victoria*. London: Allen and Unwin, 1972.

Ågren, Maria. 'Introduction'. In *Making a Living, Making a Difference: Gender and Work in Early Modern European Society*, edited by Maria Ågren, 1–23. Oxford: Oxford University Press, 2017.

Aldrich, Winifred. 'The Impact of Fashion on the Cutting Practices for the Woman's Tailored Jacket 1800–1927'. *Textile History* 34 (2003): 134–70.

Alexander, Kimberly S. *Treasures Afoot: Shoe Stories from the Georgian Era*. Baltimore: Johns Hopkins University Press, 2018.

Anishanslin, Zara. *Portrait of a Woman in Silk: Hidden Histories of the British Atlantic World*. London: Yale University Press, 2016.

Arnold, Janet. 'The Classical Influence on the Cut, Construction and Decoration of Women's Dress c.1785–1820'. *Costume* 4 (1970): 17–23.

Arnold, Janet. *Patterns of Fashion 1: Englishwomen's Dresses and Their Construction. c.1660–1860*. London: Macmillan, 1972.

Arnold, Janet. 'The Dressmaker's Craft'. *Costume* 7 (1973): 29–40.

Arnold, Janet. 'The Lady's Economical Assistant of 1808'. In *The Culture of Sewing: Gender, Consumption, and Home Dressmaking*, edited by Barbara Burman, 223–34. Oxford: Berg, 1999.

Ashmore, Sonia. *Muslin*. London: V & A Publishing, 2012.

Baillio, Joseph, Katherine Baetjer and Paul Lang. *Vigee Le Brun*. London: Yale University Press, 2016.

Batchelor, Jennie. *Dress, Distress and Desire: Clothing and the Female Body in Eighteenth-Century Literature*. London: Palgrave Macmillan, 2005.

Batchelor, Jennie. *Women's Work: Labour, Gender and Authorship, 1750–1830*. Manchester: Manchester University Press, 2010.

Batchelor, Jennie and Manushag N. Powell, eds. *Women's Periodicals and Print Culture in Britain, 1690–1820s: The Long Eighteenth Century*. Edinburgh: Edinburgh University Press, 2018.

Baumgarten, Linda. *What Clothes Reveal: The Language of Clothing in Colonial and Federal America, the Colonial Williamsburg Collection*. The Colonial Williamsburg Foundation. London: Yale University Press, 2002.

Beaudry, Mary C. *Findings: The Material Culture of Needlework and Sewing*. London: Yale University Press, 2006.

Beauvoir, Simone de. *The Second Sex*. New York: Vintage Books, 1974.

Bellamy, Liz. 'It-Narrators and Circulation: Defining a Subgenre'. In *The Secret Life of Things: Animals, Objects, and It-Narratives in Eighteenth-Century England*, edited by Mark Blackwell, 117–46. Lewisburg: Bucknell University Press, 2007.

Benson, John and Laura Ugolini, eds. *A Nation of Shopkeepers: Five Centuries of British Retailing*. London: I.B. Tauris, 2003.

Berg, Maxine. *The Age of Manufactures, 1700–1870*. London: Routledge, 1985.

Berg, Maxine. *Luxury and Pleasure in Eighteenth-Century Britain*. Oxford: Oxford University Press, 2005.

Berg, Maxine. 'The Genesis of "Useful Knowledge"'. *History of Science* 45, no. 2 (2007): 123–33.

Berg, Maxine and Elizabeth Eger, eds. *Luxury in the Eighteenth Century: Debates, Desires and Delectable Goods*. Basingstoke: Palgrave, 2003.

Bermingham, Ann. *Learning to Draw: Studies in the Cultural History of a Polite and Useful Art*. London: University of Yale Press, 2000.

Berry, Helen. 'Polite Consumption: Shopping in Eighteenth-Century England'. *Transactions of the Royal Historical Society* 6, no. 12 (2002): 375–94.

Blackwell, Bonnie. 'Corkscrews and Courtesans: Sex and Death in Circulation Novels'. In *The Secret Life of Things: Animals, Objects, and It-Narratives in Eighteenth-Century England*, edited by Mark Blackwell, 265–91. Lewisburg: Bucknell University Press, 2007.

Blackwell, Caitlin. '"The Feather'd Fair in a Fright": The Emblem of the Feather in Graphic Satire of 1776'. *Journal for Eighteenth-Century Studies* 36, no. 3 (2013): 353–76.

Blackwell, Mark. 'Introduction: The It-Narrative and Eighteenth-Century Thing Theory'. In *The Secret Life of Things: Animals, Objects, and It-Narratives in Eighteenth-Century England*, edited by Mark Blackwell, 9–14. Lewisburg: Bucknell University Press, 2007a.

Blackwell, Mark. 'Hackwork: It-Narratives and Iteration'. In *The Secret Life of Things: Animals, Objects, and It-Narratives in Eighteenth-Century England*, edited by Mark Blackwell, 187–217. Lewisburg: Bucknell University Press, 2007b.

Brewer, John and Roy Porter, eds. *Consumption and the World of Goods*. London: Routledge, 1993.

Brooks, Mary M. 'Decay, Conservation, and the Making of Meaning through Museum Objects'. In *Ways of Making and Knowing: The Material Culture of Empirical Knowledge*, edited by Pamela Smith, Amy Meyers and Harold J. Cook, 377–404. New York: Bard Graduate Centre, 2014.

Browne, Clare and Jennifer Earden, eds. *Samplers from the V&A Museum*. London: V&A Publishing, 1999.

Buck, Anne. *Dress in Eighteenth-Century England*. London: Batsford, 1979.

Burman, Barbara, ed. *The Culture of Sewing: Gender, Consumption, and Home Dressmaking*. Oxford: Berg, 1999.

Burman, Barbara, ed. '"A Linnen Pockett, a Prayer Book & Five Keys": Approaches to a History of Women's Tie-on Pockets'. In *Textiles and Text: Re-Establishing the Links between Archival and Object-Based Research*, edited by Maria Hayward and Elizabeth Kramer, 157–63. London: Archetype Publications, 2007.

Burman, Barbara and Ariane Fennetaux. *The Pocket: A Hidden History of Women's Lives 1660–1900*. London: Yale University Press, 2019.

Cahn, Susan. *Industry of Devotion: The Transformation of Women's Work in England 1500–1660*. New York: Columbia University Press, 1987.

Chrisman-Campbell, Kimberly. 'The Face of Fashion: Milliners in Eighteenth-Century Visual Culture'. *Journal for Eighteenth Century Studies* 25, no. 2 (2008): 157–71.

Chrisman-Campbell, Kimberly. *Fashion Victims: Dress at the Court of Louis XVI and Marie-Antoinette*. London: Yale University Press, 2015.

Collins, Jessica. 'Jane Holt, Milliner, and Other Women in Business: Apprentices, Freewomen and Mistresses in The Clothworkers' Company, 1606–1800'. *Textile History* 44 (2013): 72–94.

Cornforth, John. *English Interiors, 1799–1848: The Quest for Comfort*. London: Barrie & Jenkins, 1978.

Cox, Nancy. *The Complete Tradesman: A Study of Retailing, 1550–1820*. London: Routledge. 2016.

Cressy, David. *Literacy & the Social Order*. Cambridge: Cambridge University Press, 1980.

Crowley, John. *The Invention of Comfort. Sensibilities and Design in Early-Modern Britain and Early America*. Baltimore: Johns Hopkins University Press, 2001.

Crowston, Clare Haru. *Fabricating Women: The Seamstresses of Old Regime France 1675–1791*. London: Duke University Press, 2001.

Davidson, Hilary. 'Reconstructing Jane Austen's Silk Pelisse'. *Costume* 49, no. 2 (2015): 198–223.

Davidson, Hilary. 'Jane Austen's Pelisse Coat'. In *Jane Austen: Writer in the World*, edited by Kathryn Sutherland, 56–75. Oxford: The Bodleian Library, 2017.

Davidson, Hilary. *Dress in the Age of Jane Austen: Regency Fashion*. London: Yale University Press, 2019.

DeJean, Joan. *The Age of Comfort. When Paris Discovered Casual and the Modern Home Began*. London: Bloomsbury, 2009.

De Vries, Jan. *The Industrious Revolution Consumer Behaviour and the Household Economy, 1650–Present*. Cambridge: Cambridge University Press, 2008.

Dyer, Serena. 'Shopping and the Senses: Retail, Browsing and Consumption in Eighteenth-Century England'. *History Compass* 12, no. 9 (2014): 694–703.

Dyer, Serena. 'Trained to Consume: Dress and the Female Consumer in England 1720–1820'. PhD Thesis, University of Warwick, 2016.

Dyer, Serena. 'Training the Child Consumer: Play, Toys and Learning to Shop in Eighteenth-Century Britain'. In *Childhood by Design: Toys and the Material Culture of Childhood, 1700–Present*, edited by Megan Brandow-Faller, 31–46. London: Bloomsbury, 2018.

Dyer, Serena. '"Magnificent as Well as Singular": Hester Thrale's Polynesian Court Dress of 1781'. In *Fashion and Authorship: Literary Production and Cultural Style from the Eighteenth to the Twenty- First Century*, edited by Gerald Egan, 43–62. London: Palgrave Macmillan, 2020.

Efenbein, Andrew. 'Lesbian Aestheticism on the Eighteenth-Century Stage'. *Eighteenth-Century Life* 25, no. 1 (2001): 1–16.

Eger, Elizabeth. 'Paper Trails and Eloquent Objects: Bluestocking Friendship and Material Culture'. *Parergon* 26, no. 2 (2009): 109–38.

Eger, Elizabeth. *Bluestockings: Women of Reason from Enlightenment to Romanticism*. London: Palgrave Macmillan, 2010.

Eger, Elizabeth and Lucy Peltz. *Brilliant Women: Eighteenth-Century Bluestockings*. London: National Portrait Gallery, 2008.

Ehrman, Edwina. *The Judith Hayle Samplers*. London: Needleprint, 2007.

Elliot, Kamilla. *Portraiture and British Gothic Fiction: The Rise of Picture Identification*. Baltimore: Johns Hopkins University Press, 2012.

Engel, Laura. *Women, Performance, and the Material of Memory: The Archival Tourist, 1780–1915*. Basingstoke: Palgrave Macmillan: 2019.

Evans, Chris and Alun Withey. 'An Enlightenment in Steel?: Innovation in the Steel Trades of Eighteenth-Century Britain'. *Technology and Culture* 53, no. 3 (2012): 533–60.

Faiers, Jonathan. 'Dress Thinking: Disciplines and Indisciplinarity'. In *Dress History: New Directions in Theory and Practice*, edited by Charlotte Nicklas and Annebella Pollen, 15–32. London: Bloomsbury, 2015.

Feeser, Andrea. *Red, White, and Black Make Blue: Indigo in the Fabric of Colonial South Carolina Life*. Athens: University of Georgia Press, 2013.

Fennetaux, Ariane. 'Female Crafts: Women and *Bricolage* in Late Georgian Britain, 1750–1820'. In *Women and Things, 1750–1950: Gendered Material Strategies*, edited by Maureen Daly Goggin and Beth Fowkes Tobin, 91–108. Farnham: Ashgate, 2009.

Fennetaux, Ariane. 'Transitional Pandoras: Dolls in the Long Eighteenth Century'. In *Childhood by Design: Toys and the Material Culture of Childhood*, edited by Megan Brandow Faller, 47–66. London: Bloomsbury, 2018.

Fennetaux, Ariane, Amélie Junqua and Sophie Vasset, eds. *The Afterlife of Used Things: Recycling in the Long Eighteenth Century*. London: Routledge, 2015.

Fergus, Jan. *Provincial Readers in Eighteenth-Century England*. Oxford: Oxford University Press, 2006.

Finn, Margot. *The Character of Credit: Personal Debt in English Culture, 1740–1914*. Cambridge: Cambridge University Press, 2003.

Fowler, Christine. 'Robert Mansbridge, a Rural Tailor and His Customers 1811–1815'. *Textile History* 28, no. 1 (1997): 29–38.

Freedgood, Elaine. *The Ideas in Things: Fugitive Meaning in the Victorian Novel*. Chicago: University of Chicago Press, 2006.

Freud, Sigmund Freud and Josef Breuer. *Studies on Hysteria, 1883–1885*. In *The Complete Psychological Works of Sigmund Freud*, 24 vols. edited by James Strachey et al. London: The Hogarth Press, 1953–74.

Frye, Susan. *Pens and Needles: Women's Textualities in Early Modern England*. Philadelphia: University of Pennsylvania Press, 2010.

Garcin, Emmanuelle. 'La Restauration des Matériaux Textiles: La Fibre du Métier'. *Patrimoines, Revue de l'Institut National du Patrimoine* 6 (2010): 78–85.

Garcin, Emmanuelle. 'Le Costume comme source historique: Pour une philologie de la matière'. In *Fabrique de l'habit: Artisans, techniques et production du vêtement*, (fin du Moyen Âge-XVIIIe siècle), edited by Astrid Castres and Tiphaine Gaumy. Paris: Publications de l'École nationale des chartes, forthcoming, 2019.

Gerritsen, Anne and Giorgio Riello. 'Introduction: Writing Material Culture History'. In *Writing Material Culture History*, edited by Anne Gerritsen and Giorgio Riello, 1–13. London: Bloomsbury, 2015.

Geuter, Ruth. 'Embroidered Biblical Narratives and Their Social Contexts'. In *English Embroidery from the Metropolitan Museum of Art, 1580–1700*, edited by Melinda Watt and Andrew Morrall, 57–77. London: Yale University Press, 2008.

Gibson, James Jerome. *The Ecological Approach to Visual Perception*. Dallas: Houghton Mifflin, 1979.

Ginsburg, Madeleine. 'The Tailoring and Dressmaking Trades, 1700–1850'. *Costume* 6 (1972): 64–71.

Girouard, Mark. *Life in the English Country House*. London: Yale University Press, 1978.

Goggin, Maureen Daly. 'One English Woman's Story in Silken Ink: Filling in the Missing Strands in Elizabeth Parker's circa 1830 Sampler'. *Sampler and Antique Needlework Quarterly* 8, no. 4 (December 2002): 8–49.

Goggin, Maureen Daly. 'Stitching a Life in "Pen of Steele and Silken Inke": Elizabeth Parker's circa 1830 Sampler'. In *Women and the Material Culture of Needlework and Textiles, 1750–1950*, edited by Maureen Daly Goggin and Beth Fowkes Tobin, 31–49. Farnham: Ashgate, 2009.

Goggin, Maureen Daly. 'The Extra-Ordinary Powers of Red in Eighteenth- and Nineteenth-Century Needlework'. In *The Materiality of Color: The Production,*

Circulation, and Application of Dyes and Pigments, 1400–1800, edited by Andrea Feeser, Maureen Daly Goggin and Beth Fowkes Tobin, 29–43. Abingdon: Ashgate 2012.

Goggin, Maureen Daly and Beth Fowkes Tobin, eds. *Material Women, 1750–1950: Consuming Desires and Collecting Practices*. Aldershot: Ashgate, 2009.

Goggin, Maureen Daly and Beth Fowkes Tobin, eds. *Women and the Material Culture of Needlework and Textiles, 1750–1950*. Aldershot: Ashgate, 2009.

Goggin, Maureen Daly and Beth Fowkes Tobin, eds. *Women and Things, 1750–1950*. Aldershot: Ashgate, 2009.

Goggin, Maureen Daly and Beth Fowkes Tobin, eds. *Women and the Material Culture of Death*. Aldershot: Ashgate, 2010.

Greig, Hannah, Jane Hamlett and Leonie Hannan. 'Introduction: Gender and Material Culture'. In *Gender and Material Culture in Britain since 1600*, edited by Hannah Greig, Jane Hamlett and Leonie Hannan, 1–14. London: Palgrave, 2016.

Guest, Harriet. *Small Change: Women, Learning, Patriotism, 1750–1810*. Chicago: University of Chicago Press, 2000.

Hackel, Heidi Brayman. *Reading Material in Early Modern England: Print, Gender, and Literacy*. Cambridge: Cambridge University Press, 2007.

Hafter, Daryl, ed. *European Women and Preindustrial Craft*. Bloomington: Indiana University Press, 1995.

Hayden, Ruth. *Mrs Delany, Her Life and Her Flowers*. London: British Museum Publication, 1992.

Hedges, Elaine. 'The Needle or the Pen: The Literary Rediscovery of Women's Textile Work'. In *Tradition and the Talents of Women*, edited by Florence Hower, 338–64. Urbana: University of Illinois, 1991.

Hesketh, Sally. 'Needlework in the Lives of the Brontë Sisters'. *Brontë Society Transactions* 22, no. 1 (1997): 72–85.

Hilaire-Pérez, Liliane. 'Technology as a Public Culture in the Eighteenth Century: The Artisans' Legacy'. *History of Science* 45, no. 2 (2007): 135–53.

Hill, Bridget. *Women, Work and Sexual Politics in Eighteenth-Century England*. Montreal: McGill Queens University Press, 1993.

Hivet, Christine. 'Needlework and the Rights of Women in England at the End of the Eighteenth Century'. In *The Invisible Woman: Aspects of Women's Work in Eighteenth-Century Britain*, edited by Isabelle Baudino, Jacques Carré and Cécile Révauger, 37–46. Aldershot: Ashgate, 2004.

Hodder, Ian. *Entangled: An Archaeology of the Relationships between Humans and Things*. Chichester: Wiley-Blackwell, 2012.

Humphrey, Carol A. *Samplers*. Cambridge: Cambridge University Press, 1997.

Humphrey, Carol A. *Sampled Lives: Samplers from the Fitzwilliam Museum*. Cambridge: Fitzwilliam Museum Publications, 2017.

Hunter, J. Paul. *Before Novels: The Cultural Contexts of Eighteenth-Century English Fiction*. New York: Norton, 1990.

Ingold, Tim. *Making: Anthropology, Archaeology, Art and Architecture*. London: Routledge, 2013.

Jefferies, Janis. 'Crocheted Strategies: Women Crafting Their Own Communities'. *Textile* 14, no. 1 (2016): 14–35.

Jenkins, Eugenia Zuroski. *A Taste for China: English Subjectivity and the Prehistory of Orientalism*. Oxford: Oxford University Press, 2013.

Johnson, Ellen Kennedy. 'Alterations: Gender and Needlework in the Late Georgian Arts and Letters'. PhD Dissertation, Arizona State University, 2009.

Jones, Anna Rosalind and Peter Stallybrass. *Renaissance Clothing and the Materials of Memory*. Cambridge: Cambridge University Press, 2000.

Jones, Jennifer M. *Sexing La Mode: Gender, Fashion and Commercial Culture in Old Regime France*. Oxford: Berg, 2004.

Kettle, Alice and Jane McKeating, eds. *Hand Stitch: Perspectives*. London: Bloomsbury, 2011.

King, Kathryn. 'Of Needles and Pens and Women's Work'. *Tulsa Studies in Women's Literature* 14, no. 1 (1995): 77–93.

Kokoli, Alexandra M. 'Undoing "Homeliness" in Feminist Art: *Feministo: Portrait of the Artist as a Housewife* (1975–7)'. *N. paradoxa* 13 (2006): 75–83.

Kopytoff, Igor. 'The Cultural Biography of Things: Commoditization as Process'. In *The Social Life of Things*, edited by Arjun Appadurai, 64–84. Cambridge: Cambridge University Press, 1986.

Kowaleski-Wallace, Elizabeth. *Consuming Subjects: Women, Shopping, and Business in the Eighteenth Century*. New York: Columbia University Press, 1997.

Kowaleski-Wallace, Elizabeth. 'Representing Corporeal "Truth" in the Work of Anna Morandi Manzolini and Madame Tussaud'. In *Women and the Material Culture of Death*, edited by Maureen Daly Goggin and Beth Fowkes Tobin, 283–309. Aldershot: Ashgate: 2013.

Laird, Mark and Alicia Weisberg-Roberts, eds. *Mrs Delany and Her Circle*. London: Yale University Press, 2009.

Lambert, Miles. '"Sent from Town:" Commissioning Clothing in Britain during the Long Eighteenth Century'. *Costume* 43 (2009): 66–84.

Lane, Joan. *Apprenticeship In England, 1600–1914*. London: UCL Press, 2009.

Lazaro, David E. and Patricia Campbell Warner. 'All-Over Pleated Bodice: Dressmaking in Transition, 1780–1805'. *Dress* 31 (2004): 15–24.

Lembke, Janet. *Virgil's Georgics: A New Verse Translation*. London: Yale University Press, 2005.

Lemire, Beverly. 'Developing Consumerism and the Ready-Made Clothing Trade in Britain, 1750–1800'. *Textile History* 15, no. 1 (1984): 21–44.

Lemire, Beverly. 'Redressing the History of the Clothing Trade in England: Ready Made Clothing, Guilds, and Women Workers, 1650–1800'. *Dress* 21 (1994): 61–74.

Lévi-Strauss, Claude. *The Savage Mind*. London: Weidenfield and Nicolson, 1966.

Lieb, Laurie Yager. '"The Works of Women Are Symbolical": Needlework in the Eighteenth Century'. *Eighteenth-Century Life* 10, no. 2 (1986): 28–44.

Llewellyn, Nigel. 'Elizabeth Parker's Sampler: Memory, Suicide and the Presence of the Artist'. In *Material Memories: Design and Evocation*, edited by Marius Kwint, Christopher Breward and Jeremy Aynsley, 59–71. Oxford: Berg, 1999.

Lowengard, Sarah. *The Creation of Colour in Eighteenth-Century Europe*. New York: Columbia University Press, 2006.

Lupton, Christina. 'Giving Power to the Medium: Recovering the 1750s'. *The Eighteenth Century: Theory and Interpretation* 52, nos. 3–4 (2011): 289–302.

Lupton, Christina. *Reading and the Making of Time in the Eighteenth Century*. Baltimore: Johns Hopkins University Press, 2018.

Lynch, Deidre Shauna. 'Counter Publics: Shopping and Women's Sociability'. In *Romantic Sociability: Social Networks and Literary Culture in Britain, 1770–1840*, edited by Gillian Russell and Clara Tuite, 211–36. Cambridge: Cambridge University Press, 2006.

Macheski, Cecilia. 'Penelope's Daughters: Images of Needlework in Eighteenth-Century Literature'. In *Fetter'd or Free? British Women Novelists, 1670–1815*, edited by Mary Anne Schofield and Cecilia Macheski, 85–100. Athens: Ohio University Press, 1986.

Mack, Ruth. 'Hogarth's Practical Aesthetics'. In *Mind, Body, Motion, Matter: Eighteenth-Century British and French Literary Perspectives*, edited by Mary Helen McMurran, 21–46. Toronto: University of Toronto Press, 2016.

Mackie, Erin. *Market à la Mode: Fashion, Commodity, and Gender in* The Tatler *and* The Spectator. Baltimore: Johns Hopkins University Press, 1997.

Maeder, Edward. Editor of *An Elegant Art: Fashion & Fantasy in the Eighteenth Century*. New York: Los Angeles County Museum of Art Collection of Costumes and Textiles, 1983.

Marsh, Gail. *18th Century Embroidery Techniques*. Lewes: Guild of Master Craftsman Publications, 2006.

McCreery, Cindy. *The Satirical Gaze*. Oxford: Oxford University Press, 2004.

McKendrick, Neil, John Brewer and John Harold Plumb. *The Birth of a Consumer Society: The Commercialization of Eighteenth-Century England*. Bloomington: Indiana University Press, 1982.

McNeil, Peter. *Pretty Gentlemen: Macaroni Men and the Eighteenth-Century Fashion World*. London: Yale University Press, 2018.

McNeil, Peter and Patrik Steorn. 'The Medium of Print and the Rise of Fashion in the West'. *Konsthistorisk tidskrift/Journal of Art History* 82, no. 3 (2013): 135–56.

Mida, Ingrid and Alexandra Kim. *The Dress Detective: A Practical Guide to Object-Based Research in Fashion*. London: Bloomsbury, 2015.

Mitch, David. 'Education and Skill of the British Labour Force'. In *The Cambridge Economic History of Modern Britain, Vol. I: Industrialisation, 1700–1860*, edited by Roderick Floud and Paul Johnson, 332–56. Cambridge: Cambridge University Press, 2004.

O'Malley, Andrew. *The Making of the Modern Child: Children's Literature and Childhood in the Late Eighteenth Century*. Abingdon: Routledge, 2011.

Panzanelli, Roberta. 'The Body in Wax, the Body of Wax'. In *Ephemeral Bodies: Wax Sculpture and the Human Figure*, edited by Roberta Panzanelli, 1–12. Los Angeles: Getty Research Institute, 2008.

Park, Julie. *The Self and It: Novel Objects in Eighteenth-Century England*. Stanford: Stanford University Press, 2010.

Parker, Rozsika. *The Subversive Stitch: Embroidery and the Feminine*. New York: Routledge, 1984.

Parker, Rozsika and Griselda Pollack, eds. *Framing Feminism: Art and the Women's Movement, 1970–1985*. London: Pandora, 1987.

Paulson, Ronald. *Hogarth*, 3 vols. New Brunswick: Rutgers University Press, 1991–3.

Peers, Laura. 'Material Culture, Identity, and Colonial Society in the Canadian Fur Trade'. In *Women and Things, 1750–1950: Gendered Material Strategies*, edited by Maureen Daly Goggin and Beth Fowkes Tobin, 55–74. Aldershot: Ashgate, 2009.

Percoco, Cassidy. *Regency Women's Dress: Techniques and Patterns 1800–1830*. London: Batsford, 2015.

Pristash, Heather, Inez Schaechterle and Sue Carter Wood. 'The Needle as Pen: Intentionality, Needlework, and the Production of Alternate Discourses of Power'. In *Women and the Material Culture of Needlework and Textiles, 1750–1950*, edited by Maureen Daly Goggin and Beth Fowkes Tobin, 13–29. Aldershot: Ashgate, 2009.

Rana, Leena A. 'Stories behind the Stitches: Schoolgirl Samplers of the Eighteenth and Nineteenth Centuries'. *Cloth and Culture* 12, no. 2 (2014): 158–79.

Rasmussen, Pernilla. 'Creating Fashion: Tailors' and Seamstresses' Work with Cutting and Construction Techniques in Women's Dress, *c*.1750–1830'. In *Fashionable Encounters: Perspectives and Trends in Textile and Dress in the Early Modern Nordic World*, edited by Tove Engelhardt Mathiassen et al. 50–71. Oxford: Oxbow Books, 2014.

Retford, Kate. *The Art of Domestic Life: Family Portraiture in Eighteenth-Century England*. London: Yale University Press, 2005.

Retford, Kate. *The Conversation Piece: Making Modern Art in Eighteenth-Century Britain*. London: Yale University Press, 2017.

Ribeiro, Aileen. *Dress in Eighteenth-Century Europe, 1715–1789*. London: Yale University Press, 1985.

Richardson, Catherine. 'Written Texts and the Performance of Materiality'. In *Writing Material Culture History*, edited by Anne Gerritsen and Giorgio Riello, 43–58. London: Bloomsbury, 2015.

Richardson, Robbie. *The Savage and Modern Self: North American Indians in Eighteenth-Century British Literature and Culture*. Toronto: University of Toronto Press, 2018.

Richmond, Vivienne. 'Stitching the Self: Eliza Kenniff's Drawers and the Materialization of Identity in Late-Nineteenth-Century London'. In *Women and Things, 1750–1950: Gendered Material Strategies*, 43–54. Farnham: Ashgate, 2009.

Richmond, Vivienne. 'Stitching Women: Unpicking Histories of Victorian Clothes'. In *Gender and Material Culture in Britain since 1600*, edited by Hannah Greig, Jane Hamlett and Leonie Hannan, 90–103. Basingstoke: Palgrave, 2016.

Riello, Giorgio. *A Foot in the Past: Consumers, Producers and Footwear in the Long Eighteenth Century*. Oxford: The Pasold Research Fund, Oxford University Press, 2006.

Sanderson, Elizabeth C. *Women and Work in Eighteenth-Century Edinburgh*. Abingdon: Macmillan, 1996.

Sanderson, Elizabeth C. 'The New Dresses: A Look at How Mantuamaking Became Established in Scotland'. *Costume* 35 (2001): 14–23.

Sennett, Richard. *The Craftsman*. Harmondsworth: Penguin, 2009.

Schmahmann, Brenda. 'Intertextual Textiles: Parodies and Quotations in Cloth'. *TEXTILE* 15, no. 4 (2017): 336–43.

Schofield, R. S. 'Dimensions of Illiteracy, 1750–1850'. *Explorations in Economic History* 10, no. 4 (1973): 437–54.

Scobie, Ruth. '"To Dress a Room for Montagu": Pacific Cosmopolitanism and Elizabeth Montagu's Feather Hangings'. *Lumen* 33 (2014): 123–37.

Scobie, Ruth. '"Bunny! O! Bunny!": The Burney Family in Oceania'. *Eighteenth-Century Life* 42, no. 2 (2018): 56–72.

Sharpe, Pamela, ed. *Women's Work: The English Experience 1600–1914*. Oxford: Oxford University Press, 1998.

Shawcross, Rebecca. *Shoes: An Illustrated History*. London: Bloomsbury, 2014.

Sheriff, Mary D. *The Exceptional Woman: Elisabeth Vigee-Lebrun and the Cultural Politics of Art*. Chicago and London: University of Chicago Press: 1996.

Shimbo, Akiko. *Furniture-Makers and Consumers in England, 1754–1851: Design as Interaction*. London: Routledge, 2016.

Silver, Sean. 'Afterward: What Do We Mean by "Material"?' *Eighteenth-Century Fiction* 31, no. 1 (2018): 213–30.

Smith, Chloe Wigston. *Women, Work, and Clothes in the Eighteenth-Century Novel*. Cambridge: Cambridge University Press, 2013.

Smith, Chloe Wigston. 'Gender and the Material Turn'. In *Women's Writing, 1660–1830: Feminisms and Futures,* edited by Jennie Batchelor and Gillian Dow, 149–68. London: Palgrave Macmillan, 2016.

Smith, Chloe Wigston. 'Fast Fashion: Style, Text, and Image in Late Eighteenth-Century Women's Periodicals'. In *Women's Periodicals and Print Culture in Britain 1690–1820s,* edited by Jennie Batchelor and Manushag N. Powell, 440–57. Edinburgh: Edinburgh University Press, 2018.

Smith, Kate. 'Sensing Design and Workmanship: The Haptic Skills of Shoppers in Eighteenth-Century London'. *Journal of Design History* 25 (2012): 1–10.

Smith, Kate. 'In Her Hands: Materializing Distinction in Georgian Britain'. *Cultural and Social History* 11 no. 4 (2014), 489–506.

Smith, Kate. *Material Goods, Moving Hands: Perceiving Production in England, 1700–1830.* Manchester: Manchester University Press, 2014.

Smith, Kate and Leonie Hannan. 'Return and Repetition: Methods for Material Culture Studies'. *The Journal of Interdisciplinary History* 48, no. 1 (2017): 1–17.

Smith, Pamela H. 'Making as Knowing: Craft as Natural Philosophy'. In *Ways of Making and Knowing: The Material Culture of Empirical Knowledge,* edited by Pamela H. Smith, Amy R. W. Meyers and Harold J. Cook, 17–47. New York: Bard Graduate Centre, 2014.

Smith, Pamela H., Amy R. W. Meyers and Harold J. Cook. *Ways of Making and Knowing: The Material Culture of Empirical Knowledge.* New York: Bard Graduate Center, 2017.

Solkin, David H. *Painting Out of the Ordinary: Modernity and the Art of Everyday Life in Early Nineteenth-Century Britain.* London: Yale University Press, 2008.

Stabile, Susan M. *Memory's Daughters: The Material Culture of Remembrance in Eighteenth-Century America.* Ithaca: Cornell University Press, 2004.

Steedman, Carolyn. 'Englishness, Clothes and Little Things'. In *The Englishness of English Dress,* edited by Christopher Breward, Becky Conekin and Caroline Cox, 29–44. Oxford: Berg, 2002.

Stein, Sarah Abrevaya. *Plumes: Ostrich Feathers, Jews, and a Lost World of Global Commerce.* London: Yale University Press, 2010.

Sterckx, Marjan. 'Pride and Prejudice: Eighteenth-Century Women Sculptors and Their Material Practices'. In *Women and Material Culture, 1660–1830,* edited by Jennie Batchelor and Cora Kaplan, 86–102. Basingstoke: Palgrave Macmillan, 2007.

Stobart, Jon. 'The Shopping Streets of Provincial England, 1650–1840'. In *The Landscape of Consumption: Shopping Streets and Cultures in Western Europe, 1600–1900,* edited by Jan Hein Furnée and Clé Lesger, 16–36. London: Palgrave Macmillan, 2014.

Stobart, Jon. 'Luxury and Country House Sales in England, c.1760–1830'. In *The Afterlife of Used Things: Recycling in the Long Eighteenth Century,* edited by Ariane Fennetaux, A. Amélie Junqua and Sophie Vasset, 25–37. London: Routledge, 2015.

Styles, John. *The Dress of the People: Everyday Fashion in Eighteenth-Century England.* London: Yale University Press, 2007.

Styles, John and Amanda Vickery, eds. *Gender, Taste and Material Culture in Britain and North America, 1700–1830.* London: Yale University Press, 2006.

Styles, John. *Threads of Feeling: The London Foundling Hospital's Textile Tokens, 1740–1770.* London: Foundling Museum, 2010.

Suarez, Michael. 'Introduction'. In *The Cambridge History of the Book in Britain, Vol. V, 1695–1830,* edited by Michael Suarez, S.J. and Michael L. Turner, 1–36. Cambridge: Cambridge University Press, 2009.

Swenson, Rivka. 'Secret History and It-Narrative'. In *The Secret History in Literature, 1660–1820,* edited by Rebecca Bullard and Rachel Carnell, 117–33. Cambridge: Cambridge University Press, 2017.

Syson, Luke. 'Polychrome and Its Discontents: A History'. In *Like Life: Sculpture, Color, and the Body*, edited by Luke Syson, Sheena Wagstaff, Emerson Bowyer and Brinda Kumar, 14–41. New York: Metropolitan Museum of Art, 2018.

Taylor, Jane. '"Important Trifles": Jane Austen, The Fashion Magazine, and the Inter-Textual Consumer Experience'. *History of Retailing and Consumption* 2, no. 2 (2016): 113–28.

Thatcher Ulrich, Laurel. 'Of Pens and Needles: Sources in Early American Women's History'. *The Journal of American History* 77, no. 1 (1990): 200–7.

Thomas, Nicholas. *Entangled Objects: Exchange, Material Culture, and Colonialism in the Pacific*. London: Harvard University Press, 1991.

Thunder, Moira. 'Object Lesson: Designs and Clients for Embroidered Dress, 1782–94'. *Textile History*, 37, no. 1(2006): 82–9.

Tobin, Amy. '"I'll Show You Mine, If You Should Me Yours": Collaboration, Consciousness-Raising and Feminist-Influenced Art in the 1970s'. *Tate Papers* 25 (2016), https://www.tate.org.uk/research/publications/tate-papers/25/i-show-you-mine-if-you-show-me-yours, accessed 22 May 2020.

Tobin, Beth Fowkes. 'Women, Decorative Arts, and Taxidermy'. In *Women and the Material Culture of Death*, edited by Maureen Daly Goggin and Beth Fowkes Tobin, 311–30. Aldershot: Ashgate, 2010.

Tobin, Beth Fowkes. *The Duchess's Shells: Natural History Collecting in the Age of Cook's Voyages*. London: Yale University Press, 2014.

Tobin, Beth Fowkes. 'Bluestockings and the Culture of Natural History'. In *Bluestockings Now: The Evolution of a Social Role*, edited by Deborah Heller, 55–71. Abingdon: Ashgate, 2015.

Tuite, Clara. '"Comedy, Too Fatal Emblem": Anne Damer and Occult Theatricality'. *Eighteenth-Century Fiction* 27, no. 3–4 (2015): 557–96.

Tyner, Judith A. *Stitching the World: Embroidered Maps and Women's Geographical Education*. Aldershot: Ashgate, 2015.

Unsworth, Rebecca. 'Hands Deep in History: Pockets in Men and Women's Dress in Western Europe'. *Costume* 51, no. 2 (2017): 148–70.

Van Horn, Jennifer. *The Power of Objects in Eighteenth-Century British America*. Chapel Hill: University of North Carolina Press, 2017.

Vickery, Amanda. *The Gentleman's Daughter*. London: Yale University Press, 1998.

Vickery, Amanda. *Behind Closed Doors: At Home in Georgian England*. London: Yale University Press, 2009.

Vickery, Amanda. 'The Theory and Practice of Female Accomplishment'. In *Mrs Delany and Her Circle*, edited by Mark Laird and Alicia Weisberg-Roberts, 94–103. London: Yale University Press, 2009.

Walford, Jonathan. *The Seductive Shoe: Four Centuries of Fashion Footwear*. New York: Steward, Tabori and Chang, 2007.

Walsh, Claire. 'Shop Design and the Display of Goods in Eighteenth Century London'. *Journal of Design History* 8 (1995): 157–76.

Walsh, Claire. 'Social Meaning and Social Space in the Shopping Galleries of Early Modern London'. In *A Nation of Shopkeepers: Five Centuries of British Retailing*, edited by John Benson and Laura Ugolini, 52–79. London: I.B. Tauris, 2003.

Walsh, Claire. 'Shops, Shopping, and the Art of Decision Making'. In *Gender, Taste and Material Culture in Britain and North America, 1700–1830*, edited by John Styles and Amanda Vickery, 151–87. London: Yale University Press, 2006.

Waugh, Norah. *The Cut of Women's Clothes, 1600–1930*. New York: Theatre Arts Books, 1968.

Weatherill, Lorna. *Consumer Behaviour and Material Culture in Britain, 1660–1760*. London: Routledge, 1996.

White, Jonathan. 'A World of Goods? The "Consumption Turn" and Eighteenth-Century British History'. *Cultural and Social History* 3, no. 1 (2006): 93–104.

Wilcox, David. 'Cut and Construction of a Late Eighteenth-Century Coat'. *Costume* 33 (1999): 95–7.

Wilson, Carol Shiner, 'Lost Needles, Tangled Threads: Stitchery, Domesticity, and the Artistic Enterprise in Barbauld, Edgeworth, Taylor, and Lamb'. In *Re-Visioning Romanticism: British Women Writers, 1776–1837*, edited by Carol Shiner Wilson and Joel Haefner, 167–90. Philadelphia: The University of Pennsylvania Press, 1994.

Wyld, Helen. *Embroidered Stories: Scottish Samplers*. Edinburgh: National Museums Scotland, Edinburgh, 2018.

Yarrington, Alison. 'A Female Pygmalion: Anne Seymour Damer, Allan Cunningham and the Writing of a Woman Sculptor's Life'. *Sculpture Journal* 1 (1997): 32–44.

Zuroski, Eugenia and Michael Yonan. 'Material Fictions: A Dialogue as Introduction'. *Eighteenth-Century Fiction* 31, no. 1 (2018): 1–18.

Index